AMERICAN SEACOAST, ANTI-AIRCRAFT, AND RAILWAY ARTILLERY 1886-1955

Glen M. Williford

Redoubt Press, McLean, Virginia 2026

Library of Congress Cataloging-in-Publication Data

Library of Congress Control Number: 2026938544
Williford, Glen M.
American Seacoast, Anti-Aircraft, and Railway Artillery 1886-1955
— 1st Edition
p. cm.
includes index

ISBN 978-1-7323916-8-0
1. American seacoast defenses—history—20th century. 2. U.S. Army Artillery
3. Illustrated guide to U.S. Army Artillery
I—title

Front Cover Image: 10-inch seacoast gun on disappearing carriage at Battery Gillmore, Fort Hamilton NY

Rear Cover Images clockwise from upper left: 3-inch AA firing drill in Puerto Rico in 1940; 14-inch railway gun at Fort MacArthur; postcard of a 6-inch M1900 coasr artillery gun; 12-inch mortars on railway carrages ready to fire; 155mm GPF gun at Fort Tilden
Frontspiece Page image: Seacoast 8-inch gun on barbette carriage
emplaced at the United States Military Academy, West Point, New York

Cover design by Mark Berhow

Redoubt Press is a division of McGovern Publishing, which along with Three Sisters Press, publishes books on subjects of historical interest. Under the Redoubt Press label, we have published ***The American Defences of the Panama Canal*** by Terrance McGovern, ***The Concrete Battleship - Fort Drum, El Fraile Island, Manila Bay,*** by Francis J. Allen, ***A Legacy in Brick and Stone,*** by John R. Weaver II, ***Pacific Rampart, A History of Corregidor and the Harbor Defenses of Manila and Subic Bays*** by Glen M. Williford, and ***Pacific Fortress, a History of the Seacoast Defenses of Hawaii*** by Glen M. Williford. Under the Three Sisters Press label, we have published ***The Chesapeake Bay at War! – The Coastal Defenses of Chesapeake Bay During World War Two*** by Terrance McGovern, ***Seacoast Cannon Coloring Book*** by Brian B. Chin, and ***The Delaware Bay at War! – The Coastal Defenses of Delaware Bay During World War Two*** by Terrance McGovern. McGovern Publishing is interested in new titles, especially those dealing with fortifications, please contact Terry McGovern at 703/538-5403 or at tcmcgovern@att.net if you have a title that you are seeking to have published.

Visit www.mcgovernpublishing.com

Redoubt Press
A Division of McGovern Publishing
1700 Oak Lane, McLean, Virginia 22101 USA
tcmcgovern@att.net

Contents

Preface and Acknowledgments

This volume is the companion to ***American Breech loading Mobile Artillery 1875-1953*** (ISBN 978-0-7643-5049-8). Like that work, it draws primarily on historical technical publications and documentary resources -- and on the cooperation of the many people and institutions that helped make those resources accessible. The pages that follow are an attempt to acknowledge that debt in full.

Principal Correspondents

Four people deserve special recognition, and I want to name them here. All have been long-time correspondents of mine, principally in locating, analyzing, and sharing information about American coast defense guns and mounts that is otherwise difficult to find. Two of them have recently passed. Alex Holder of Scituate, Massachusetts, and Bolling Smith of Accokeek, Maryland, each contributed enormously to this work and to the research that preceded it. David Kirchner of Germantown, Virginia, and Tom Batha of Rouses Point, New York, remain active partners in the ongoing pursuit of American artillery history.

Sources

The technical content of this book draws on two distinct bodies of material. The first is the body of official technical publications -- army and navy manuals, pamphlets, and related documents -- that were originally produced to accompany the weapons themselves. Written as instructions for use and maintenance, these publications contain detailed descriptions of what the guns were, how they were constructed, their major components, and their service performance. Surviving copies are scattered unevenly across military libraries and archival collections, but all but a very few titles were eventually tracked down. Taken together, they provided the technical descriptions and performance data that form the backbone of this work.

The second body of material is considerably harder to come by. The narrative context -- why a given type was acquired in the first place, where and how many were produced, how they were actually deployed and used in combat -- has largely gone unreported in secondary sources. Most of it had to be extracted directly from the correspondence files of the Army Ordnance Department and the Navy Bureau of Ordnance. These records exist only as original paper files, held at the two National Archives facilities in the Washington, D.C. area: the downtown depository and the facility at College Park, Maryland. I owe a considerable debt to the professional staff at the military records divisions of both institutions, and I want to acknowledge that debt plainly here.

Illustrations

The illustrations in this work come from several sources, and I want to account for them as clearly as I can. The gun plans are drawn from official U.S. Army and Navy ordnance technical publications. In later years these technical drawings were generally replaced by clear photographs with background detail removed; those photographs are similarly from official publications. In a handful of cases the drawing used comes from a publication produced by the private contractor or supplier; all such publications are listed in the bibliography.

Period photographs showing the guns in service come primarily from public domain holdings at the National Archives and Records Administration, Still Photos Division, Archives II, College Park, Maryland. I want to thank the staff there for their patience with my requests over the years -- in particular Holly Reed and Ruth Beamon, who have been consistently helpful on a great many occasions. A number of photographs were drawn from National Park Service holdings as well. Of particular value was the collection held by the Sandy Hook unit of the Gateway National Recreation Area, and I am grateful to them for making available their photographs from Sandy Hook's years as the Army's proving ground.

Most of the images were selected from the author's digital collection, either scanned from original photographs or obtained from other digital collections.

Several photos that were selected were of such a low resolution that they would not meet printing requirements. These were enhanced to a higher resolution using a commercial program that uses AI. The reader should be aware that some of the fine details in the enhanced photographs might be artifacts of the enhancement process. The enhanced photos were carefully compared to the original digital images and only those that revealed little discernable changes between the printed images were used.

Memorial Guns and Surviving Examples

A small group of us tracks down and documents memorial artillery guns still standing across the United States. All of us owe a debt to Richard Pope, who started the WW1 Memorial Survey and whose persistent efforts to recruit partners in finding, photographing, and recording these guns made the work considerably easier for those of us who followed. Most of the photographs of surviving guns in this volume are from my own visits to their locations. In a number of cases, however, the location itself and advice on how to get there came from fellow searchers. I want to thank Tom Batha, Charles Bugajsky, Roger Davis, C.B. Drennon, Mike Fiorini, Alan Hardey, and Ivan Hlavacek for their help in this regard.

Museum Collections

American military museums are generally generous in permitting photography of their collections and in allowing those photographs to be reproduced in publications, provided they are properly credited. I have tried to identify each institution accurately in the caption of every such photograph in this book. Two collections deserve particular mention here. The Air Defense Artillery Training Support Facility at Fort Sill, Oklahoma, and the Ordnance Training Facility at Fort Lee -- which now houses much of the former U.S. Army Ordnance Museum collection from Aberdeen, Maryland -- both made their holdings available to me, and I am grateful for it.

Postcards

In a few cases, period postcards represent the only surviving photographic record of certain guns. I want to thank Karl W. Schmidt for making his collection available and for granting permission to reproduce those images here.

A Final Note—

Despite every effort to get the facts right and the words straight, errors of interpretation and transcription will have found their way into this book. There is no good excuse for failing to catch them before publication. Any that remain are entirely my own.

Glen M. Williford
Zionsville, Indiana
January 2026

Introduction

American seacoast, anti-aircraft, and railway artillery have received less attention in print than their mobile field artillery counterparts. The technical details of these weapons -- what they were, how they were built, how many were produced and by whom -- have relatively little coverage in the historical literature. Most seacoast guns and all but one type of railway artillery never saw combat, and so they are missing from the battle histories and combat narratives where mobile artillery appears prominently. That does not mean these weapons did not serve. They did, extensively, in defensive roles within the United States and its possessions -- roles that simply never required them to fire in anger. Readers and historians today have few ready references that describe what was, in fact, a significant and sustained military undertaking. This book exists to fill that gap.

This is not a definitive history of American seacoast, anti-aircraft, or railway artillery. For individual gun types there are technical manuals and sometimes secondary histories that go into far greater depth than anything presented here. What this book offers instead is an illustrated identification guide -- a directory of the various types of American fixed artillery used in the modern era, organized to allow quick reference and reliable identification.

Scope

The most difficult challenge in compiling this work was deciding what to include and what to leave out. Some of those decisions were already made by the publication of the companion volume on mobile artillery. That book covered the period from the introduction of modern breechloading artillery in 1875 through the end of the Korean War in 1953. This volume follows the same temporal boundaries -- 1886 to 1955 -- and addresses a complementary subject: fixed artillery used by U.S. military services.

In practice, that has meant Army seacoast artillery -- a term I use synonymously with coast artillery throughout this work. Railway artillery (even though technically not "fixed") followed naturally as an extension of that scope. With the exception of the First World War generation, American-produced railway guns were intended for seacoast service. They are mobile in the sense that they can be relocated by rail, but they were generally designed to fire only from semi-permanent emplacements, and again with one exception they were exclusively manned by Coast Artillery Corps units and personnel.

A seacoast gun, once emplaced, needs only itself, a trained crew, and a supply of ammunition to be effective. The concerns that dominate field artillery -- standardization of types to simplify supply, interchangeable parts to ease maintenance, consistent performance across guns of the same type -- matter much less in coastal defense. As a result, a wide variety of guns from different sources found their way into temporary batteries at one time or another. I have tried to include every type of seacoast gun employed in a fixed emplacement by American forces, regardless of how many were produced or how briefly they served. On the other hand, instances where field artillery pieces were positioned along the coast for potential use against naval targets did not seem to qualify for inclusion here. In any case, the technical details of those gun types are already covered in the companion volume on mobile artillery.

The decision to include anti-aircraft artillery was less straightforward. Some production anti-aircraft guns of the First and Second World Wars were as fully mobile as their field artillery counterparts. Others were always intended for permanent emplacement. Responsibility for air defense was initially assigned to the Coast Artillery Corps and only split off into its own organization much later. Lighter anti-aircraft guns were used by all branches -- the Marine Corps, the Army Air Forces, even the U.S. Air Force during the Korean War. I decided that anti-aircraft artillery belonged in this volume rather than in the earlier one on mobile artillery, primarily because of the Coast Artillery connection and the fact that many types were designed for fixed defensive positions. Navy shipboard mounted anti-aircraft was left for coverage elsewhere.

Reasonable arguments can be made for exceptions to any of these decisions. At some point, however, a line has to be drawn, often influenced by the practical need to keep a work of this kind to a manageable size. These are the lines I drew.

Organization

The book is organized into three sections: seacoast artillery, anti-aircraft artillery, and railway artillery. Within each section, chapters follow in roughly chronological order by general category or generation of guns. Each chapter begins with an introductory page, followed in most cases by a two or three-page spread for each individual gun type.

The first of those two pages is text. It provides a brief history of the gun's development and use, a technical description with attention to innovative or distinguishing features, notes on how many were produced and by which arsenal or manufacturer, and comments on service use, combat experience if any, and when the type was retired. I also note how many examples survive and sometimes where they can be found. The second page contains a diagram of the gun -- the scale varies depending on the available plan and the space on the page -- and a table of the gun's major characteristics. Where possible, I have included period photographs showing the weapon in service and a photograph of a surviving example.

In some cases the available documentation does not support this format exactly. Where that is true, I have used additional photographs or made other adjustments as needed. Production quantities, or the breakdown of production by manufacturers, were sometimes unavailable or the available figures seemed unreliable. For temporary coast defense guns, I have focused on the number acquired for emplacement rather than the initial production run for other Army or Navy uses.

Sources

A technical guide like this one does not lend itself to the usual practice of detailed footnoting. So many technical facts and specifications appear on every page that referencing even the most important ones would be impractical. Because most of the statistical data comes from a relatively small number of official service publications, and because information about production sources and quantities comes from specific archival files, I have chosen to use a descriptive "Sources Consulted" section as a bibliography rather than attempting individual citations throughout the text.

Information about the number and location of surviving guns has been compiled personally over many years and does not exist in any single published or verifiable source. I present it here as accurately as I can, with the understanding that the situation changes as guns are moved, scrapped, or rediscovered.

Photographs and plans are not individually cited in the text. Plans are taken from service technical manuals and drawings. For the sake of clarity, most have been cleaned of identifying labels and detail notes, which would not be legible at the reproduced size. Period photographs are from public domain archival sources. The majority come from the National Archives, Still Photos Branch, College Park, Maryland. Additional sources include the Naval History and Heritage Command in Washington, D.C., the U.S. Army Heritage and Education Center at Carlisle Barracks, Pennsylvania, and the National Park Service, Gateway National Recreation Area. All photographs of surviving examples were taken by me or by personal associates.

Designations

I have tried to use the actual designations for marks, models, and bore sizes as they were employed by the services at the time. During the First World War the U.S. Army changed most -- though not all -- of its gun bore sizes from English inches to metric measurement. I have not attempted to express designations in both systems throughout the book. What the service used at the time of that particular weapon is what appears here. I have similarly adhered to the U.S. system for range (yards) and weight (pounds).

For some earlier guns, particularly those acquired from British firms and early naval types, the designation "Pounder" for bore size was sometimes retained. In those cases, the physical measurement is provided either in the text or in the table of gun characteristics.

Seacoast Artillery

The U.S. Army had a long history in developing and deploying an entirely separate series of guns for seacoast employment. It was based on considerably different criteria from field artillery—guns emplaced on permanent carriages in fixed fortifications did not have restrictions on weight that mobile guns require. For much of the 19th and early 20th centuries the Army's Ordnance Department spent as much effort developing seacoast guns as for field army usage; and certainly, expenditures for such exceeded that for the mobile forces during the 35 years between 1882 and 1917.

Most of the Army guns and carriages listed in this work were the direct result of the work of the Board of Ordnance and Fortification. This board was created by the Congressional Fortification Act of 1888, and lasted almost fifty years until disbanded in 1930. It was an interdepartmental board with members from the Army's Ordnance, Engineer, and Artillery (and later Coast Artillery) corps and at times even included a civilian member. The board was charged with identifying and specifying artillery needs of the Army, and then in developing and evaluating candidates to fill these specifications. While not dealing with infantry equipment or small arms, the three major types of artillery discussed in this book, along with field artillery were all the responsibility of this board. It had its own budget for purchasing both domestic and foreign guns to evaluate and for running trials. Within the Army, gun designing was done by the Ordnance Department, often by technical specialists assigned at the various Army arsenals. There were also instances when a private design was adopted and ordered into production. Production was overseen by the Ordnance Department, and accomplished either under contract with private domestic or foreign firms, or by shop orders with government-owned arsenals. The Navy had a separate, though similar in many respects, system of development and acquisition.

American coast defenses can be closely associated with "systems" or generations roughly twenty years apart starting with the early start of the nation. Each of these generations, usually established by a major government initiative, featured common elements of fortification and a specifically developed system of armament. In the breechloading era this was true for the first modern system (the Endicott System of 1886-1905), the follow-on generation prompted by the acquisition of overseas territories (the Taft System of 1905-1914), the First World War era (the Board of Review System of 1915-1930) and the final Second World War system (the 1940 Program of 1940-1949).

Each of these four modern systems used a combination of heavy ordnance (8, 10, 12, 14, or 16-inch guns). The first two systems primarily used special disappearing carriages, where the gun retracted behind a protective parapet after firing. For high sites barbette carriages were used, where the gun projected permanently above the level of a protective frontal parapet. In the last two generations, only barbette carriages were used, accompanied with either protective casemates of concrete or being widely dispersed and carefully camouflaged for protection. Intermediate guns (5 or 6-inch in bore) were used in each system, also on barbette or disappearing carriages, and even on a hybrid type known as the balanced pillar. Lighter, rapid-fire guns to cover minefields and landing beaches were of 3-inch, 4.7-inch, and finally of 90mm size. These also were on either barbette (usually called parapet for this size) or another type of balanced pillar mount. Finally, an essential element of American coast defense was the high-angle, plunging fire weapon. The early systems used 12-inch mortars (some of which persisted to the very end of coast artillery), but the use of 16-inch howitzers and even 16-inch guns firing at high angle provided the same effect in the later years.

Also, during each major conflict, a variety of guns (of both naval and army origin) were mounted as temporary defenses in minor harbors and to supplement guns in major locations. Use of these weapons was made extensively in the U.S. during both the Spanish American and Second World Wars. The Army, Navy, and Marines erected batteries in insular and foreign locations during both world wars. And finally, a unique set of Navy turrets removed from warships were used to augment the defenses in Oahu during the war.

Coast Defense 3-inch Guns

The original Endicott Board for American seacoast defense did not perceive the need for lighter guns. However, in the mid-1890s the ordnance and engineer departments began to discuss characteristics for smaller weapons. The primary driver seems to have been the increasing familiarity and practice of submarine mining for defense. Once placed, minefields needed to be protected against an enemy's sweeping efforts. Additionally, the emergence of powerful (and with enough range to endanger the American coast) torpedo boats and destroyers, capable of "sneaking" into a harbor and destroying capital ships with torpedoes needed countermeasures. In 1896 the technical services agreed that a gun firing about a 12-14 pound projectile in a rapid-fire mode was of ideal size. This gun, which eventually emerged having a 3-inch bore size, would augment intermediate guns of 5-6 inch bore that were simultaneously being developed. In January 1897 a formal development project for a coast defense 3-inch gun was undertaken.

The first two major types of 3-inch guns were actually developed by private firms, though both had been actively supplying guns and mounts for the U.S. Navy. The balanced pillar mount was also called a masking parapet mount, and both the 3-inch and 5-inch types of this carriage at times were referred to with either term. As a result of the Spanish American War emergency, the ordnance department ordered 120 3-inch guns on a patented masking parapet mount from the Driggs-Seabury Gun and Ammunition Company of Derby, Connecticut. Delivered in 1899-1900, they were quickly emplaced in concrete batteries throughout the continental U.S. While the gun was quite acceptable, the mounting proved inadequate; it never worked very well and was not easily fixed. They served only up to the end of the First World War.

Sixty of a second pedestal type were ordered from Bethlehem Steel. The Model 1902 gun was not dissimilar from the Driggs-Seabury model, but the carriage was a much simpler and straightforward pedestal mount. It proved much more successful and some served through the 1940s.

For special situations an embrasure type mount was also developed. Using a standard Model 1898 gun tube, the gun was to be coupled with a specific carriage to allow it to be fired from an enclosed embrasure. Work done by Driggs-Seabury and Watertown Arsenal eventually led to an acceptable carriage, however only two were ever produced and emplaced.

For the third type, the Ordnance Department depended on its own internal development. The Model 1903 3-inch gun and pedestal was the most successful of the 3-inch types and served up until the end of the coast artillery service. Longer at 55-calibers, the gun was of nickel steel and high performance. The carriage was a logical improvement on the successful Bethlehem type pedestal from the M1902. Between 1904 and 1908, 108 tubes and 107 carriages were produced at the government arsenals of Watervliet and Watertown. They completed the domestic Endicott emplacements and armed all the insular possessions for the Taft generation. Many were re-located in 1940-42 as they performed the anti-torpedo boat role pending the availability of the new 90mm gun and mount. Almost all the guns survived until the late 1940s, when they were declared obsolete and scrapped.

Relatively little 3-inch seacoast gun materiel still remains. The 3-inch M1898 guns were offered as memorial pieces in 1929-30 (though by then the carriages had been scrapped), and quite a few barrels were shipped to local communities. Over fifteen tubes still exist. The well-used M1902s and M1903s weren't so fortunate. Only by oversight does a pair of M1902 survive, and just three M1903s, abandoned in the Philippines during the war and repatriated for park display purposes remain. There are also a pair of "practice" Model 1911 guns once used in coast artillery militia armories on display.

3-inch M1903 gun and carriage at Fort Casey, WA

3-inch (15-pounder) Gun Model 1898 On Balanced Pillar Mount Model 1898

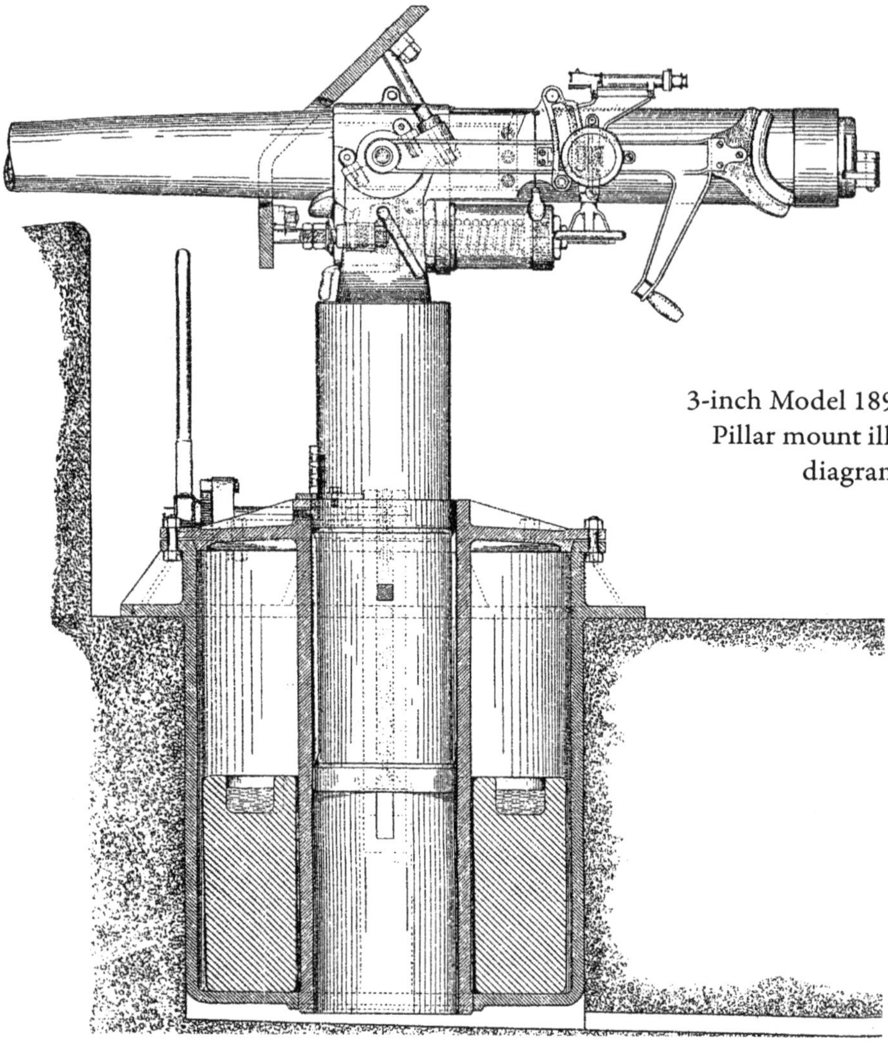

3-inch Model 1898 Balanced Pillar mount illustration diagram.

With the rush to secure light rapid-fire guns at the start of the Spanish American War, the Ordnance Department accepted an offer by the private Driggs Seabury Gun and Ammunition Company for a large quantity of its 15-pounder (3-inch) gun and a unique, patented masking parapet carriage. Eighty guns were contracted in mid-1898, followed by another 40 the next year. Driggs had difficulty meeting its deliver dates, but all were delivered by the end of 1900.

The tube closely resembled the 50-caliber type Driggs was delivering to the Navy. It was made of steel, built-up with tube, jacket and hoop. The breech-block was of interrupted-screw type, with six sectors. It was hinged on the right side for a block carrier and opened by a single handle. Rifling was of 24 grooves, increasing from 1 in 50 to 1 in 25 until 9-inches from the muzzle. The gun had a small protective shield with one bend.

Innovation was in the carriage. The gun sat in a recoil sleeve, with the hydropneumatic recoil cylinder slung underneath. The sleeve was held in a yoke sitting atop an inner cylindrical pedestal. In turn this cylinder could be raised and lowered within an outer cylinder sunk into the platform's concrete foundation. The raising action was by chain mechanism and featured a counterweight. It was not a true disappearing mechanism, simply a raising mechanism that could elevate the pedestal above the parapet for action but be withdrawn when not in use—not raised or lowered after every shot. The gun could achieve an elevation from -5 to +12-degrees.

As completed the guns were quickly mounted. The engineers designed a simple emplacement, usually for two guns (but the actual range was 1-4). Platforms had excavations for the retracting cylinder and a niche on the side wall to accommodate the barrel when lowered. Usual spacing was 29-feet between gun centers, but some had as little as 21-feet. A traverse magazine between the guns held the stored fixed ammunition, which was carried to the guns by hand. Forty-nine batteries were built between 1898-1903 to hold 111 guns (two batteries were abandoned early and never armed). All continental coasts received them, usually one or two batteries in each major fort, often on the flanks for minefield coverage.

It seems the carriage never really worked well. The mechanism did not support the gun well enough; with frequent breaking of parts and irregular effects on aiming. In 1904 changes were made to the wheels and gearing for traverse. In 1914-17 the breechblock was changed to eight sectors, a range drum and telescope added. Finally during 1915-1919 the carriages were modified with the pedestal permanently fixed in the upright position, the elevation mechanism removed, and the lower well filled-in with concrete. At least 102 of the 111 emplaced M1898 carriages were so modified and became designated the M1898M1.

3-inch (15-pounder) Gun Model 1898 On 15-pounder Barbette Carriage Model 1898

Gun Characteristics
Bore: 3-inch
Construction: Gun Steel, built-up
Tube length overall: 154.5-inches
Nominal caliber: 50 caliber
Tube and breech weight: 1,782-pounds
Usual carriage: Balanced Pillar Mount Model 1898
Weight of gun and carriage: 14,122-pounds
Maximum elevation: 12-degrees
Loading: Fixed
Shell weight: 15-pounds
Maximum muzzle velocity: 2,600-fps
Maximum range: 7,849-yards

Builders: Guns and carriages by Driggs-Seabury Gun and Ammunition Co.

Number built: 120

A 3-inch (15-pounder) on a Model 1898 carriage in a typical Endicott System emplacement.

Never highly thought of, they were not candidates for relocation during the First World War. The tubes were ordered removed in 1919 and put into storage pending possible future re-use, the carriages all scrapped or buried in place by order of April 10, 1920. The tubes stayed in storage for another ten years.

Thought was given that these excess tubes might be useful as anti-aircraft guns. A project explored the concept, but the lack of an automatic breech mechanism and the rifling twist eventually persuaded otherwise. Two guns were converted into experimental AA guns Model 1920E1. The remaining tubes were finally declared obsolete and made available for scrapping or donation in 1929-1931. Consequently at least 15 tubes still exist as monument guns. Four balanced pillar carriages were recovered from emplacement fill at Fort Taylor in Florida and are in the process of restoration. Hopefully they will be married to appropriate tubes and put on public display someday.

(Below) Battery Irwin with four 3-inch guns on M1898MI carriages at Fort Monroe for an officer training class in April 1918.

(Above) a 3-inch gun on the modified M1898MI carriage

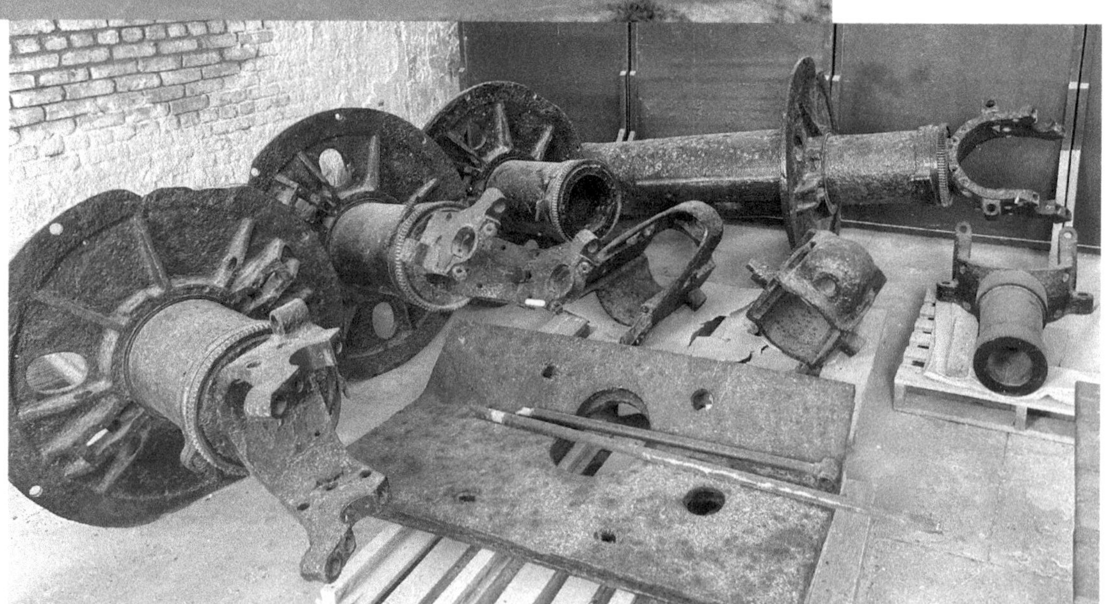

While there are many surviving gun tubes, very few carriages survive of the 3-inch balanced pillar mount. These parts were found in the filled-in moat of Fort Taylor in Key West, Florida.

3-inch (15-pounder) Gun Model 1902
On 3-inch 15-pounder Barbette Carriage Model 1902

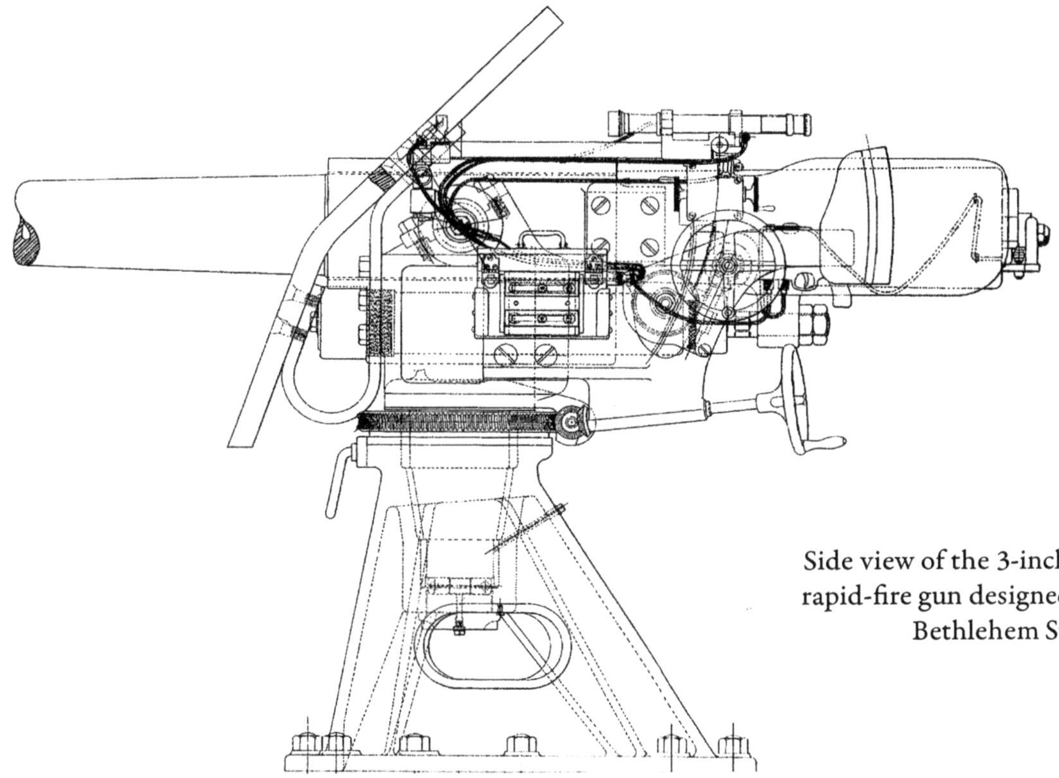

Side view of the 3-inch Model 1902 rapid-fire gun designed and built by Bethlehem Steel.

The second 3-inch gun type purchased for seacoast defense was a commercial model offered by Bethlehem Steel. The gun tube was quite similar to that sold by Driggs; it was a 50-caliber, built-up steel gun with tube, jacket and breech bushing. It had the same chamber size and used the same ammunition as the M1898 gun. The interrupted-screw breech was cylindrical but had the eight sectors from the beginning. The operating lever was mounted lower on the block; the block carrier was still on the right side. It

3-inch model 1902s under manufacture at Bethlehem Steel. Note distinctive shape of the mounting pedestal.

had the same rifling of 24 grooves as its predecessor. It had a 2-inch thick, bent shield made of nickel steel.

The pedestal carriage was much simpler than the previous masking parapet type. Meant to always stay in a deployed, firing position, its simplicity was its major asset. The gun was held in a cradle which had trunnions. The recoil cylinder was on the underside of the cradle. The cradle in turn was held in a yoke from which it could be elevated with hand wheel from -10 to +15-degrees. A telescopic sight was on the left side, with a corresponding cut-out in the shield. The yoke rode on ball bearings allowing a 360-degree traverse on a concentric cone pedestal. This cast steel pedestal was anchored to the platform with ten unevenly-spaced bolts. The slightly later M1903 pedestal had twelve evenly-spaced bolts—which meant that its carriages could fit on M1902 emplacements but not the other way around.

Sixty guns and carriages were ordered from Bethlehem in October 1902, with deliveries complete in 1904. Unlike many other guns, the guns and carriages always were installed with matched serial numbers, though there was nothing inherently that demanded this. In 1914 the guns had their breech blocks replaced by eccentric blocks, thereafter they were known as the M1902M1. They were marked in the usual Army way with gun information on the muzzle face and a separate, attached carriage plate.

There were 26 emplacements constructed for these guns, of from one to six guns each, though as usual just two was the most common (22 of the 26). These were started also in 1902-1904 and soon completed. They were simply built where next needed according to the priorities within the Endicott system, with no major concentration in any particular harbor, though all were in domestic U.S. locations. The design of course eliminated the deep wells and niches for the masking parapet type, and they were more spacious with gun centers of 42-feet. In later years many received attached or adjacent battery commander's stations and others had adjacent coincidence rangefinder positions.

As emplaced the guns were an important part of the defenses, often covering the flanks of forts, smaller channels, and minefields. They were kept through the First World War. By the Second World War only seven batteries had been disarmed—all in discontinued harbor defenses. A number were relocated in 1940-42 to temporary positions still within the defended harbor system; usually to just simple concrete platforms with earth-covered magazines. The Model 1902 3-inch gun and pedestal were finally declared obsolete in 1946 and authorized for disposal.

Almost all of these guns were scrapped early postwar. Two guns and pedestals of Battery Irwin at Fort Monroe (actually initially the armament of Battery Hindman, Fort Wool until they moved to Fort Monroe in 1945) were kept as display guns, where they still exist today. Interestingly these two had an enlarged box-shaped shield giving the breech overhead protection adopted during the Second World War.

3-inch (15-pounder) Gun Model 1902 On 3-inch 15-pounder Barbette Carriage Model 1902

Gun Characteristics
Bore: 3-inch
Construction: Gun Steel, built-up
Tube length overall: 154.35-inches
Nominal caliber: 50 caliber
Tube and breech weight: 1,925-pounds
Usual carriage: 3-inch 15-pounder Barbette Carriage Model 1902
Weight of gun and carriage: 6,468-pounds
Maximum elevation: 15-degrees
Loading: Fixed
Shell weight: 15-pounds
Maximum muzzle velocity: 2,600-fps
Maximum range: 8,771-yards

Builders: Guns and Carriages Bethlehem Steel Co.

Number built: 60

A Model 1902 3-inch gun in its emplacement in the defenses of Narragansett Bay.

The only surviving 3-inch Model 1902 guns and carriages are now at Battery Irwin (which was originally designed for M1898 Masking Parapet carriages) at Fort Monroe Virginia. These guns still have their "new" style of shield installed during the Second World War.

15-pounder (3-inch) Gun Model 1898
On 15-pounder Casemate Mount Model 1902

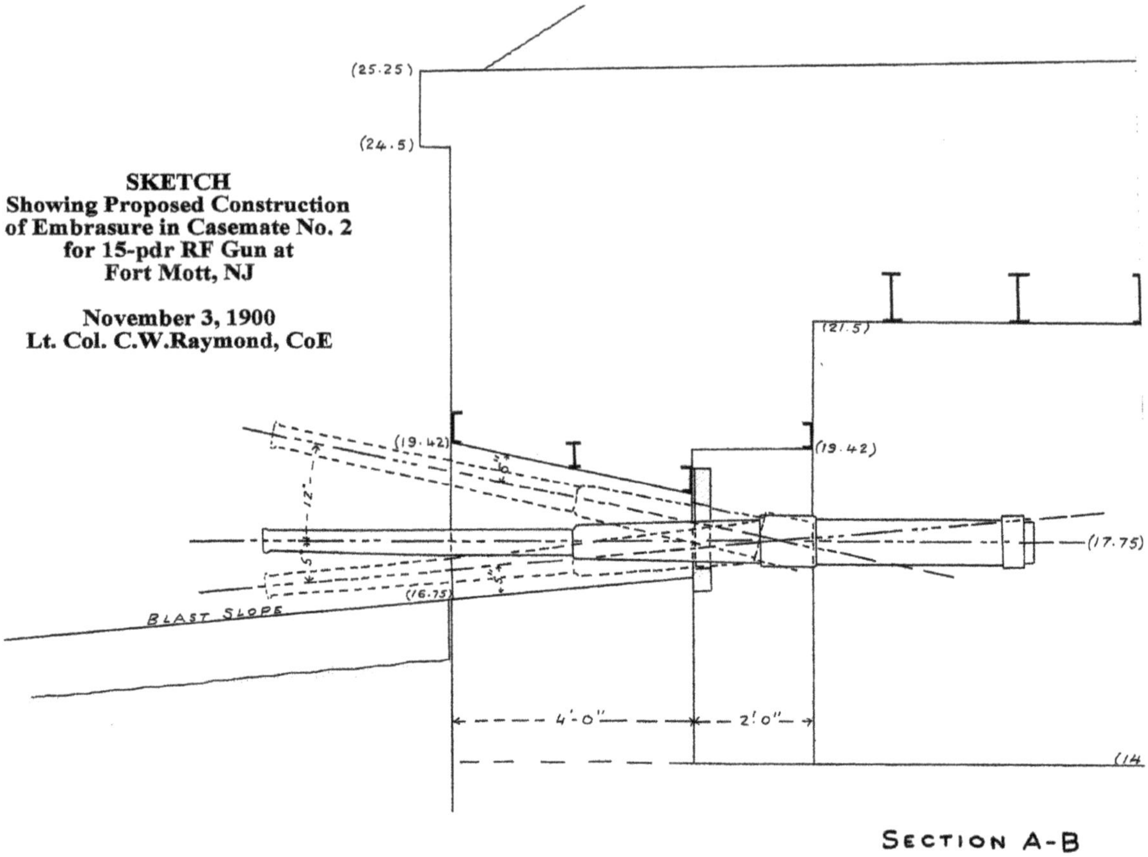

Ordnance Department sketch for the placement of the special 3-inch casemate mount ultimately emplaced at Battery Edwards, Fort Mott of the Delaware defenses. Unfortunately there are no known photographs of the gun or mount, either in an arsenal or at this battery.

Early intentions envisioned the use of 3-inch guns (and in fact, 5-inch guns too) emplaced in completely covered batteries, firing through a narrow embrasure opening. The Army Ordnance Department approached contractor Driggs-Seabury to produce a specialized carriage for such an emplacement carrying the relatively successful 15-pounder, 3-inch Model 1898 gun and breech.

Work was ordered on April 15, 1898 with funding originating with the National Defense Act preceding the Spanish American War. As mentioned above, the gun was the standard 50-caliber 15-pounder (3-inch) 1898 type already being made by this company. In fact, guns serial No. 62 and 63 were taken from production to be matched to new carriages. All of the gun characteristics described previously still applied to the expected performance. It fired the same ammunition, at the same muzzle velocity of 2,600-fps. The 15-degree maximum elevation could deliver a range of 8,771-yards.

For some reason the embrasure mount proved very difficult to fabricate. At least Driggs-Seabury certainly found that to be the case. Initially getting the forgings proved problematic—Driggs was a fabricator, not a forger, they did not have in-house capability to make their own steel forgings. In particular, the rounded shield proved difficult to make, and subcontractor Bethlehem Steel had to admit its inability to make an acceptable forging. Two carriages were eventually made and delivered in 1903, but they were rejected by the Department. Eventually Watertown Arsenal had to be engaged in making the carriages, apparently though able to use some of the pieces from the Driggs-Seabury effort. Finalized as the Model 1902 Embrasure Mount, they were delivered late that year. The carriages carried information plates designating them as carriages No. 1 and 2. No drawings, photographs, or textual descriptions of these carriages have come to light to allow a fuller description of their pertinent characteristics. The casemates taking the guns allowed for a 60-degree field of fire.

An emplacement was constructed for mounting of these two guns and carriages. Work was done at Fort Mott in the Delaware River defenses in 1899-1900. Two former brick magazines at the northern end of the 1870s gun battery were selected for conversion. This work had been part of a major system of batteries built after the Civil War to take 15-inch Rodman carriages on barbette carriages. On the flanks of each pair of guns was a buried magazine made of brick, with a flat floor and arched ceiling. Two such adjacent magazines had their front earthen cover removed and were opened with an embrasure facing seaward. Inside the casemate the carriage was to be held by foundation bolts and fired through the embrasure opening behind a steel shield. Named Battery Edwards, the emplacement was transferred to coast artillery troops on January 16, 1902.

This type of emplacement was subsequently never reproduced. No additional embrasure carriages were ordered and no additional such emplacements built (though conceptually similar 6-inch embrasures were part of Fort Drum some fifteen years later). Apparently never considered truly successful, the guns of Battery Edwards were declared obsolete and ordered removed in March 1920. The two tubes were shipped to Watervliet Arsenal in October 1920 and maintained with similar Model 1898 tubes as reserve material for several years, but the embrasure mounts were soon scrapped. The interesting emplacement itself still exists and can be accessed at Fort Mott State Park in New Jersey.

15-pounder (3-inch) Gun Model 1898 On 15-pounder Casemate Mount Model 1902

Gun Characteristics
Bore: 3-inch
Construction: Gun Steel, built-up
Tube length overall: 154.5-inches
Nominal caliber: 50 caliber
Tube and breech weight: 1,782-pounds
Usual carriage: 15-pounder casemate mount
Weight of gun and carriage: n.a.
Maximum elevation: 15-degrees
Loading: Fixed
Shell weight: 15-pounds
Maximum muzzle velocity: 2,600-fps
Maximum range: n.a.

Builders: Guns: Driggs-Seabury
Carriages: Driggs-Seabury/Watertown Arsenal

Number built: 2

The front exterior of Battery Edwards, Fort Mott

3-inch (15-pounder) Gun Model 1903
On 3-inch 15-pounder Barbette Carriage Model 1903

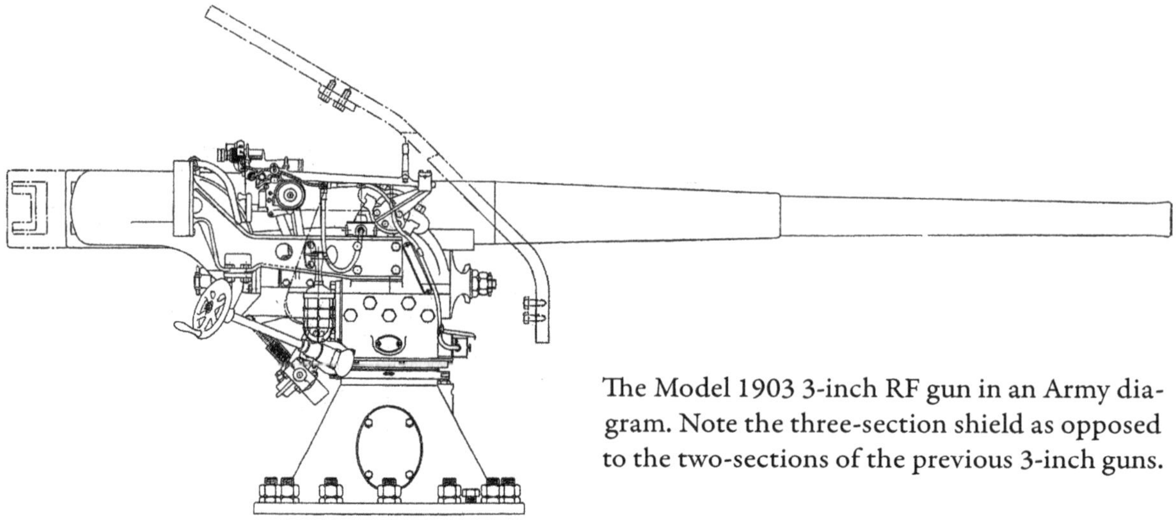

The Model 1903 3-inch RF gun in an Army diagram. Note the three-section shield as opposed to the two-sections of the previous 3-inch guns.

After purchasing commercial RF 3-inch guns, the Ordnance Department finally prepared and produced the final gun acquired in this category. The tube for the Model of 1903 was lengthened to 55 calibers. For the first time with seacoast artillery, a strong nickel steel alloy was used in the gun tube, jacket and breech bushing. A larger 296-cubic inch powder chamber was adopted, the gun used fixed ammunition with a greater powder charge than the previous types. The tube had 24 grooves and the familiar rifling pattern of 1/50 increasing to 1/25. The breech block was still of the interrupted-screw type, it was cylindrical in shape with eight sectors. Attempts were made to develop a semi-automatic breech mechanism, but they all failed and the gun remained a manually-operated type. It could be fired either by firing lever or lanyard.

The model 1903 pedestal was of simple design, in general design similar to the type used on the Model 1902 carriage. The pedestal was of cast steel in the form of a frustum of a cone. The base was drilled for twelve foundation bolts but could also use the ten bolts of a Model 1902 platform. The carriage used a double elevating screw for elevation vs. the previous rack and pinion style. The gun recoiled through guide rails, with a hydropneumatic recoil cylinder mounted below the gun. The upper carriage mechanism was

A 1903 3-inch gun and pedestal photographed in 1916 in an arsenal setting.

held in a yoke with two vertical cheeks and trunnion beds. The gun had two shoulder guards, versus just a single one on the previous types.

In total 108 guns were fabricated. Starting in 1904 and ending in 1916 seven separate orders were placed at Watervliet Arsenal for these guns. They were all marked in the usual fashion on the muzzle end and carried serial numbers No. 1 through No. 108. Most of the 107 Model 1903 carriages were made at Watertown Arsenal between 1904 and 1916, but runs were also made by the American and British Manufacturing Company (No. 22-44) and the Detrick and Harvey Machinery Company (No. 83-95). They were marked with an oval plate mounted on the side of the upper carriage.

A new type of emplacement was designed by the Corp of Engineers. Compared to other Endicott and Taft types, they were highly standardized. The emplacements had gun centers of 62-feet. Most were just for two guns with two magazines and a storeroom in a lower level traverse between the guns. Allowance was made for standing platforms on either side of the stairs leading to the gun platform for gun crews to manually pass ammunition. In the continental U.S. 34 battery emplacements were built in 1902-1917, and 13 more in overseas locations. Most were for two guns, but they ranged from 1-6 guns. In later years many were given battery commander's stations between the guns above the traverse, and frequently they also got a separate coincidence rangefinder building on one flank.

3-inch (15-pounder) Gun Model 1903 On 3-inch 15-pounder Barbette Carriage Model 1903

Gun Characteristics
Bore: 3-inch
Construction: Nickel Steel, built-up
Tube length overall: 170-inches
Nominal caliber: 55 caliber
Tube and breech weight: 2690-pounds
Usual carriage: 3-inch 15-pounder Barbette Carriage Model 1903
Weight of gun and carriage: 6,000-pounds
Maximum elevation: 16-degrees
Loading: Fixed
Shell weight: 15-pounds
Maximum muzzle velocity: 2,600-fps
Maximum range: 8,771-yards

Builders: Guns: Watervliet Arsenal
Carriages: Watertown Arsenal.
American and British Manufacturing,
Detrick and Harvey Machinery Co.

Number built: 108 guns, 107 carriages

The 1903s had a long service life. This gun of Battery James, Fort Mills, Manila Bay was in active service in the late 1930s.

These were effective guns and a valued element of the defense. The Model 1903 guns were kept in place during the First World War and interwar years, generally removed only when a harbor defense was totally abandoned. A few were re-located in the 1918 period in the Philippines to arm searchlight-protection batteries, and others were moved in 1941-42 within harbor defenses to new locations. In 1939 the carriages were authorized to be updated with new lighting, and 72 new box shields were produced for remaining guns in 1943. They were not declared obsolete until 1946, when removed and scrapped. None were designated for memorial or museum contribution. However, three guns did survive at Fort Wint in Subic Bay after being abandoned during the war and have been brought back at put on display in state parks in Washington state.

1903 3-inch at the bastion point of El Morro, San Juan, Puerto Rico in 1945. Note the WWII fire control station on the parapet directly above the gun.

The few surviving 3-inch M1903 guns and mounts were those remaining abandoned at Fort Wint, Subic Bay, Philippines. This gun was relocated in the 1960s for display purposes to the Washington state defenses of Puget Sound.

Coast Defense 4.7 and 5-inch Intermediate RF Guns

Guns of 6-inch size for coast defense were projected from the beginning of the Endicott era, but understandably initial ordnance efforts were concentrated on designing and preparing for production the guns and mortars of larger size. In 1896 the first project for a new 6-inch gun was begun, which would result in the Model 1897 gun on Model 1898 disappearing carriage. While a valuable size for engaging enemy cruisers and raiders, in American service this size of coast defense gun was not generally considered rapid-fire. It still used bagged powder, as the 100-pound projectile with cartridge case was too heavy for one-man loading. To get the desirable attributes of rapid or quick-firing, the Army need to drop to a gun size of about 5-inch. What was needed was a relatively small gun that could quickly reload, and be quickly traversed against relatively close, rapidly moving enemy ships. A new category for guns of 4.7-5 inch size thus emerged in the late 1890s.

In 1895 the Army's ordnance department began developing its own 5-inch RF gun. Despite one trial gun bursting and debates over what type of breech mechanism to use, the gun was accepted by the end of 1896. A rather complicated balanced-pillar carriage (allowing the gun to be withdrawn into a less-noticeable position between engagements) was completed and ordered the following year.

Next considered were wire-wound guns offered by the Brown Segmental Wire-wound Gun Company. Promising guns made under the patents of Mr. J.H. Brown, the guns offered considerable advantages in longitudinal strength without the high costs of forged, built-up construction. The firm had a long-term project with the Ordnance Department to build a 10-inch wire-wound gun, but then also offered to make 25 each 5-inch and 6-inch RF guns. In May 1898 a contract was signed for 50 guns. But they were never made. The firm was not a manufacturer, it used subcontracting firms, and great difficulty was encountered with the winding machines. The contract was eventually annulled on January 17, 1901; but not before the engineers had already completed at least 21 emplacements for the 5-inch gun type. It took four years, but the Ordnance Department designed and proofed a new 5-inch, built-up RF gun and simple pedestal carriage (the Model 1900 gun on Model 1903 pedestal) to fit the waiting emplacements.

The other gun in this category is the Armstrong 4.7-inch RF gun. Inquiries made by the War Department in March of 1898 at the beginning of the Spanish American War to quickly secure guns of foreign manufacture for coast defensive yielded two orders for a total of 34 British 4.7-inch guns (of three different calibers—apparently what was physically available in the firm's warehouses and works) and eight 6-inch guns. Even though they can also be categorized as temporary batteries, both of these types of guns were subsequently fully integrated into the harbor defense system. The 4.7-inch type is described in this section, and the 6-inch in its corresponding 6-inch chapter.

All of these guns served through the First World War, though some 5-inch were removed for use on wheeled field artillery carriages, some guns of 5 and 4.7-inch size were removed for temporary use on transport ships, and others were quickly re-located in 1917-1918. All were declared obsolete in 1919-1922. In the final generation of coast artillery, the need for lighter RF guns was adequately covered with the new 6-inch barbette and 90mm mounts.

Finally, a static land-defense 4.7-inch howitzer was developed in 1913 to specifically arm the new forts in Panama. On account of its bore size, it is also covered in this section.

A 4.7 inch battery at Fort Pickens

4.7-inch Armstrong Gun On 4.7-inch Pedestal Mount

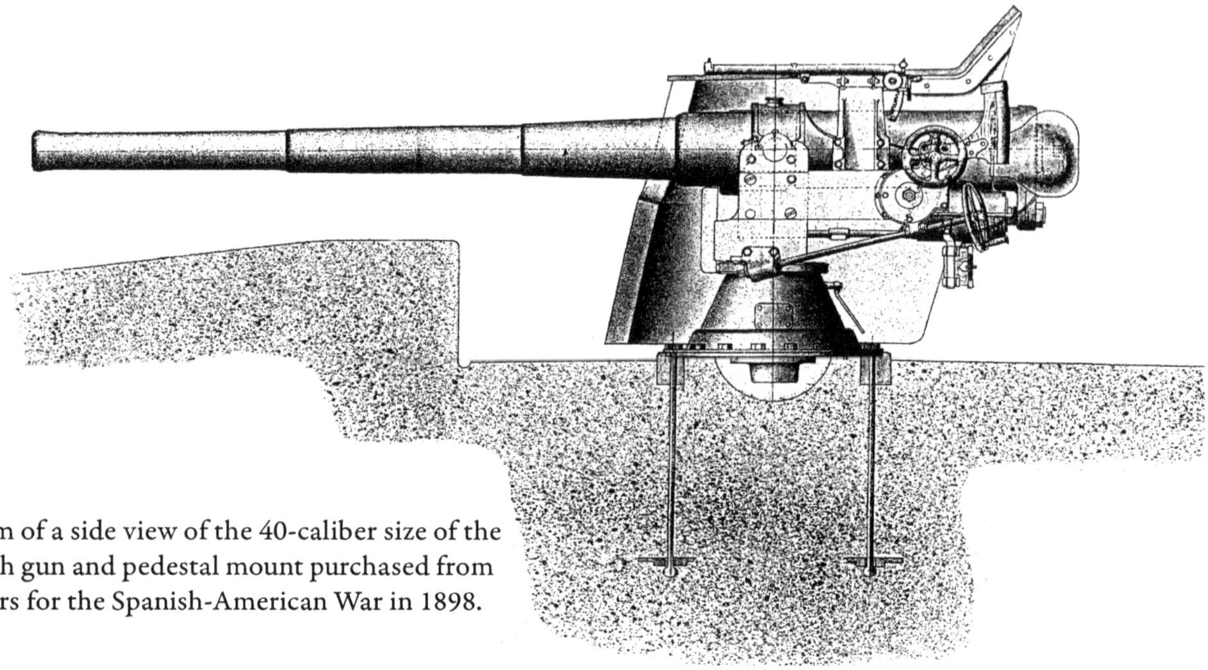

Diagram of a side view of the 40-caliber size of the 4.7-inch gun and pedestal mount purchased from Vickers for the Spanish-American War in 1898.

In early March of 1898, the Army's Ordnance Department placed inquiries with agents of foreign armament manufacturers for light, rapid-fire guns. It perceived a critical shortage of seacoast defense guns offering protection against fast-moving light enemy ships, such as the powerful ocean-going destroyers of the Spanish Navy. While the 5-inch balanced pillar mount was just going into installations, no rapid or quick firing American guns were even on the drawing boards. Positive responses came from two sourc-

The 4.7-inch guns, usually in pairs, were quickly dispersed to many forts in different defenses to improve rapid-fire capabilities. This shows one in its temporary location at Fort Delaware on the Delaware River in 1898.

es. The American Ordnance Company had located and could offer 12 (later increased to 14) 4.72-inch guns of Armstrong pattern apparently available from that firm and available as soon as May 1898. Vickers Sons & Maxim had 20 4.72-inch of the same Armstrong-type pattern also available.

These 34 guns were quickly ordered (along with 300 complete rounds each), paid for with funds from the war emergency National Defense Act of March 9, 1898. As early as April 11th the first 12 such guns were delivered to the U.S. aboard new navy cruiser USS New Orleans also purchased in England by the Navy. All the guns were received by the end of the summer of 1898. They came in a mixture of sizes. There were 17 of 40 caliber, 13 of 45 caliber, and 4 of 50 caliber length. The different lengths meant different weights, muzzle velocity, and range, but structurally they were designed the same. They did not have the same cased powder charges but otherwise fired the same 45-pound shell.

The gun was a standard British and export model of the naval 4.7-inch, deck pedestal-mounted RF gun firing semi-fixed, cased powder ammunition. It was steel, built-up tube without trunnions that fit into a cradle. The breech ring held a 2-step interrupted screw breech block with forward truncated conical

4.7-inch Armstrong Gun (40 caliber)
On 4.7-inch Pedestal Mount

Gun Characteristics
Bore size: 4.7-inches
Construction: Steel, built-up
Tube length overall: 198.1-inches
Nominal caliber: 40 caliber
Tube and breech weight: 4,648-pounds
Usual carriage: Armstrong 4.7-inch Pedestal Mount
Weight of gun and carriage: 19,124-pounds
Maximum elevation: 20-degrees
Loading: Semi-fixed cased
Shell weight: 45-pounds
Maximum muzzle velocity: 2,150-fps
Maximum range: 9,669-yards

Builders: W. G. Armstrong & Co. (Elswick Ordnance Co.)

Number built: 34 acquired by U.S. Army for coast defense

Another one of the emplaced Armstrong 4.7-inch guns, this one at Battery Drum, Fort Strong, Boston.

shape and six sections. A single-motion operating lever hinged on the right could open or shut the breech. The simple pedestal was composed of a base anchor supporting a Y-shaped yoke with trunnion bearings. The bronze cradle was attached to a single hydraulic recoil cylinder and two spring counter recoil cylinders mounted underneath the tube. A complete outer curved shield was 4.5-inch thick in front, with 2-inches on the side and 1-inch on the hood. The gun fired 45-pound shells of common, steel, or shrapnel type.

As soon as purchase was confirmed the engineers began preparing emplacements. Using an allocation of $6000 per emplacement, these ranged from separate concrete batteries with gun platforms and adjacent magazines to simple buried gun blocks on existing old fort parapets. After construction in 1898-99, several emplacements had to be rebuilt during the next few years. Seventeen batteries of from 1-4 guns were emplaced, all at East or Gulf coast forts. In subsequent years they saw considerable relocation. Eight guns and pedestals were removed in 1913 and sent to Oahu, where six were emplaced in new, fully casemated batteries. Then in 1917 several were taken out for use as armament on troop transports and others were re-located to temporary emplacements, including three to San Juan, Puerto Rico.

The Armstrong 4.7-inch gun was declared obsolete in June 1919 and slated for elimination. The weapon (both gun and carriage with shield) was considered suitable for donation as a memorial gun. At least 29 of the 34 guns were donated or allocated for use by the Army as post ornaments (such as all eight still on Oahu) in the years 1921-1926. Many of these were in turn donated for the scrap drives of the Second World War. Still, at least eight of the 4.7-inch guns survive. One is mounted on public display in a suitable emplacement at Battery Bingham, Fort Moultrie. Charleston, South Carolina.

In the 1920s the 4.7-inch gun was declared obsolete and offered as civic memorial guns. More than a few survived the scrap drives on the Second World War and can be seen today. This gun is on public display in Dravosburg, Pennsylvania.

4.7-inch Howitzer Model 1913 On Pedestal Mount Model 1915

The small 4.7-inch howitzer Model 1913 was specially designed to provide fire support for the Panama Canal and be emplaced on dispersed fixed concrete emplacements. This is an arsenal photograph of the gun and Model 1915 pedestal mount.

This was a new howitzer and mount specifically designed to serve as land defense artillery for seacoast forts. A concern of defense planners right before the First World War was providing adequate protection against hostile raiding parties that might neutralize a defense, particularly in the insular possessions of the Philippines, Hawaii, and Panama where a large defensive land army would not be available. One of the solutions for Panama was the development and emplacement of a new defensive howitzer on a fixed, all-around firing pedestal to cover key beaches and landing locations.

The howitzer was a relatively simple, 22.5-caliber wired-wound tube. This tube had 42 grooves, and rifling was from 1 in 40 increasing to 1 in 20. It had the usual Army interrupted screw breech mechanism with mechanical firing device. Recoil and counter recoil were controlled by hydrosprings located together in a single, large cylinder mounted atop the tube. A large, prominent lug projected from the breech of the tube to attach to the recoil mechanism. The tube sat in its cradle supported by a yoke fitting into the lower carriage, or pedestal. Elevation was given by a simple screw and nut attached to the bottom of the cradle, allowing -10 to +40-degrees of elevation. The top cradle carrying the pivot yoke rotated to a full 360-degrees using bevel gear and operated with a hand wheel.

The simple lower pedestal (the Model 1915) was fixed to foundation bolts pre-emplaced in concrete. The pedestal held the yoke with the cradled gun. The pedestal was provided with its own azimuth circle.

Photograph of the 4.7-inch M1913 howitzer on one of its simple emplacements in Panama from the early 1920s.

There were ten foundation bolts, in a 37-inch diameter circle. This was identical to the foundation arrangements for the seacoast 3-inch Model 1903 gun, the emplacement being capable of carrying either type of weapon. However, as there were no such guns of this type ever assigned to the Panama defenses, the utility was never realized. As mounted the howitzer had a large steel shield meant to protect the firing crew. On the other hand, the emplacement itself was entirely open with no concrete or earth protection whatsoever.

Watervliet Arsenal manufactured 21 howitzers in 1914. The first pilot was delivered on December 20th, all the rest were completed by June 20, 1915. The 21 carriages followed pace. In March of 1916 17 completed units were shipped to Panama. Twenty-six individual gun blocks, on existing fort reservations both on the Atlantic and Pacific coast, were constructed in 1917. The completed emplacements were transferred to coast artillery troops in April of 1918. Four armored railroad cars were also constructed to take this same howitzer also in Panama (see separate listing for them in this volume's railroad artillery section).

These guns didn't last long. They were used for a few years in Panama, though probably left in warehouses for protective storage rather than being emplaced in the open outside. Along with many other pre-war guns, they were declared obsolete on June 15, 1919—just three years since their arrival. The gun tubes were scrapped in accordance with orders in 1922; however, the carriages and emplacements found a new life. The gun's role was taken by adapting at first a number of Model 1897 75mm field guns to the pedestal (in 1926, becoming known thus as the Mount Pedestal M1). Then in 1934 these guns in turn were replaced by Model 1916 75mm field guns (the modified pedestal becoming the M2 type). It seems by the early 1940s these too were taken out of service and the pedestals scrapped. All that remains today in Panama is a few of the stark concrete pads which once were intended for this rather unique weapon.

4.7-inch Howitzer Model 1913 On Pedestal Mount Model 1915

Gun Characteristics
Bore size: 4.7-inches
Construction: Alloy steel, wire-wound
Tube length overall: 111.4-inches
Nominal caliber: 22.5-calibers
Tube and breech weight: 2,100-pounds
Usual carriage: Pedestal Mount Model 1915
Weight of gun and carriage: n.a.
Maximum elevation: 40-degrees
Loading: Fixed
Shell weight: 60-pounds
Maximum muzzle velocity: 1,300-fps
Maximum range: 10,000-yards

Builders: Gun: Watervliet Arsenal
Carriage: Watertown Arsenal

Number built: 21

The 4.7-inch M1913 howitzer didn't last long, being declared obsolete in the mid-20s. However, the need for guns on the emplacements continued. Initially replaced with 75mm Model 1897 field guns (the French 75), those were replaced in the 1930s with the 75mm M1916 field gun. This photo shows one still in service in July 1942 in Panama.

5-inch RF Gun Model 1897
On 5-inch Balanced Pillar Mount Model 1896

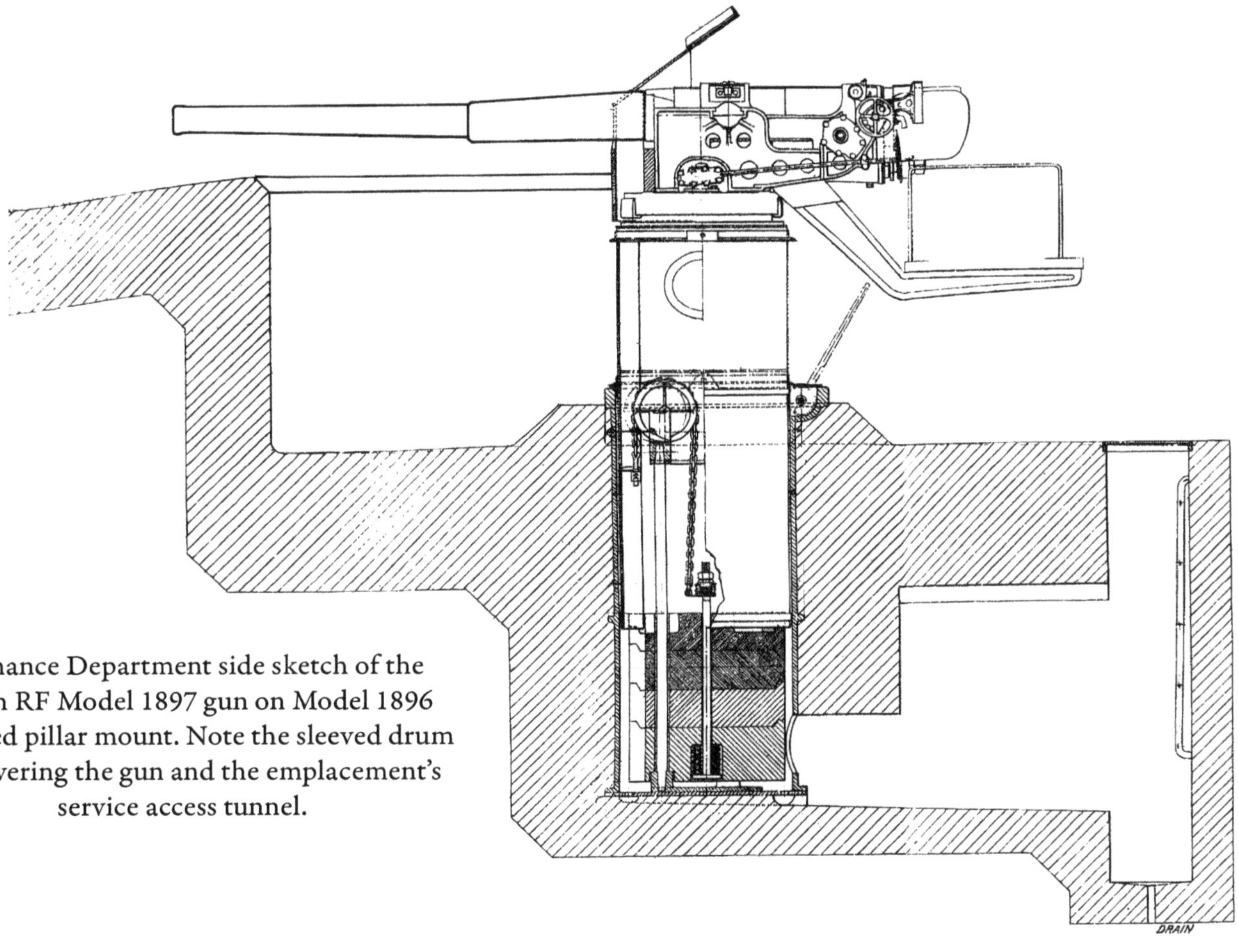

Ordnance Department side sketch of the 5-inch RF Model 1897 gun on Model 1896 balanced pillar mount. Note the sleeved drum for lowering the gun and the emplacement's service access tunnel.

The first coast defense 5-inch gun developed was a conventional steel built-up design. The Model 1897 gun was composed of a tube, surrounding jacket fixed by a locking hoop. The jacket projected beyond the posterior end of the tube and held the breech mechanism. That was a normal interrupted-screw conical block. For the block the hinge was on the right side, and the handle on the left. The powder chamber was conical in shape. There were 30 grooves and rifling increased from 1 in 50 to 1 in 25. The gun fired by a standard combination electric and friction primer. It used separate loading to fire projectiles of 58-pounds weight as common and shrapnel rounds.

What really distinguished this weapon was its carriage. An Armstrong 4.7-inch balanced pillar gun and mount had recently been evaluated by the Ordnance Department. Using that knowledge the department designed its own version. The upper carriage was of conventional design. The tube, which was without trunnions, was held in a slide. Below the tube were the hydraulic recoil cylinder and two cylinders carrying counter recoil springs. The cradle was held by its trunnions on a yoke sitting on a pillar of 49.5-inches in diameter. In turn this pillar moved within a 10-foot long cylinder extending into the concrete emplacement. Using chains and pulleys and a 22,400-pound counterweight the upper carriage with gun could be elevated 42.6-inches to a locked firing position. Two men could raise the gun using ratchets. When not in use it could be lowered out of sight, the emplacement had a niche on the side to receive the lowered barrel. While it did not depress between firings (like a true disappearing carriage), it could be lowered out of sight before and after an engagement. Needless to say, this was a complicated and expensive mount that ultimately proved to not offer enough advantages to offset its problems. While the Model 1898 masking parapet 3-inch gun had similar capabilities, the design was otherwise never duplicated. The gun was well-protected with a 3-inch nickel steel inner shield and a thinner outer and top shield.

Thirty-five guns were produced in 1898-1901 by Watervliet Arsenal and Bethlehem Steel. Just 32 carriages were contracted to Watertown, Bethlehem, and Detrick & Harvey; delivery was finished by the end of 1900. The Model 1897 tube was only used on this carriage. The 32 emplacements (in batteries of one to three guns each) were spread widely among forts and built from 1897-1900. They were somewhat more complicated than pedestal emplacements, needing deep pits, and a passage under the platform to allow access to the counterweight pit. Ammunition was stored in below-ground magazines in the traverse between the platforms.

The guns and their emplacements served through the pre-war period. Most (28) were removed in 1917 to use the tubes in wheeled field artillery mounts for France, another two were relocated to Gray's Harbor, Washington. On July 3, 1919 both guns and carriages were declared obsolete, the carriages were scrapped in-place in 1920, though the gun tubes were kept for a while in storage for potential future use. In fact, six of the tubes were converted at Watervliet Arsenal in the early 1920s to an experimental 3.3-inch gun. The remaining inventory of 5-inch Model 1897 tubes were authorized for scrapping or donation for memorials in 1930. Numerous tubes were distributed as memorial guns, and quite a few still exist in city parks and war monuments throughout the country. None of the Model 1896 balanced pillar carriages managed to survive.

5-inch RF Gun Model 1897
On 5-inch Barbette Carriage Model 1896
Balanced Pillar Mount

Gun Characteristics
Bore size: 5-inch
Construction: Steel, built-up
Tube length overall: 231.5-inches
Nominal caliber: 45 caliber
Tube and breech weight: 7,583-pounds
Usual carriage: 5-inch Barbette Carriage Model 1896
Weight of gun and carriage: 56,392-pounds
Maximum elevation: 15-degrees
Loading: Separate
Shell weight: 58-pounds
Maximum muzzle velocity: 2,600-fps
Maximum range: 10,431-yards

Builders: Guns: Watervliet Arsenal (10),
Bethlehem Steel (25)
Carriages: Watertown Arsenal (8),
Bethlehem Steel (12),
Detrick & Harvey (12)

Number built: 35 Guns, 32 Carriages

One of the 5-inch M1897 guns on balanced pillar mount in its emplacement. It is in a depressed, or lowered position.

The 5-inch M1897 displayed at a public exposition in 1904. This time the gun is in elevated position—note the difficulty in passing the relatively heavy ammunition by hand.

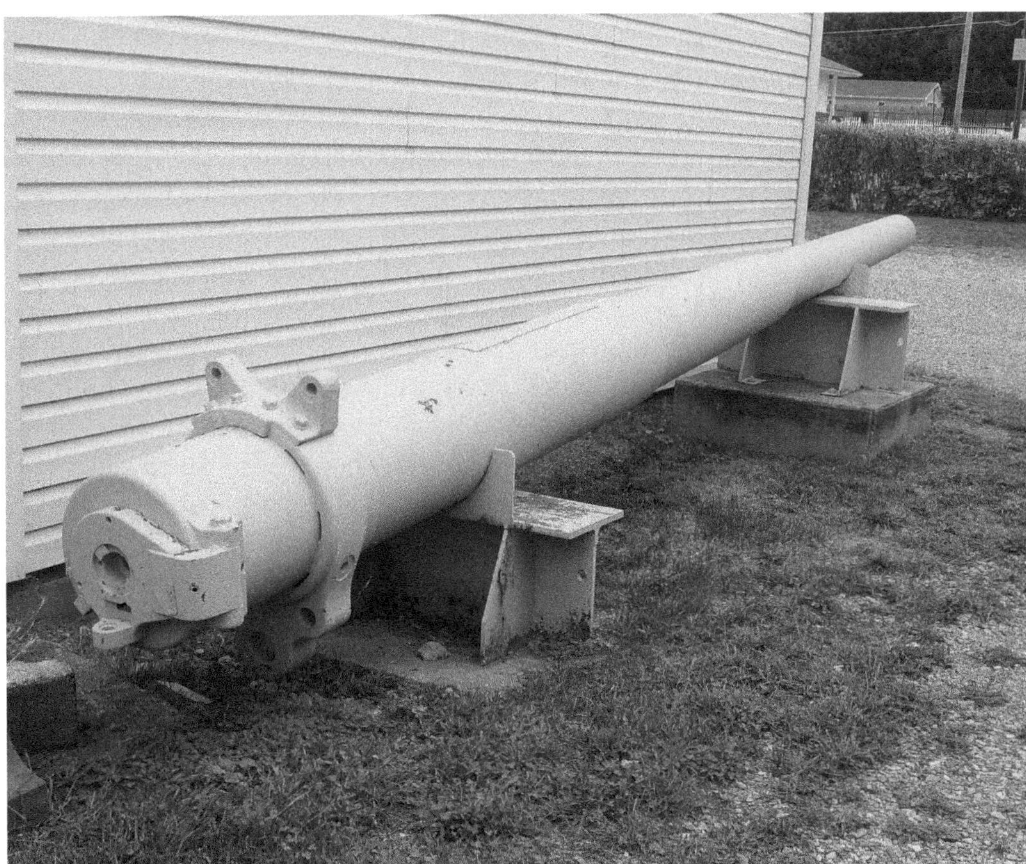

5-inch guns for coast defense were declared obsolete and dismounted in 1921. While no complete carriages survive, numerous gun tubes were donated for civic memorials and veteran's organizations. This tube is neglected outside a small town museum in Maryland.

5-inch RF Gun Model 1900 On 5-inch Barbette Carriage Model 1903

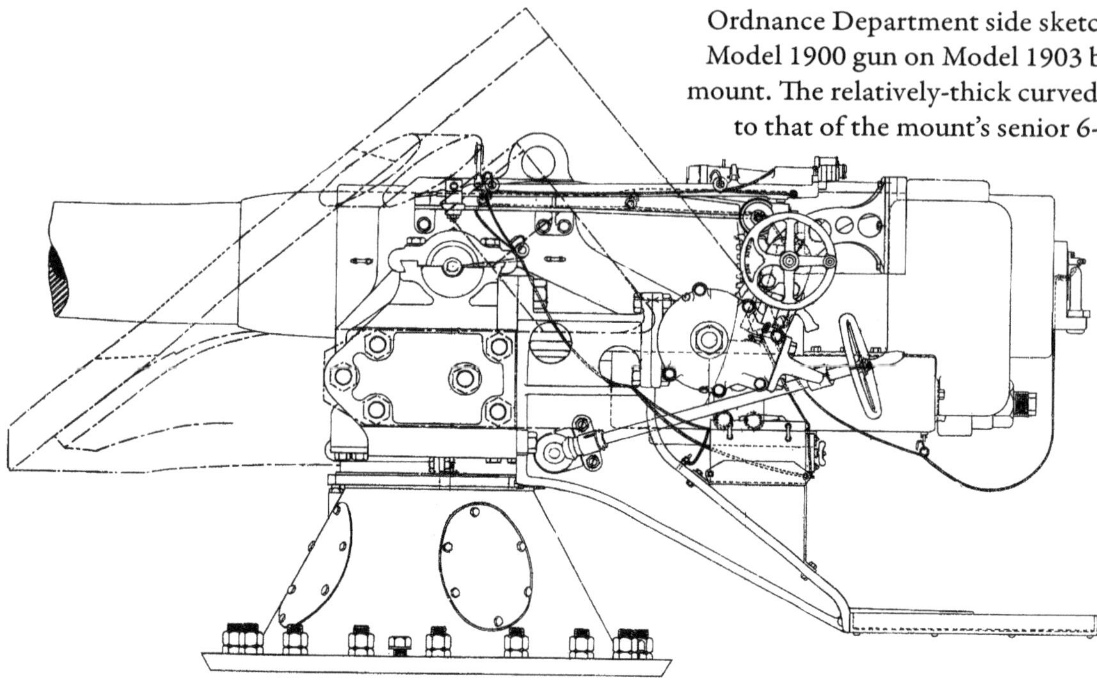

Ordnance Department side sketch of the 5-inch RF Model 1900 gun on Model 1903 barbette (pedestal) mount. The relatively-thick curved shield was identical to that of the mount's senior 6-inch sister type.

This gun and carriage were specifically produced to substitute for Brown wire-wound 5-inch guns. Twenty-five of the latter had been ordered in 1898. They were generally projected less expensive than built-up guns of the same size, and with a muzzle velocity of 2,600-fps and a 55-pound shell. However the company failed to deliver the guns, the Ordnance Department cancelled the contract in January 1901. Unfortunately, an appropriation had already been made for the intended emplacements. The Corps of Engineers reported that 21 emplacements were actually completed. The Army subsequently ordered conventional 5-inch built-up guns Model 1900 it had already developed. A new carriage Model 1903, basically a scaled-down 6-inch Model 1900 barbette carriage, was developed just for this application. Exactly 21 guns and mounts were ordered to fill the ready emplacements.

The gun was a conventional steel, built up tube of 50 calibers (an increase of 5 vs. the previous Model 1897). It had a breech bushing holding a Stockett-type interrupted screw breech mechanism. It was hinged on the right, and closed lever was on the left. The barrel had 30 grooves and used rifling increasing from 1 in 50 to 1 in 25. The cylindrical chamber accepted separate-loaded, bagged ammunition. It fired a 58-pound common, armored-piercing, or shrapnel round. The breech was similar to that of the M1897 except for having a bronze, automatic loading tray that raised when the block carrier swung out. The carriage was cone-shaped pedestal with a revolving pivot yoke. The yoke had two vertical cheeks with trunnion beds

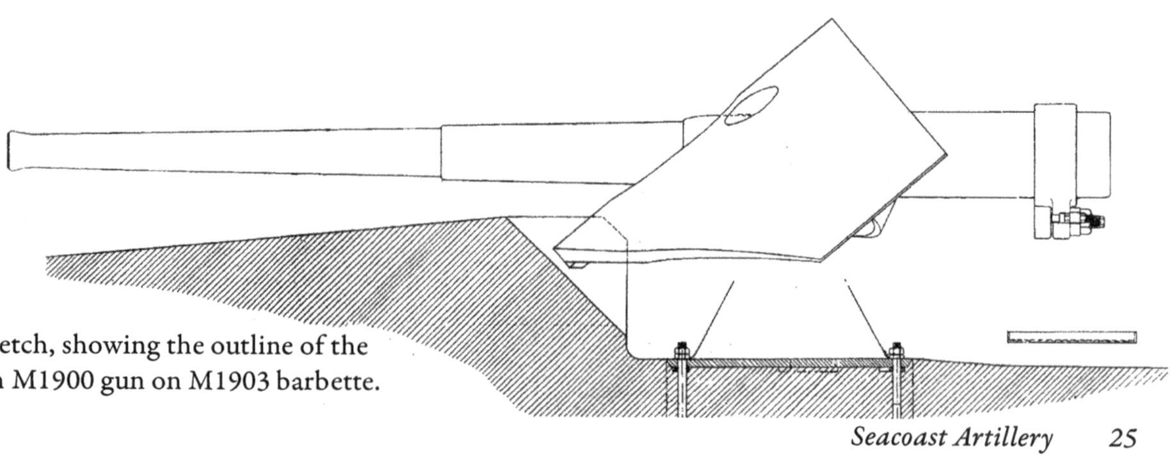

Another sketch, showing the outline of the entire 5-inch M1900 gun on M1903 barbette.

for the cradle holding the barrel. The barrel recoiled within the cradle and was braked by an underslung recoil cylinder and two counter recoil spring cylinders. Crew was protected with a 40-degree inclined shield 4.5-inches thick attached to the yoke—it was identical to that used on the 6-inch Model 1900 barbette gun. The carriage could attain elevation from -5 to +15-degrees.

The guns were all made by Watervliet Arsenal; they were numbered No. 1-21 and marked on the muzzle face in the usual manner. Carriage bids were accepted in early 1904. Watertown Arsenal built carriages Model 1903 No. 1-11, and the Morgan Engineering Company was awarded No. 12-21. They carried manufacturing information and serial numbers on a carriage plate on the yoke. The emplacements, built in 1898-1900, were relatively simple. Gun blocks with hold-down bolts were surrounded with a low parapet. On the side (or between multiple mounts) was a magazine space 13 by 22-foot in size. All ammunition was hand-carried, no hoists or cranes were required. The emplacements did not require the loading platform space as the similar 6-inch barbette gun type. There were 11 5-inch pedestal batteries built, ten dual and one single emplacement—though one battery never received its armament.

5-inch RF Gun Model 1900 On 5-inch Gun Barbette Carriage Model 1903

Gun Characteristics
Bore size: 5-inch
Construction: Steel, built-up
Tube length overall: 258.5-inches
Nominal caliber: 50 caliber
Tube and breech weight: 11,120-pounds
Usual carriage: 5-inch Barbette Carriage Model 1903
Weight of gun and carriage: 21,573.5-pounds
Maximum elevation: 15-degrees
Loading: Separate
Shell weight: 58-pounds
Maximum muzzle velocity: 2,600-fps
Maximum range: 10,431-yards

Builders: Gun: Watervliet Arsenal (No. 1-21)
Carriage: Watertown Arsenal (No. 1-11),
Morgan Engineering (No. 12-21)

Number built: 21 guns, 21 carriages

Ordnance Department photograph of a 5-inch gun on a M1903 barbette mount.

The relatively simple gun and carriage proved successful, there were no major problems encountered during service. In 1910 two guns were relocated to Hawaii, and then in 1917 ten more were moved to new, temporary locations to protect passages or anchorages otherwise left defenseless. On July 3, 1919 both the 5-inch Model 1900 gun and Model 1903 carriage were declared obsolete and approved for disposal. They simply were too few in number to keep actively maintained and the carriages lacked elevation and range for modern combat conditions. All the carriages were subsequently scrapped; the guns remained in storage pending possible future use. In 1930 these too were authorized for scrapping or donation. While two tubes survived for several years on ornamental display in Hawaii, they too were eventually eliminated. Consequently, there are no surviving examples of the gun or carriage.

Photograph used by Morgan Engineering in a brochure showing the 5-inch gun on the barbette M1903 mount the firm manufactured.

The Rapid-fire 5-inch M1900 gun on M1903 barbette carriage in its emplacement of Battery Fraser, Fort Slocum, New York about 1910.

Coast Defense Intermediate 6-inch Guns

Of the Endicott and Taft Systems

The Endicott Board had called for only four 6-inch guns, obviously not seeing much utility for guns of this size. However, the following decade proved that 6-inch guns were useful in supplementing the fire from larger guns, countering secondary batteries of capital ships, and opposing enemy cruisers. The Ordnance Board in 1896 called for three hundred 6-inch RF guns. Consequently from 1897 to 1910 a large number of 6-inch guns on both disappearing and barbette carriages were produced and added to the seacoast defenses. The very first modern 6-inch gun of any type developed by the U.S. Army was the Model 1897 gun. It was of standard built-up construction with a conventional Stockett breech mechanism. This gun was used exclusively on the Model 1898 disappearing carriage. The latter pulled heavily from the designs of larger disappearing carriages and introduced no new features to that type of mount. Thirty guns and 29 carriages were fabricated; all 29 were emplaced in coast defense posts in the continental U.S.

During the Spanish American War eight British-type rapid-fire 6-inch guns were acquired by purchase and emplaced in selected harbors. They were never deemed truly successful and were eliminated and removed soon after the First World War.

Next came an effort to produce a new rapid-fire barbette gun and mount. Initially hope was placed in the wire-wound gun offered by the Brown Wire-wound Segmental Gun Company. A contract for 25 guns of this type was signed in 1898, but in 1901 this was annulled when the firm failed to begin manufacturing the gun. Subsequently the Ordnance Department developed its own, conventional built-up 6-in/50-caliber gun on barbette carriage. This became the Model 1900 gun and carriage, destined to serve into the post-World War II period.

The second disappearing 6-inch gun, and the most successful, was the Model 1903 6-inch gun on the Model 1903 disappearing carriage. The tube was a progressive development of the steel, built-up type with Stockett breech mechanism. The disappearing carriage offered several incremental improvements over the first M1898 type. This gun and carriage were built and emplaced in large numbers throughout the continental U.S. The Model 1905 gun and M1905 disappearing carriage were similar designs; in fact the two guns could (and were) used interchangeably on the two carriages. The M1905s were mostly used in Taft system fortifications in the insular possessions and final generation of continental sites.

Finally, the Model 1908 gun was developed, using and realizing the advantages promised in wire wrapped fabrication, although it had the dimensions and powder chamber size of the first M1897s. Some of these tubes were used in the final disappearing carriage batteries and a very few served on unique barbette and casemate mounts in the Canal Zone and Philippines.

Many of the 6-inch guns were gradually removed from their emplacements. In the First World War over 100 Model 1900, 1903, and 1905 guns were taken for wheeled field artillery roles. Some barbette mounts were relocated to new or temporary emplacements in 1917-1918. The Model 1900 barbettes were returned to emplacements after the war, but the guns from the disappearing mounts were eventually modified for use on new barbette carriages of the 1940 Program. Several additional disappearing 6-inch guns were eliminated by the closure of several harbor defenses in the 1920s. Still, there were numerous guns of this type still in service until deleted in 1942-44. However, the modified 6-inch guns used in the Second World War batteries, and more than a few Model 1900 barbettes were still giving valuable service at the end of American coast artillery in 1948-49. Most of these guns were never considered for memorial donation or preservation, the few that continue to survive did so mainly through oversight.

A 6-inch gun on a disappearing carriage

6-inch RF Gun Model 1897 and 1897M1 On Disappearing Carriage LF Model 1898

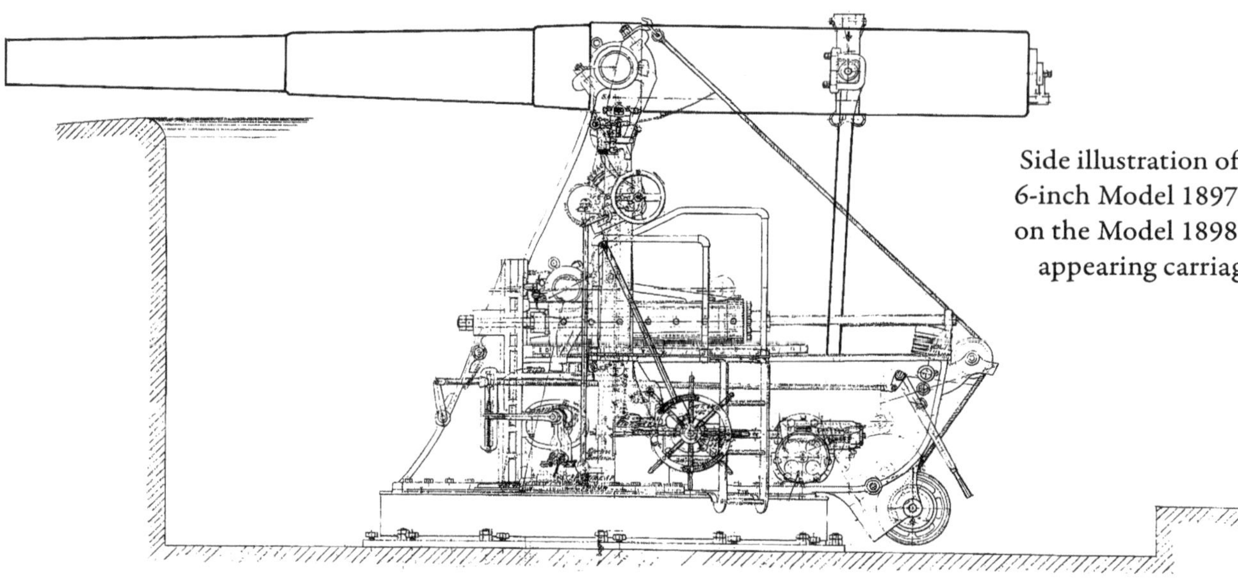

Side illustration of the 6-inch Model 1897 gun on the Model 1898 disappearing carriage.

This was the first modern 6-inch gun project undertaken by the U.S. Army. The 45-caliber tube was of built-up steel construction. It had a tube, jacket and seven hoops. A Stockett-type cylindrical interrupted screw breech was fitted, with eight sections. The breech was opened with a lever parallel to the breech face connected to the block carrier hinge on the right side. The tube was rifled with 36 grooves, and rifling increased from 1 in 50 to 1 in 25. Thirty gun tubes were produced. The first prototype was ordered in April 1898, and the next 29 in three batches from July 1898 to August 1900. Later guns had a slightly different configuration of the powder chamber, and all earlier ones were similarly modified so that guns No. 2-30 were changed to the designation M1897M1. All tubes were made at Watervliet Arsenal and delivered in 1899-1902. In normal fashion they were stamped with manufacturing and serial detail on the face of the muzzle.

The center-pintle, limited-fire disappearing carriage was typical in design of this type of mechanism. The primary parts were the base ring and racer with individual traversing rollers (20), a chassis, top carriage with recoil rollers and two hydraulic recoil cyl-

Morgan Engineering (Alliance, Ohio) photograph of one of their Model 1898 disappearing carriages used with a Model 1897 gun in retracted position.

inders, gun levers, crosshead, lead counterweight, elevating arms, working platform and sighting platform. Both cast iron and cast steel were used for carriage parts. The mechanical action of the recoil raised the counterweight; the recoiling gun was stopped by this effort and the resistance of the recoil cylinders. The gun always returned to the same level to ease loading from hand trucks. The tube moved back into firing position through the action of the rollers on the inclined chassis top. The gun was raised once again to firing position by releasing the pawls that dropped the counterweight. The latter weighed 19,100-pounds and was made of lead plates.

All 29 Model 1898 disappearing carriages were ordered during 1898, deliveries coming in 1899-1900. They were made by Watertown Arsenal (20), Philadelphia Engineering Works (6) and Morgan Engineering (3). The guns were assigned a mission of contending with similar-sized enemy guns, of repelling boat attacks, and sweeping mine fields. A board in October 1900 determined after firing tests that while the carriage was mechanically satisfactory, in the interest of economy and efficiency no other quantities of this type of carriage should be acquired.

Despite a rather short production run, the guns and carriages were all emplaced. The Corps of Engineers designed a suitable emplacement plan, featuring separate adjacent gun pits with magazines located in traverses at a lower level. There was one step down

6-inch RF Gun Model 1897 and 1897M1 On 6-inch Disappearing Carriage LF Model 1898

Gun Characteristics
Bore size: 6-inch
Construction: Steel, built-up
Tube length overall: 277.85-inches
Nominal caliber: 44.5 caliber
Tube and breech weight: 16,216-pounds
Usual carriage: 6-inch Disappearing Carriage LF Model 1898
Weight of gun and carriage: 65,406-pounds
Maximum elevation: 15-degrees
Loading: Separate
Shell weight: 106-pound
Maximum muzzle velocity: 2,600-fps
Maximum range: 11,799-yards

Builders: Guns Watervliet Arsenal
Carriages: Watertown Arsenal, Morgan Engineering, Philadelphia Engineering Works

Number built: 30 guns, 29 carriages

An emplaced 6-inch Model 1897MI gun on its Model 1898 limited-fire disappearing carriage at Battery Mills, Fort Wadsworth, NY.

from the loading platform to the working platform. Twenty-nine emplacements, using all the available disappearing carriages were built. Usually these were in dual batteries, though some single and triple batteries were built. The batteries were well-spread among harbors and forts; no fort received more than a single battery of this type gun and carriage. They were only emplaced in domestic, continental U.S. locations. Once placed, they were not relocated; the gun was not one of the 6-inch types chosen for field gun adaptation.

Beginning in the mid-1920s, with the discontinuance of several minor harbor defenses, the guns and carriages began to be removed. One was taken out in 1918, others in 1925-1927. More than a few served up until the Second World War. With the completion of the modern 1940 Program armament (including an entire new generation of 6-inch guns), the final survivors were eliminated in 1943-1945. Both the gun tubes and carriages were scrapped; no other usage being found for them. Also, none were designated as suitable for donation or use as ornaments or memorials. Consequently, while many emplacements survive, no original gun or carriage does.

A 6-inch M97/98 disappearing gun being raised into firing position.

No 6-inch Model 1897 gun tubes or Model 1898 disappearing carriages survive. However, a fairly accurate reproduction has been fabricated and is on public display at Battery Pratt, Fort Stevens in the Defenses of the Columbia River in Oregon.

6-inch Armstrong Gun
On Armstrong 6-inch Barbette Carriage

Manufacturer illustration of the 6-inch British gun. Curiously these eight guns were purchased through an agent from Vickers & Sons, but most of the surviving records indicate that they are actually Armstrong guns.

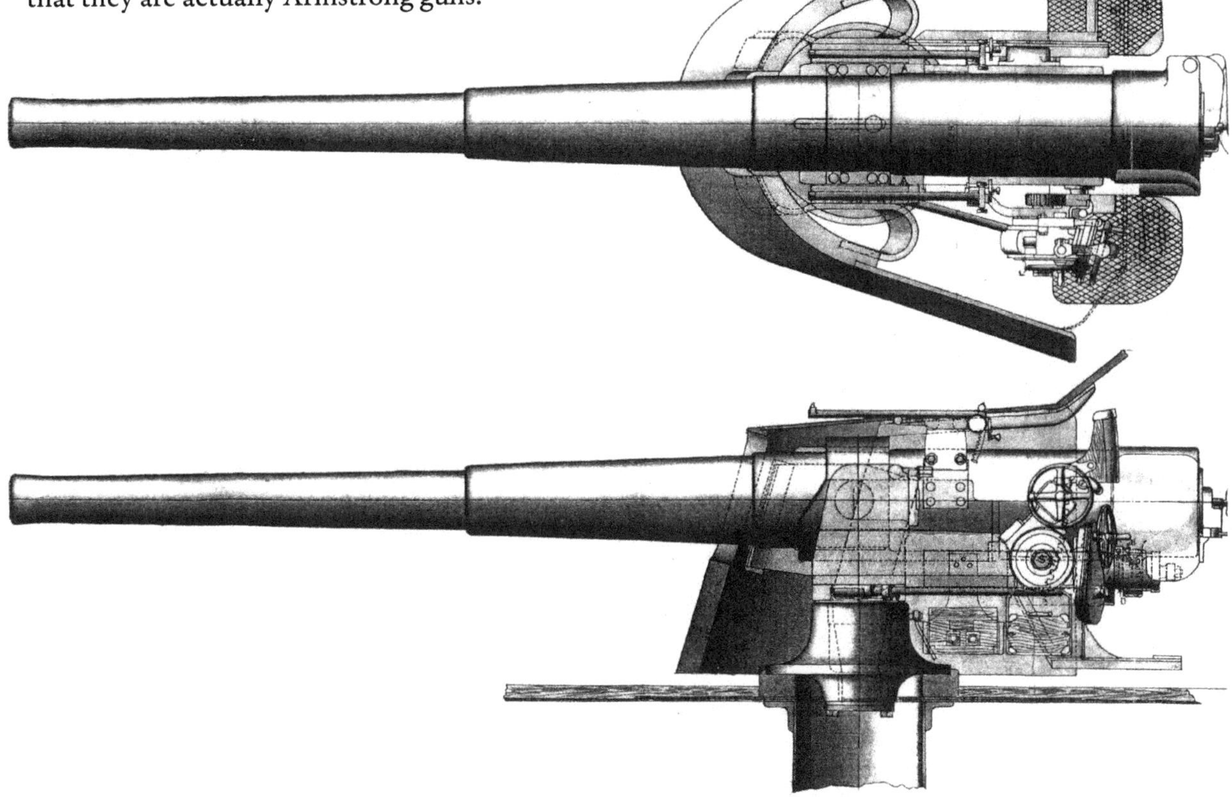

American efforts to import rapid-fire guns at the outbreak of the Spanish-American War led to the purchase of 42 guns placed for British-made weapons. Vickers Sons & Maxim advised the availability of eight 6-inch RF guns, deliverable in twelve weeks. Eight tubes and carriages, originally manufactured by Armstrong, were subsequently ordered. Acceptance of the offer was made on April 8, 1898 and the guns and carriages were delivered on August 19, 1898. The exact explanation for why Vickers had a supply of Armstrong guns is not known, though it was not unknown for warship contractors to occasionally be requested or contracted to arm ships with guns from a competitive firm.

The guns were typical 40-caliber, quick-firing, steel built-up guns. They had no trunnion and were held by a cradle on the carriage. It had an interrupted-screw (2-step) type breech, and electrical pistol firing gear. The tube had 24 grooves and was rifled from 0 to 1 in 30 at the muzzle. The carriage was a naval-type pedestal mount. It consisted of forged steel cradle, pivot yoke and pedestal. The recoil cylinder, using a concentric recoil system, was bolted to the underside of the carriage with two spring counter recoil spring cylinders. On the left side of the carriage was the worm gear for hand elevation. Gun elevation was possible from -7 to +16 degrees. The crew was protected by a multi-pieced shield, the front of which was 3.25-inches thick. There were slots in the shield to allow for sights on each side of the gun. The gun was quick-firing, meaning it used separate, cased powder. A 100-pound armored-piercing or common shot could be fired at 1,850-fps using smokeless powder.

Anticipating the delivery of this ordnance, the Corps of Engineers began construction of emplacements on April 27, 1898 using funds from the National Defense Act of March 9. The engineers were already informed that the gun used 24 hold-down bolts arranged in a 34-inch circle. For emplacements an elevated, stepped platform (these guns did not need high parapets like disappearing guns) held the carriage. The usual magazines were in the lower levels of the protected traverse. No lifts were needed for the ammunition. Six original battery emplacements, holding either one or two guns were built, each in a different harbor defense.

There was a considerable number of relocations for these guns between1904-1907. Three eventually went to a new, modern emplacement at Fort Adams (Battery Bankhead). However soon after completion (in 1913) these three guns and carriages plus one other were sent to Oahu. There two new, casemated batteries (Battery Adair and Boyd) were built on Ford Island in the middle of Pearl Harbor. They were part of the island's land defenses—providing defense of the coast artillery posts and naval bases from rearward, land attack. The choice to use these (by then) older Armstrong guns was dictated by their relatively low muzzle velocity and hence ability to fire different types of shrapnel rounds. Completed from April 1916 to February of 1917, these too did not last long.

6-inch Rapid-fire Gun, Armstrong
On 6-inch Barbette Carriage, Armstrong

Gun Characteristics
Bore size: 6-inch
Construction: Steel, built-up
Tube length overall: 249-inches
Nominal caliber: 40 caliber
Tube and breech weight: 14,784-pounds
Usual carriage: Barbette Carriage Armstrong or Vickers
Weight of gun and carriage: 34,484-pounds
Maximum elevation: 16 degrees
Loading: Separate, cased
Shell weight: 100-pounds
Maximum muzzle velocity: 1,850-fps
Maximum range: 10,185 yards

Builders: Elswick (Armstrong) pattern purchased from Vickers

Number built: 8 guns and carriages acquired by U.S. Army

During the Spanish-American War the eight purchased British 6-inch LF guns were distributed to various east coast forts. This photograph shows a mount at Fort Wadsworth, New York. It was mounted in a modified old 1870s emplacement and named Battery Barbour.

Both the guns and carriages were declared obsolete in June 1919. Two guns in the U.S. were scrapped, but the Hawaiian guns were not finally disposed of until after 1925. The remaining two guns were abandoned as too difficult to recover from Egmont Key in Tampa Harbor (Fort Dade). They sat there rusting for sixty more years until recovered, restored, and put on display at Fort Desoto State Park in St. Petersburg. They are the only survivors of this gun from U.S. service, though similar ordnance from the same time and manufacturer exist internationally.

Another one of the 6-inch purchased guns. This gun and a sister were mounted in Battery Burchsted at Fort Dade, Tampa Bay.

The 6-inch guns of the previous illustrated Battery Burchsted survived scrapping in the 1920s. Abandoned in place, they were eventually moved, restored, and put on display at a park in nearby Fort DeSoto, St. Petersburg, Florida.

6-inch RF Gun Model 1900
On 6-inch Barbette Carriage Model 1900

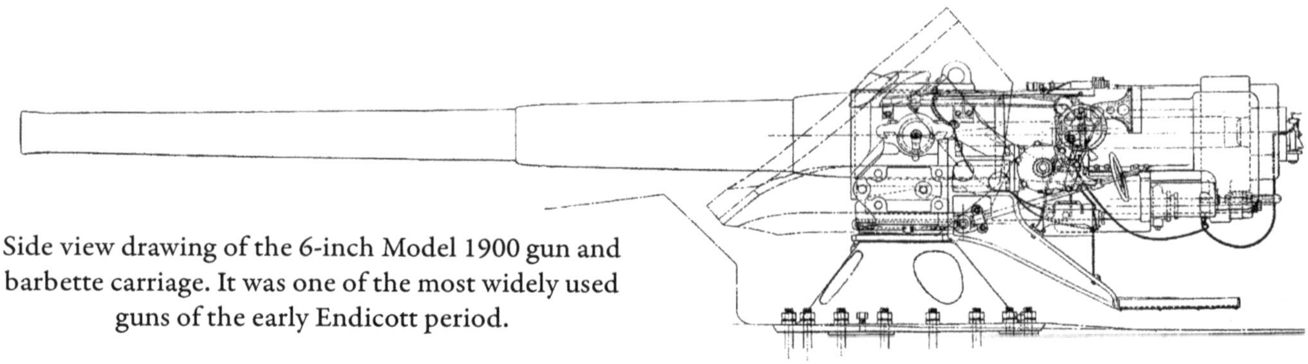

Side view drawing of the 6-inch Model 1900 gun and barbette carriage. It was one of the most widely used guns of the early Endicott period.

Following the failure of the Brown Wire-wound Segmental Gun Company to deliver 6-inch guns, the department developed its own new weapon. It borrowed heavily from the previous Model 1897 gun. The gun was built-up but had just two "A" hoops. It had a full 50-caliber length. The barrel had 36 grooves, and rifling was 1 in 50 increasing to 1 in 25. The powder chamber was larger in diameter but shorter. The thread for the breechblock was cut in a bushing rather than the jacket. The block was of the Stockett type, of eight sectors. However, the breech block itself was made of relatively mild gun steel, just after issue (in 1904) these blocks had to be replaced with new ones of nickel steel. The tube had two recoil slides to fit the cradle; that is except for two tubes (No. 45 and 46) that were given trunnion bands to fit the Model 1903 disappearing carriage.

The barbette carriage was a new type for the Army. A steel pedestal held a pivot yoke, gunner's platform and shield. The yoke's two vertical cheeks held a cradle to mount the tube. Underneath the tube were a single hydraulic recoil cylinder and two spring counter recoil cylinders. This simple arrangement provided a gun always in firing position that could rapidly traverse and elevate—perfect for an anti-boat and minefield coverage assignment. The carriage was equipped with a 3-inch telescopic sight. A large, curved 4.5-inch thick steel shield, attached by two ribbon-spring supports, provided protection from splinters and light shells.

Tubes were ordered in 1900-1901. After 46, one more was added each in 1905 and 1909. All 48 were made at Watervliet Arsenal. The first ten carriages were ordered in May 1902; the 45th and last was ordered in April 1903. Some were made at Watertown Arsenal, but small runs came from Rock Island Arsenal and Watervliet Arsenal, along with 13 from Builders Iron Foundry.

The army had some difficulties in deciding on an emplacement type for this gun. Initially it used a platform that had two guns side by side. One battery was actually armed like this while several other emplacements were built but not yet armed of this type. Quickly it was found that the guns interfered with each other, and the platforms were reduced to holding just a single gun. An intermediate type emplacement with single platforms and lower, adjacent magazines

Most of the Model 1900 6-inch guns were used with the barbette carriage. However two guns were given a trunnion band and mounted on Model 1903 disappearing carriages—this photograph is of one of these mounts on display at a 1904 exposition.

was finally adopted. Finally, a more spacious type similar to that used for the disappearing guns was utilized. Emplacements for 44 guns were built using the Model 1900 barbette, and one two-gun battery using the Model 1900 6-inch gun on Model 1903 disappearing carriage.

The gun and carriage proved highly successful. The simplicity of the carriage and rapid traverse clearly made it preferred to the later but more complicated 6-inch disappearing types. These guns were extensively relocated during their long lives. In 1917 ten were pulled for use on wheeled field carriages. Modified they became guns M1900A3. After the war the guns were re-modified to their original configuration. Twelve others were relocated in 1917-20 to new sites. Some of these were temporary defenses (the Delaware and Chesapeake Capes) but some became permanent locations (Fort Tilden). Postwar, returned guns sometimes went back into original emplacements, in other cases they were remounted in emplacements that had previously held 6-inch disappearing guns. Virtually all of the guns and carriages were still in service for coast defense during the Second World War. The prototype 200-series 6-inch barbette emplacement at Fort Story, Virginia was even armed with M1900 guns and barbettes. Most were not eliminated until 1946-48. A single pair of the guns and barbette carriages managed to avoid the scrap heap, they are on display at Fort Hancock, New Jersey.

6-inch RF Gun Model 1900
On 6-inch Barbette Carriage Model 1900

Gun Characteristics
Bore size: 6-inch
Construction: Steel, built-up
Tube length overall: 310.4-inches
Nominal caliber: 50 caliber
Tube and breech weight: 19,114-pounds
Usual carriage: 6-inch Barbette Carriage Model 1900
Weight of gun and carriage: 43,914-pounds
Maximum elevation: 20-degrees
Loading: Separate
Shell weight: 106-pounds
Maximum muzzle velocity: 3,000-fps
Maximum range: 17,000-yards

Builders: Guns: Watervliet Arsenal
Carriages: Watertown Arsenal, Watervliet Arsenal, Rock Island Arsenal, Builders Iron Foundry

Number built: 48 Guns, 45 Carriages

A good view of a Model 1900 barbette gun at Fort Michie, Long Island Sound in the mid-1930s.

June 1918 photo of a typical emplacement with a Model 1900 gun and barbette carriage.

A M1900 gun and carriage at Battery Peck, Fort Hancock

Only two intact Model 1900 6-inch guns and carriages still exist. They are on display, well-interpreted, in restored condition at Battery Gunnison (New Peck), Fort Hancock, New Jersey.

6-inch RF Guns Model 1903 and 1905
On 6-inch Disappearing Carriages LF Model 1903, 1905, and 1905M1

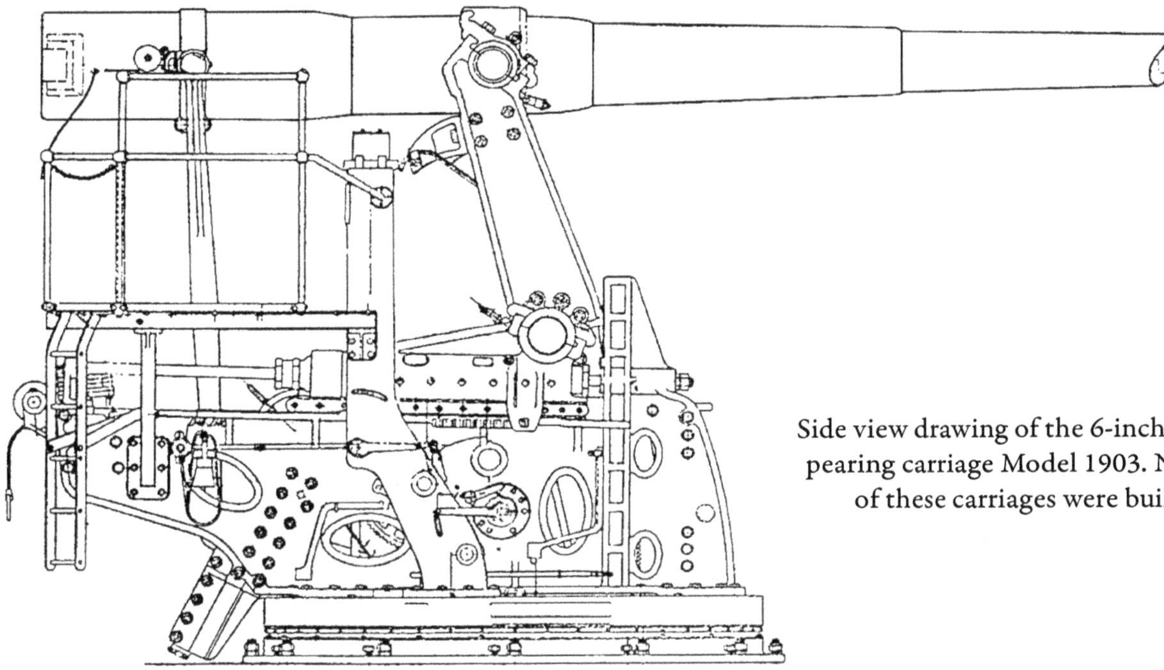

Side view drawing of the 6-inch disappearing carriage Model 1903. Ninety of these carriages were built.

A project for an improved, new 6-inch gun on disappearing carriage was begun almost as soon as the final M1897s/M1898s were emplaced. The new tube resembled the Model 1900 type. It was built-up and 50-calibers in length, though it was the first coast artillery gun tube to be made entirely of nickel steel. The breech mechanism was considerably improved, with threads cut directly into the jacket. The shape of the block was changed to a truncated ogive, patterned after the Bofors-type of patent, with twelve sections. The operating level was hinged on the right side, on the lower portion of the breech face. The slightly improved Model 1905 tube had a larger gas check seat, and the conical breechblock was changed to the Welin pattern. Otherwise, the performance characteristics were identical.

Shop photo of a gun on 6-inch disappearing carriage Model 1903 at Bethlehem Steel.

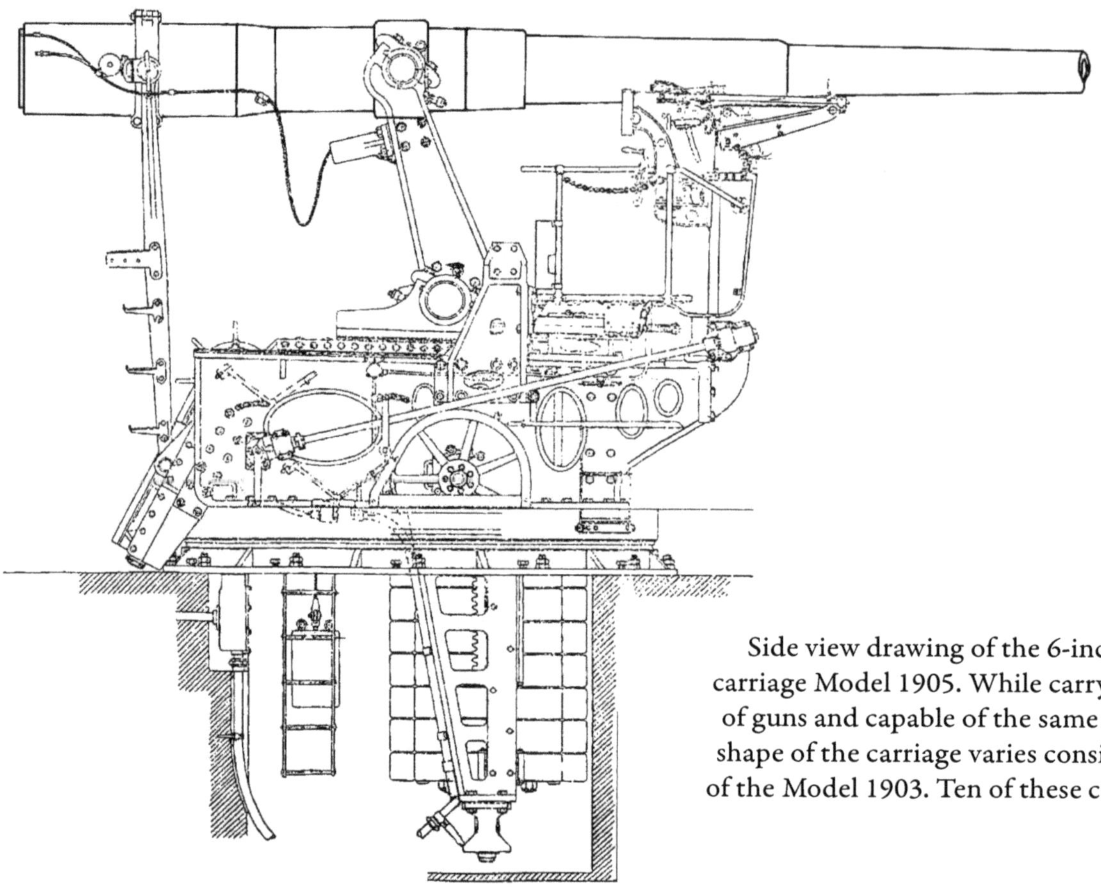

Side view drawing of the 6-inch disappearing carriage Model 1905. While carrying the same type of guns and capable of the same performance, the shape of the carriage varies considerably from that of the Model 1903. Ten of these carriages were built.

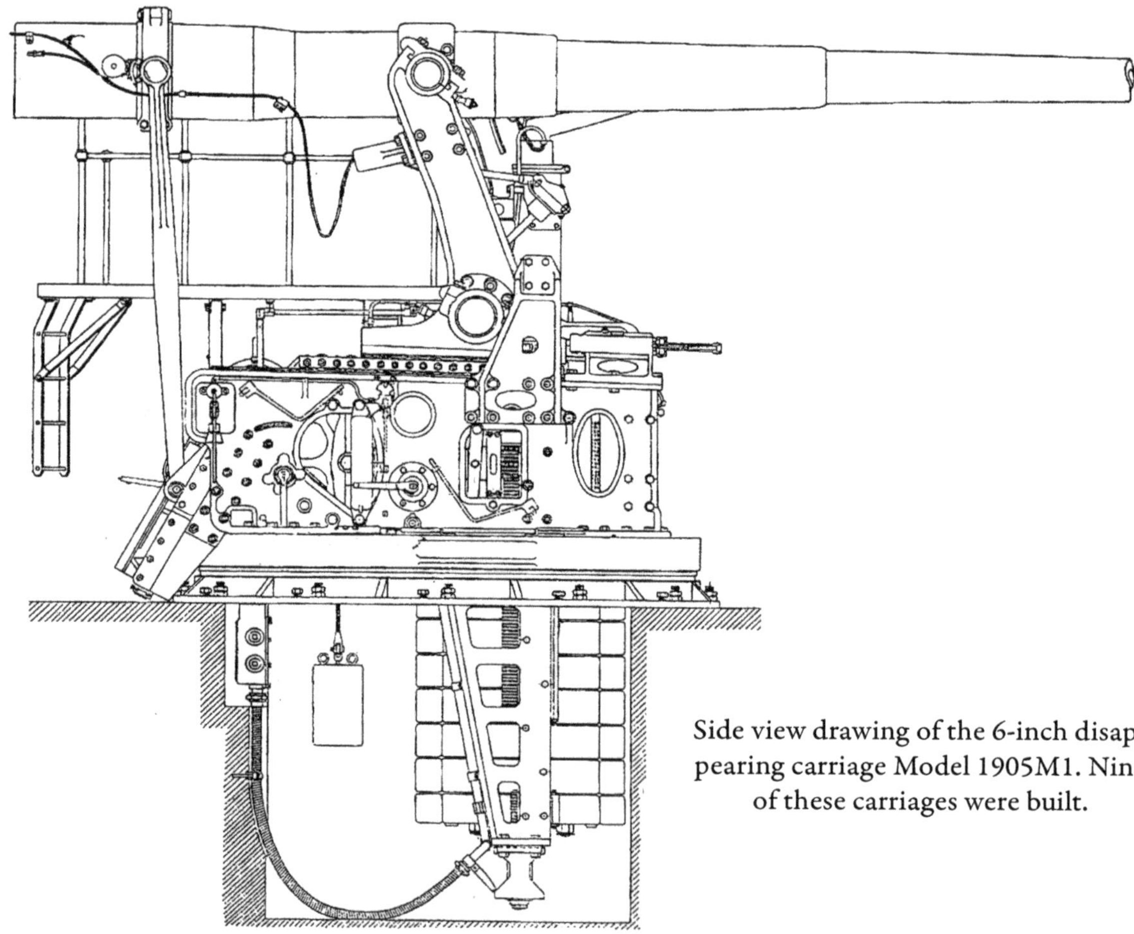

Side view drawing of the 6-inch disappearing carriage Model 1905M1. Nine of these carriages were built.

The Model 1903 carriage had the usual primary parts: the base ring and racer with traversing rollers, a chassis, top carriage with recoil rollers and hydraulic recoil cylinders, gun levers, crosshead, lead counterweight, elevating arms, working and sighting platforms. The counterweights were either of lead or cast iron. The action of the recoil raised the counterweight, which was caught by pawls. The gun moved forward into firing position through the action of the rollers on the inclined chassis top. The gun was raised again to fire by releasing the pawls that dropped the counterweight. The similar Model 1905 disappearing carriage had but a single counter recoil cylinder mounted in the center of the counterweight. The lightness of the carriage gave it a considerably different appearance than the Model 1903—including a pronounced forward projection of the front part of the chassis. An improved M1905M1 carriage had its sighting platform on the left rear, and it had a shorter chassis with a "straight" front. All carriages could deliver elevation of the gun to +15-degrees.

The guns were made in considerable numbers. Seventy Model 1903 guns were ordered from Watervliet Arsenal in 1903-04, followed by 34 Model 1905 gun from the same fabricator in 1905-06. They were accompanied by 90 carriages Model 1903 (two specially equipped to carry Model 1900 6-inch guns) ordered beginning in June 1903 and ten model 1905 disappearing carriages. Only nine M1905M1 carriages were built. As typical for this time a large number of arsenal and private contractors built the carriages. Carriages could take either gun type. Six of the later, wire-wound M1908 6-inch guns were also emplaced on disappearing carriages Model 1905M1.

Many battery emplacements of from one to four guns were built throughout the continental U.S. and Philippines. The design had spacious loading platforms and three steps from loading to working platform. An elevated half-circle concrete loading platform surrounded the rear of the gun pit. Adjacent traverse magazines fed shell carts to this circle for manual lifting, separate powder being carried up stairs on either side of the platform. The guns and loading system worked well, though the batteries were quite expensive compared to barbette batteries.

More than half of these guns in the U.S. were removed from their disappearing carriages in 1917 for use as field artillery. After the war they were not returned to their emplacements (though that was the intent at the time) but rather stored for possible future use. The unused carriages were scrapped in 1920. These tubes became the main armament of the 1940 Program 6-inch batteries. Remaining 6-inch disappearing batteries served until disarmed in 1942-44. There are several batteries of M1905 guns and carriages still remaining abandoned in the forts of Manila Bay, and two guns and carriages survived scrapping; one each is on display at National Parks at Fort Winfield Scott and Fort Pickens.

6-inch RF Guns Model 1903, 1905 On 6-inch Disappearing Carriages LF Model 1903 and 1905

Gun Characteristics
Bore size: 6-inch
Construction: Nickel steel, built-up
Tube length overall: 310.4-inches
Nominal caliber: 50 caliber
Tube and breech weight: 19,990-pounds (M1903)
Usual carriage: 6-inch Disappearing Carriage LF Model 1903 or 1905
Weight of gun and carriage: 97,690-pounds
Maximum elevation: 15 degrees
Loading: Separate
Shell weight: 108-pounds
Maximum muzzle velocity: 2,900-fps
Maximum range: 14,600-yards

Builders: Gun: M1903 & M1905 Watervliet Arsenal
Carriages: M1903 Watertown Arsenal, Morgan Engineering, Wellman-Sevier Morgan, Detrick & Harvey, Mansfield Engineering, Bethlehem Steel
Carriages: M1905 Bethlehem Steel, Watertown Arsenal

Number built: 70 M1903 guns, 34 M1905 guns 90 M1903 carriages, 10 M1905 carriages, 9 M1905M1 carriages

Good proving ground photograph of the Model 1903 6-inch disappearing carriage. Note the presence of sighting stands on both sides of the carriage.

The 6-inch gun M1905 on M1903 disappearing carriage during 1918 at Battery Schofield, a training emplacement built at the West Point Military Academy.

M1903 gun and carriage in service at Battery Tolles, Fort Worden in Puget Sound.

Only two 6-inch M1903 disappearing guns survive, but they are fortunately restored and on display. This unit is at Battery Cooper in the Fort Pickens National Park Unit near Pensacola, Florida.

6-inch Gun Model 1908 and 1908M1
On 6-inch Disappearing Carriage Model 1905M1 and 1905MII

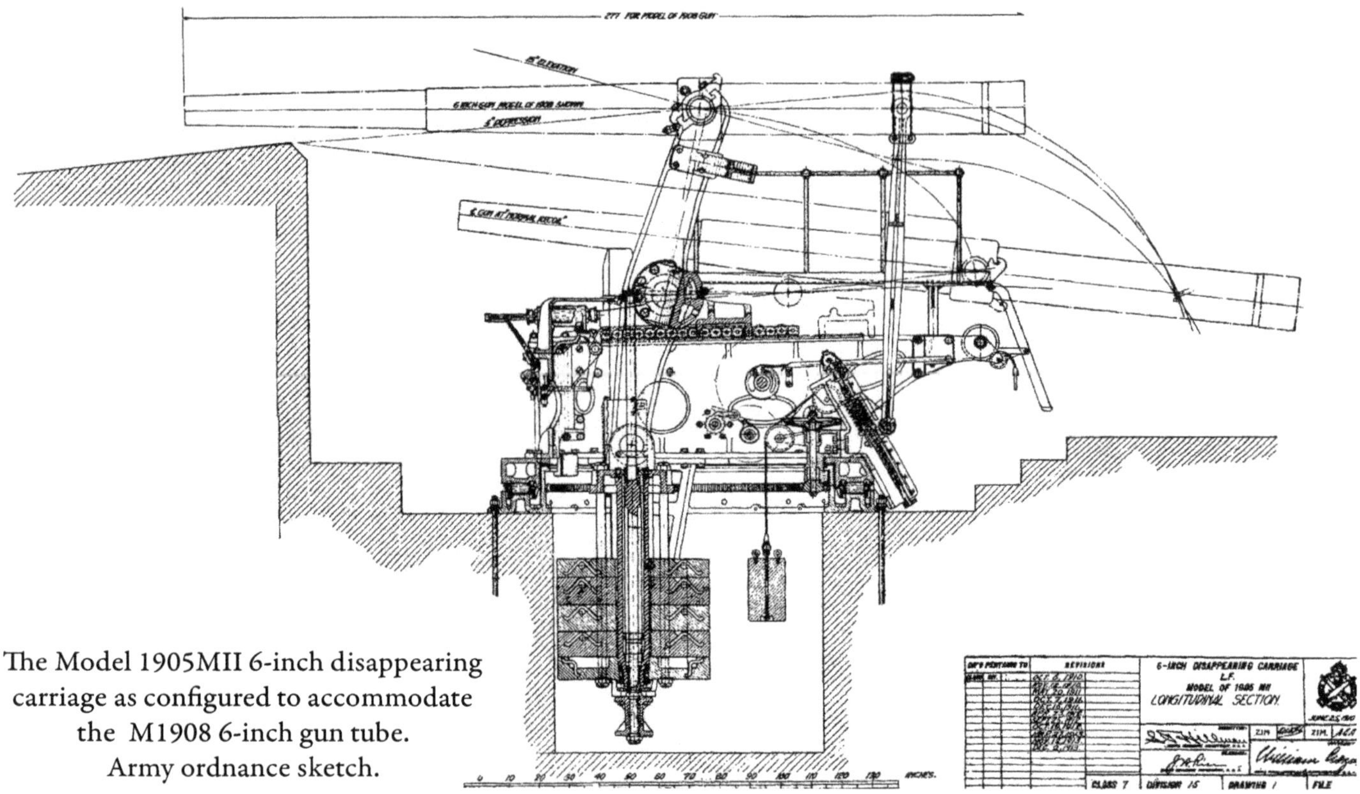

The Model 1905MII 6-inch disappearing carriage as configured to accommodate the M1908 6-inch gun tube. Army ordnance sketch.

The Ordnance Department had continued to pursue wire-wound construction for guns. Eventually they perfected the manufacturing process and designed a practical gun. Basically, it copied the first M1897 6-inch gun for overall length (44.5 caliber) and size of powder chamber. The major difference was really just the manufacturing method. Wire wound construction reduced weight and material cost by using less forged steel (the inner tube, jacket, and hoop were of gun steel). Consequently, the gun was only 17-inches in diameter over the powder chamber versus 24-inches for the previous Model 1905 6-inch gun. The tube had a shrunk-on trunnion hoop. The barrel had 54 grooves, and rifling of 1 in 50 increasing to 1 in 25. The breech mechanism was an improvement of the early type on the M1897 and M1900 6-inch guns Stockett type. The nickel steel breech block was of a straight cylindrical shape and had four slotted and four open sectors. It fired the same 6-inch C.I. and A.P. shot and shell as all the other seacoast 6-inch gun types.

A modification of the newest Model 1905 L.F. disappearing carriage was specifically designed to carry the Model 1908 gun. The lighter tube did require a different counterweight and adjustments in balance to properly work on the carriage designed for a heavier gun type. The M1905MII carriage also discontinued the use of the single hydraulic recoil cylinder embedded in the counterweight of the previous models, it was found to be too difficult to access for maintenance. The shorter carriage chassis had no sight standard in the front, with just a single sighting platform on the left rear with the traversing hand wheel. Also, this model had retraction by gears and a counter recoil system independent of the recoil system. This carriage, of which only 14 were built, exclusively was used with the Model 1908, wire-wound gun tube. All were made at Watertown Arsenal, the 14 carriages Model 1905MII were sequentially numbered with the other M1905s, being No. 20-33.

Eighteen Model 1908 6-inch tubes were ordered from 1908 to 1912; and two Model 1908M1 guns (similar, but with double tube liners) in 1909. Four Model 1908 guns were mounted on Model 1905M1 disappearing carriages at Fort Whitman. Two Model 1908M1 guns were mounted on Model 1905M1 carriages at Fort DeRussy on Oahu. The rest of the tubes were mounted on Model 1905MII disappearing carriages. All the tubes were made at Watervliet Arsenal.

The 14 M1905MII carriages were all emplaced in two-gun batteries of the latest, spacious design, designated the 1912 Type. Gun pits were separated by 170-feet between centers. A central traverse covered shell and powder rooms along with internal plotting room and latrine. A battery commander station was located atop the traverse. The seven two-gun batteries were all built in overseas possessions. One battery was in the Philippines (Fort Hughes), one in Hawaii (Fort Kamehameha) and the other five in Panama Canal forts.

These guns were mounted in their emplacements between 1914 and 1916. Being in important overseas garrisons, they were not disturbed during the dismounting and relocation of the First World War. All served until either captured in 1942 (Fort Hughes) or replaced with more modern emplacements and ordered dismantled in 1944-45. While most of the battery emplacements still exist today, the only surviving guns and carriages are the damaged, partial remains at Battery Leach at Fort Hughes, Manila Bay.

6-inch Gun Model 1908 and Model 1908M1
On 6-inch Disappearing Carriage LF Model 1905M1 and MII

Gun Characteristics
Bore size: 6-inch
Construction: Steel, wire wound
Tube length overall: 283.6-inches
Nominal caliber: 44.6 caliber
Tube and breech weight: 12,257-pounds
Usual carriage: 6-inch Disappearing Carriage LF Model 1905MII
Weight of gun and carriage: 113,148-pounds
Maximum elevation: 15-degrees
Loading: Separate
Shell weight: 106-pounds
Maximum muzzle velocity: 2,600-fps
Maximum range: 14,600-yards

Builders: Guns at Watervliet Arsenal
Carriages at Watertown Arsenal

Number built: 18 guns Model 1908 and 2 Model 1908M1 used on 6 Model 1905MI and 14 1905 MII

This is the M1908 gun on disappearing carriage M1905MII as mounted in about 1936 at Battery Leach, Fort Hughes, Manila Bay defenses.

6-inch Gun Model 1908MII
On 6-inch Barbette Carriage Model 1910

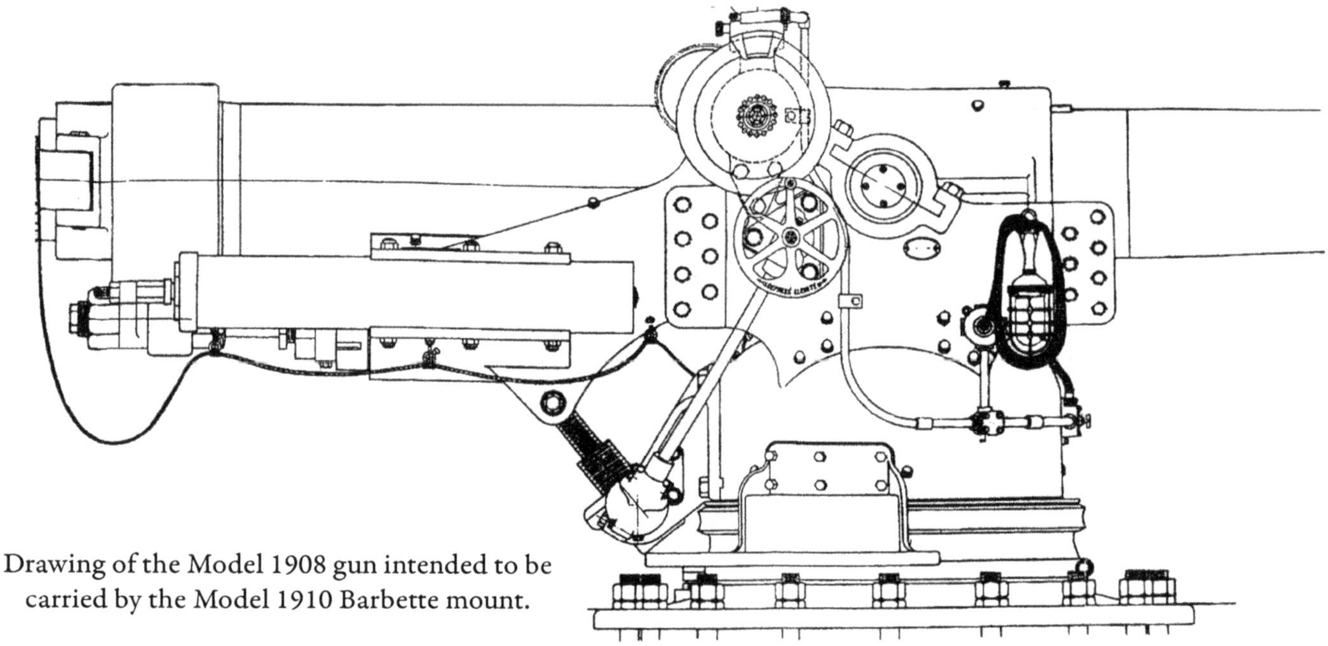

Drawing of the Model 1908 gun intended to be carried by the Model 1910 Barbette mount.

Six of the latest Model 1908 wire-wound guns were built without trunnion hoops so they could be specifically utilized in locations requiring barbette rather than disappearing carriages. They had a 44-inch long cylinder machined in the jacket to fit the cradle of the new barbette carriage. Also guns No. 1-4 (intended for Fort Drum, Manila Bay) were equipped with a gas ejection system utilizing compressed air. Otherwise, the breech and performance were similar to the earlier Model 1908 gun tubes.

The carriages were reminiscent in form to those of the Model 1900. They were a basic pedestal of cast steel, yoke and cradle, but with no shoulder guard. A recoil band on the gun connected to a single hydraulic recoil cylinder and two spring counter recoil cylinders all underneath the gun tube. Elevation was by

Watertown Arsenal photograph of the Model 1908 6-inch gun on the Model 1910 shielded barbette mount in the shop in 1915.

concentric double elevating screw, and elevation was controlled by stops for a range of -3 to +12-degrees. A gunners' platform was included alongside the gun. The first four carriages were equipped with a circular 5.3-inch thick shield designed specifically to fit the casemated openings at Fort Drum. The last two guns, intended for an open barbette emplacement in Panama, had a special semi-elliptical shield of cast steel 5.3-inch thick at the front tapering to 2-inches on the sides and 1.75-inch on the hood. In both cases the shield rotated with the gun. While the last two had a different style of shield, the gun and carriage were the exact same model. The carriages were equipped with a 3-inch telescope Model 1910.

Two emplacements were built for this gun. Four 6-inch casemated guns on El Fraile Island in Manila Bay had long been projected for that fort. They were intended to cover either the north or south mine fields. On the north flank was Battery McCrea and on the south was Battery Roberts. The final design for Fort Drum placed them towards the rear of the fort body in casemates—one above the other, a set facing north and another south. They had traversing fields of fire of 120-degrees. The relatively low power of the guns and just 12-degrees of elevation limited them to a range of 10,200-yards, but adequate for the minefield coverage envisioned. Ammunition service was by hoists from magazines located below the casemates. Guns were mounted and service began in 1917.

The second emplacement was Battery Morgan at Fort De Lesseps at the Atlantic entrance to the Panama Canal. The design was for a compact emplacement, gun centers being 96-feet and 6-inches. The center traverse covered the protected magazine that served both guns. The battery was armed in late 1916

6-inch Gun Model 1908II
On 6-inch Barbette Carriage Model 1910

Gun Characteristics
Bore size: 6-inch
Construction: Steel, wire wound
Tube length overall: 283.6-inches
Nominal caliber: 44.6 caliber
Tube and breech weight: 12,257-pounds
Usual carriage: 6-inch Barbette Carriage Model 1910
Weight of gun and carriage: 57,850-pounds
Maximum elevation: 12-degrees
Loading: Separate
Shell weight: 106-pounds
Maximum muzzle velocity: 2,600-fps
Maximum range: 10,200-yards

Builders: Guns: Watervliet Arsenal
Carriages: Watertown Arsenal

Number built: 6 guns Model 1908MII and 6 carriages Model 1910

A 1913 photo of the 6-inch gun Model 1908 on the embrasure barbette intended for Fort Drum, Manila Bay.

and transferred to coast artillery troop in early 1917. New shields for these two carriages were developed and produced in 1943, being installed late in the war. The armament was pulled and scrapped apparently after mid-1944. The emplacement still exists.

Four guns were ordered from Watervliet Arsenal in April 1910, and the last two a year later in April 1911. They were delivered for proof between 1911 and 1914. The Ordnance Department gave them a numbering sequence of their own, having serial No. 1-6. As usual they were marked on the muzzle face with year, serial, weight, and inspector's initials.

The four casemated guns at Fort Drum saw action in 1941-42. While not restored by the Japanese during occupation, they were heavily damaged during the retaking of the fort in 1945 (particularly Battery Roberts, which suffered a major secondary explosion after direct hits from American forces). Parts of the guns and carriages that were not destroyed during the war or subsequently scrapped still exist on site.

Photo of the only pair of shielded Model 1908 6-inch guns on Model 1910 barbettes at Battery Morgan, Fort De Lesseps in Panama.

Another good view of the 6-inch barbette gun at Battery Morgan, Fort De Lesseps in 1941.

A pair of 6-inch M1908 embrasure guns on the side of Fort Drum, Manila Bay about 1933.

(Below) Battery Roberts, with at least one remaining Model 1908 gun, perseveres to this day at the ruins of Fort Drum, Manila Bay.

Coast Defense 8-inch Guns

Of the Endicott and Taft Systems

The Endicott Board called for heavy guns of four sizes: 16, 12, 10, and 8-inch bores. With the exception of the 16-inch, development work started almost simultaneously on the other three. With the obvious exception of bore size, there was little difference between the guns. They all used the same built-up techniques, the same steel quality, and the same basic breech mechanism. And the development effort produced acceptable weapons that indeed had long service lives. The first 8-inch gun was the Model 1888 ordered from Watervliet Arsenal in mid-1887 and delivered two years later. Production orders started in 1890 when appropriations were received from Congress to start the acquisition of seacoast guns. Altogether 92 Model 1888 guns were produced (along with the one prototype). They had rifling of 48 lands and grooves with a twist of 1 in 250 to 1 in 25. They were made by Watervliet, Bethlehem Steel, and the West Point Foundry.

A barbette carriage was developed for the M1888 8-inch gun. It was the first of the new generation of similar carriages for all three gun sizes. Only nine were built. They were not a popular item in the coast artillery inventory, and just a few new emplacements were ever built for this carriage type. Far more successful were the two disappearing carriages developed (a third, all-around fire carriage was also designed, but none were ever ordered). The Model 1894 front-pintle and the Model 1896 center-pintle types were the lead development projects of their types. However, the engineers gradually stopped recommending new sites for 8-inch guns. The emplacements consumed about the same resources in terms of dollars and manpower and ultimately delivered less "punch" and range than their larger cousins. By the Spanish American War there was actually an excess inventory of over twenty 8-inch Model 1888 guns with no carriages that were then used in new temporary batteries emplaced on modified old 15-inch Rodman carriages. After a few new orders of the Model 1896 disappearing carriage addressed the backlogged need for carriages, the further emplacement of new 8-inch batteries ended. There was no requirement for guns of this size in the Taft or 1915 Programs.

The 8-inch Model 1888 gun turned out to be the only model for a gun of this bore size developed by the Ordnance Department for seacoast defense. Unlike its bigger sisters, there was no improved Model 1895 or high-powered Model 1900 versions. There was only one significant relocation of an 8-inch battery in the early 1900s: three guns from Fort Wadsworth's Battery Duane were moved to a new Battery Arrowsmith at Fort Hancock. In a new, heavy, intermediate role a limited number of ex-navy 8-inch, 45-caliber guns were used on new barbette carriages in the 1940 Program.

The Model 1888 8-inch gun was selected for use on a railway carriage in 1917; in this instance the compact size of the gun and payload was quite adequate for field artillery use. Fifty of the 65 mounted 8-inch guns were dismounted in late 1917, primarily taking from the inner defense lines and older barbette and Model 1894 carriage emplacements. These guns were not returned after the war, and the majority of the abandoned carriages were scrapped during 1918-1920. In the late 1930s and during the Second World War several of the old 8-inch tubes on Model 1918 railway top carriages were again emplaced in fixed emplacements as barbette mounted guns. The only two surviving Model 1888 gun tubes were each last used in service on a railroad top carriage.

8-inch gun on M1894 disappearing carriage at Battery Bowyer, Fort Morgan

8-inch Gun Model 1888
On Barbette Carriage Model 1892

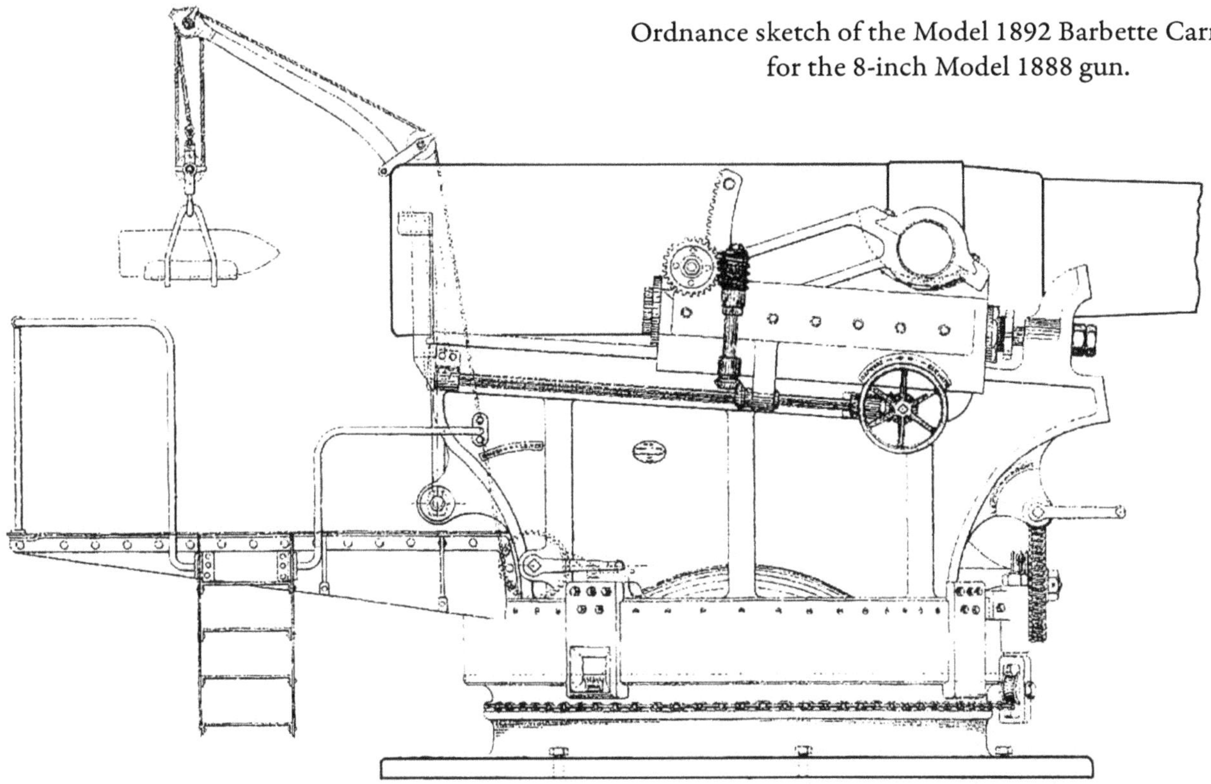

Ordnance sketch of the Model 1892 Barbette Carriage for the 8-inch Model 1888 gun.

The carriage was a center-pintle, gravity-return type. The design was also referred to as a Vavasseur style. It had a base ring of cast iron bolted to a concrete emplacement platform. The gun traversed on 20 forged steel rollers on a racer. The cast steel chassis had two side members and transoms. The top of the chassis had rails inclined 4-degrees with eight rollers on each side to allow the top carriage to return by gravity to battery. The top carriage held the gun and two hydraulic recoil cylinders. A capacious loading platform extended to the rear. Supporting this platform were triangular supports cut-out with six circular perforations, useful for photo identification. A prominent projectile hoist was attached to the chassis, as the breech height was such that it could be loaded only by lifting the shells rather than directly from carts. A rack and pinion elevation mechanism could provide -7 (useful on such carriages mounted on bluffs) to +18-degrees. A sight standard on the right side held a telescopic sight.

All of the Model 1892 barbette carriages were intended to carry 8-inch guns Model 1888. That gun had been developed in the late 1880s; orders were issued to the West Point Foundry and Watervliet in 1890 for the first 20 production guns. This gun was a 32-caliber, forged steel, built-up gun with an interrupted-screw, Welin-type breech. Later developments led to a reduction in the number of hoops for the M1888M1 and M1888MII, but their performance was otherwise identical.

The first carriage was ordered from Watertown Arsenal on June 20, 1891. This was six months before the first 12-inch barbette and 19 months before the first of 10-inch size. Seven more were ordered in 1892-93, and the final ninth carriage on June 10, 1899. The first, prototype mount had a bronze top carriage, but all subsequent carriages (and the first one later was rebuilt) had cast steel top carriages. Initially the first eight carriages used an awkward forked tong that pivoted from the rear of the chassis to lift ammunition from the platform to the breech. Seven of the carriages were modified in 1895-96. The shot tong and platform with a prominent slot to allow it to be elevated, was removed. A new loading platform was attached to the carriage with a shot crane on the right side. One carriage (no. 3) was never modified, and carriage No. 9 was built new with the changes incorporated.

The deployment of these carriages was interesting. The first carriage was a popular display item in some of the great World's Fairs in the U.S. in the 1890s. There was usually one (and sometimes two) of these 8-inch barbette carriages at Sandy Hook Proving Ground, and another for cadet training purposes at the academy in West Point. Fort Monroe received four guns and carriages placed atop the parapet of the old Third System brick fort for the Spanish American War; and another was in a concrete emplacement (Battery Barber) outside the moat on the north side of the same fort. Another was emplaced temporarily during the Spanish American War at Fort Pulaski outside Savannah. Finally, four guns were intended for permanent concrete batteries in the San Francisco defenses. Ultimately only three were so emplaced, two at Battery Duncan and one at Battery Drew.

Several batteries were removed as early as 1910-12. The few surviving emplacements were disarmed in 1917-18, their guns intended for use on railway carriages. The carriages left in place were scrapped in 1920. One or two may have survived into the 1930s before being salvaged. No examples of the 8-inch barbette carriage survive, though one Model 1888 tube is left on a Model 1888 railway top carriage in a park in Tampa, Florida.

8-inch Gun Model 1888
On 8-inch Barbette Carriage Model 1892

Gun Characteristics
Bore size: 8-inch
Construction: Steel, built-up
Tube length overall: 278.5-inches
Nominal caliber: 32 caliber
Tube and breech weight: 32,218-pounds
Usual carriage: 8-inch Barbette Carriage Model 1892
Weight of gun and carriage: 87,084-pounds
Maximum elevation: 18-degrees
Loading: Separate
Shell weight: 316-pounds (short cap)
Maximum muzzle velocity: 2,200-fps
Maximum range: 14,026-yards

Builders: Gun: Watervliet Arsenal (56),
Bethlehem Steel (25),
West Point Foundry (11)
Carriages: Watertown Arsenal (9)

Number built: 9

The 8-inch M1888 on Barbette Carriage M1892 attended to by military cadets, about 1903-1910.

Sandy Hook Proving Ground photograph of the 8-inch Model 1888 gun on the Model 1892 Barbette Carriage.

An 8-inch M1888 on M1892 barbette emplaced in the San Francisco defenses.

8-inch Gun Model 1888
On 8-inch Disappearing Carriage LF Model 1894

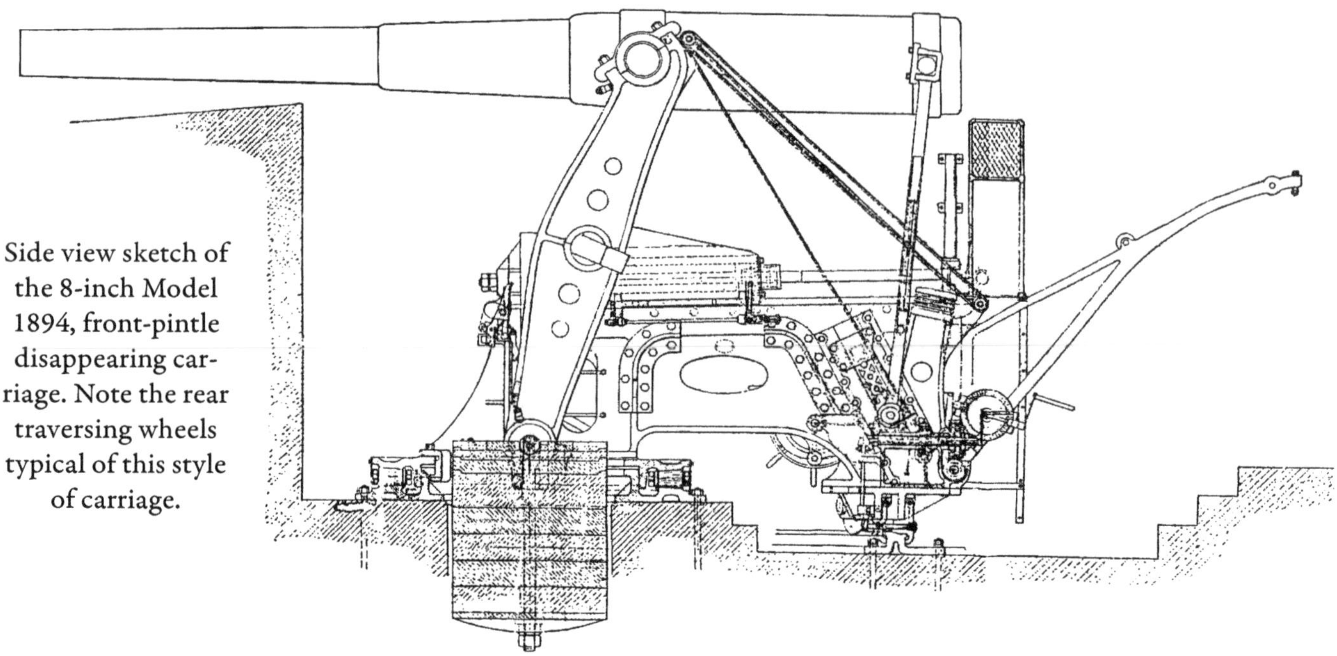

Side view sketch of the 8-inch Model 1894, front-pintle disappearing carriage. Note the rear traversing wheels typical of this style of carriage.

The first American disappearing carriages were of this front-pintle, limited-fire type. The primary parts were the base ring and racer with 20 individual conical traversing rollers, a chassis, top carriage with recoil rollers and hydraulic recoil cylinders, gun levers, crosshead, lead counterweight (of some 34,000-pounds), elevating arms, working platform and sighting platform. These parts were mostly made of cast iron or cast steel as necessary. The action of the carriage used the recoil to raise the counterweight. The gun always returned to the same level to ease loading from hand trucks. The gun moved back into firing position through the action of the rollers on the flat chassis top, just like a gravity-return carriage. The gun was raised again to firing position by releasing the pawls that dropped the counterweight. The counterweight was underneath the front part of the carriage, but the rear part was supported on wheels by a large racer ring. On this ring was a bronze azimuth circle marked in hundredths of degrees. One distinctive identification clue is the presence of four circular cutouts on each elevating gun arms, not occurring on any other type of disappearing carriage.

The carriage could accommodate any of the 8-inch model 1888 gun types, and in fact they were mounted with all three major marks. The initial single experimental 8-inch and 10-inch front-pintle carriages were ordered from Southwark Foundry on June 22, 1891. The 8-inch type was delivered two years later and extensively tested. It was not given a serial number, though later modified (in 1900) to allow it to be emplaced in a service battery at Fort Columbia. Ten production carriages were ordered from Watertown Arsenal in 1894 and 1896. The Ordnance Department then wanted to switch to the improved Model 1896 but found out that the engineers had already built or began emplacements for fifteen more M1894s. Consequently, fifteen carriages were ordered from Morgan Engineering and Providence Steam Engine Co. in late 1896. Together with the experimental carriage, 26 front-pintle disappearing carriages for 8-inch guns were produced. Starting in 1909 but taking until 1917 all were modernized to the M1891M1 standard with new traversing gear, elevating counterbalance device, wire rope retracting gear, sighting platform on the left side and electrical lighting and firing.

All the carriages were emplaced. The Corps of Engineers began designing and building emplacements for 8 and 10-inch disappearing guns as early as 1891; though funding restricted most to the mid-1890s. Soon an 1894 type emplacement was standardized. It had gun centers of 124-feet distance, and four steps between the loading and working platforms. As one of the first attempts at designing a modern emplacement, compared to later types it had narrow, independent loading platforms, and magazines were cramped and arranged at different lower levels. Ammunition service was only by crane, the use of early natural cement led to moisture problems, and the batteries were not electrified. Many had to be extensively repaired and modernized to serve in later years. In all eleven batteries were at least partially armed with the M1894 disappearing carriage, some had three or four, or even in one instance five such mounts.

Many of these batteries were disarmed in 1917-18 to release guns for railway carriages, no doubt influenced by still unresolved problems with the emplacements that carried them. A few others lasted until their defenses were discontinued in the early 1920s. With the 8-inch guns gone, the carriages were scrapped in 1918-1922. None of the carriages survived into the 1930s or 1940s, but some of the original tubes were still serving remounted on railway carriages. While many of the emplacements still exist, there are no surviving Model 1894 disappearing carriages.

8-inch Gun Model 1888
On 8-inch Disappearing Carriage LF Model 1894

Gun Characteristics
Bore size: 8-inch
Construction: Steel, built-up
Tube length overall: 278.5-inches
Nominal caliber: 32 caliber
Tube and breech weight: 32,480-pounds
Usual carriage: 8-inch Disappearing Carriage LF Model 1894
Weight of gun and carriage: 136,480-pounds
Maximum elevation: 12-degrees
Loading: Separate
Shell weight: 316-pounds (short cap)
Maximum muzzle velocity: 2,200-fps
Maximum range: 11,019-yards

Builders: Guns: Watervliet Arsenal and
Bethlehem Steel
Carriages: Southwark Foundry,
Watertown Arsenal,
Morgan Engineering,
Providence Steam Engine Co.

Number built: 26 guns Model 1888, 1888M1 and 1888MII utilized 1 experimental prototype and 25 production carriages

Scientific American etching of 8 inch M1894 disappearing carriage

Photograph by Morgan Engineering of one of its 8-inch Model 1894 Disappearing Carriages with gun.

An emplaced Model 1894 Disappearing Carriage at Battery Arrowsmith, Fort Hancock, Southern New York defenses. Observe the circular cut-outs in the elevating arms.

8-inch Gun Model 1888
On 8-inch Disappearing Carriage LF Model 1896

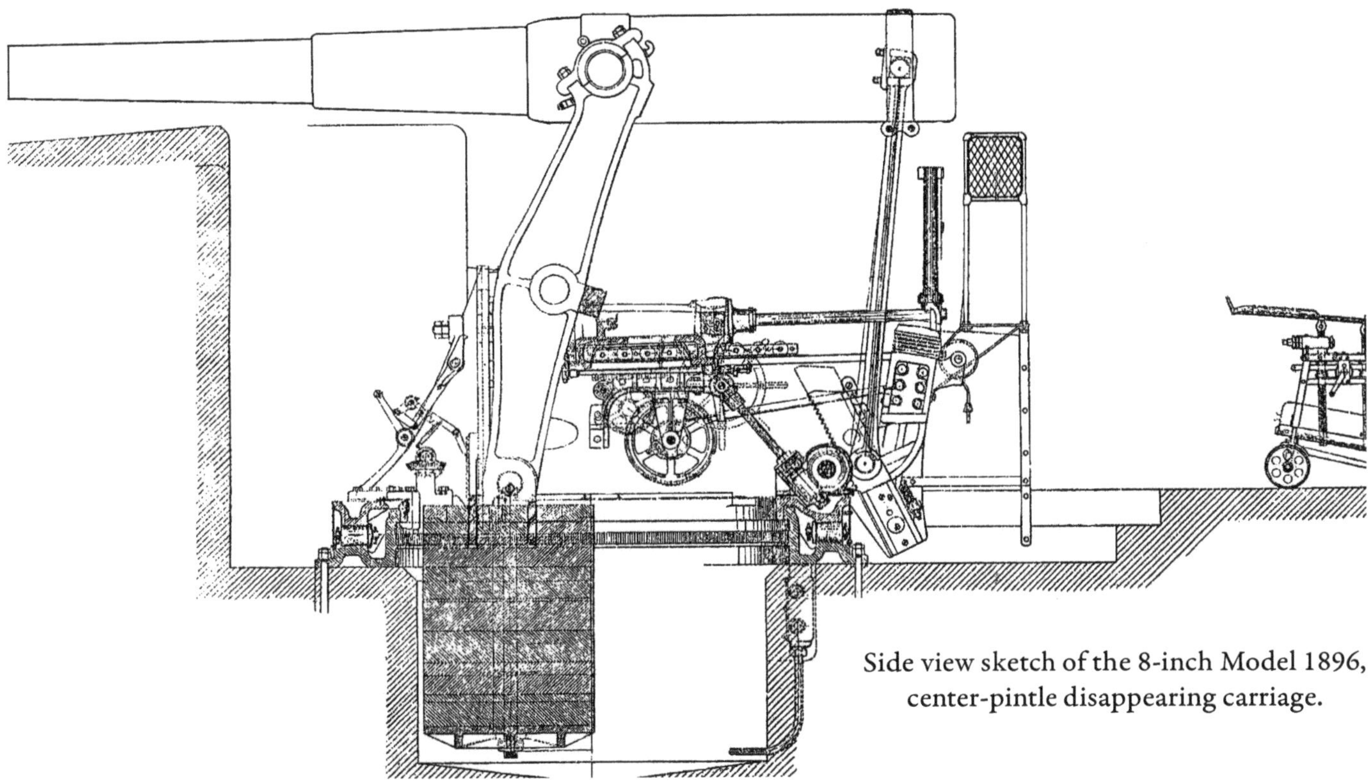

Side view sketch of the 8-inch Model 1896, center-pintle disappearing carriage.

This was the final matchup of the standard Model 1888 8-inch, built-up gun with an ordnance-designed disappearing carriage. This limited-fire, center-pintle disappearing carriage was typical in design of the Buffington-Crozier type of 1896 mechanism. The primary parts were the base ring and racer with 24 individual traversing rollers, a chassis, top carriage with recoil rollers and hydraulic recoil cylinders, gun levers, crosshead, lead counterweight (configured for 36,945-pounds weight), elevating arms, working platform and sighting platform. These parts were mostly made of cast iron or cast steel as necessary. Upon firing the action of the carriage's gun levers raised the counterweight. The recoil was halted by the action of the hydraulic pistons of the two 8-inch diameter recoil cylinders. The gun always returned to the same level to ease loading from hand trucks. It was moved back into firing position by rollers on the inclined chassis top, just like a gravity-return carriage. The gun was raised again to firing position by releasing the pawls that dropped the counterweight. Elevation from -5 to +12-degrees was possible, and traverse was limited by stops for the emplacement of 170-degrees.

No experimental carriage was thought necessary. The new carriage was strongly preferred by the Army for its lightness and ease of maneuverability. Starting in late 1896 orders were placed in small batches until June 1899 for 38 M1896 carriages. They were made either by Pond Machinery, Morgan Engineering, Walker Company, Lake Erie Engineering, or Watertown Arsenal. Unlike most of the 10 and 12-inch carriages, the 8-inch type were never modified for higher elevation—it was already realized that the gun was on the lower threshold of power and range and any expense to modify could not be justified.

A new type 1896 emplacement design was developed by the Corps of Engineers. It addressed many of the deficiencies of the 1894 emplacement, particularly the layout and size of magazines, communications between guns and levels, and the use initially of balanced-platform ammunition hoists, later changed to chain hoists. With size differences obvious from the ammunition and dimensions of the carriage, these types were virtually identical to contemporary emplacements for 10-inch disappearing guns. All 38 carriages were utilized in emplacements. They went into 19 separate batteries, most of which had two or three Model 1896 carriages—though on occasion they shared emplacements with the older 1894 disappearing type. The presence of only two steps from the loading platform to the working level distinguish them from other types for photographic recognition. Emplacements were not as heavily modified as remaining 10 and 12-inch types, but many did receive new latrines, battery commander stations, and electrification in 1910-1917.

The improved carriage and probably as importantly the improved emplacements gave some of these mounts a longer service life than its predecessor. A total of 50 8-inch guns were removed from fortifications in 1917 for use on railway carriages for France, leaving just 3 Model 1894s and 12 Model 1896s. Five more were removed in 1920, leaving just ten (all in New England) well into the Second World War. The final removals for scrapping did not come until 1942-43. Unusual for disappearing carriages, two complete sets of gun and carriage were donated for civic memorial display in Florida (the American Legion in Tampa) after the First World War. Though displayed for several years, unfortunately they were scrapped in the following war, and no Model 1896 disappearing carriages survive today.

8-inch Gun Model 1888 On 8-inch Disappearing Carriage LF Model 1896

Gun Characteristics
Bore size: 8-inch
Construction: Steel, built-up
Tube length overall: 278.5-inches
Nominal caliber: 32 caliber
Tube and breech weight: 32,480-pounds
Usual carriage: 8-inch Disappearing Carriage LF Model 1896
Weight of gun and carriage: 124,271-pounds
Maximum elevation: 12-degrees
Loading: Separate
Shell weight: 316-pounds (short cap)
Maximum muzzle velocity: 2,200-fps
Maximum range: 11,019-yards

Builders: Guns: Watervliet Arsenal,
West Point Foundry and
Bethlehem Steel
Carriages: Watertown Arsenal (3),
Pond Machinery (21),
Morgan Engineering (5),
Walker Engineering (5), and
Lake Erie Engineering (4).

Number built: 38 guns Model 1888M1 and 1888MII utilized 38 carriages

Raised gun of an 8-inch tube on Model 1896 Disappearing Carriage. Location unknown.

Completed 8-inch Model 1896 carriage and gun in lowered position at Battery Gibson, Fort Dupont, Defenses of the Delaware River.

Completing an 8-inch Model 1896 Disappearing Carriage. Note the assembled shell and powder cart also.

Very few disappearing carriages were ever provided as memorial guns to civic or veteran organizations. One known exception was the 8-inch Model 1888 gun on Model 1896 disappearing carriage donated to and put on display at Plant Park, Tampa, Florida. This postcard image from the 1930s shows the gun. It was subsequently scrapped during the Second World War.

Coast Defense 10-inch Guns

Of the Endicott and Taft Systems

This was the most common large-bore cannon built and deployed during the Endicott period. The first Model 1888 was a 34-caliber, built-up steel gun initially ordered in 1887, and delivered in 1890 (although unlike the first 12-inch, was entirely American-made). It used the conventional interrupted-screw, Welin-type breech mechanism. Since its development the Ordnance Department had successively reduced the number of necessary hoops or bands needed to reinforce the tube The Model 1888M1 and 1888M2 were progressive reductions, but otherwise gun characteristics and performance were identical. Following the type gun, one hundred seventeen 10-inch Model 1888s were fabricated. Watervliet Arsenal made sixty seven (24 M1888, 20 M1888M1, and 23 M1888MII) guns. Bethlehem Steel had been granted a "100-gun Contract" in 1891 as an incentive to make the capital investments in a new gun plant. Fifty of these guns were 10-inch Model 1888MIIs. The numerous Model 1888s were used on all 10-inch barbette carriages, the Model 1894 and a good proportion of the Model 1896 disappearing carriages.

A very similar Model 1895 gun was developed. It was slightly longer at 35 caliber, and featured an easier to operate, single handle, Stockett-type breech mechanism. Thirty-three were made at Watervliet Arsenal. It was used on Model 1896 and Model 1901 disappearing carriages. In 1900 an attempt (again in common with the 12-inch experience) was made to construct a more powerful, high-velocity 10-inch gun. However, it produced excess bore wear and had to be restricted to lower charges to produce the same muzzle velocity of the Model 1895 types. Just 14 of this 40-caliber guns were made and were used mostly on the Model 1901 disappearing carriage. All but the last four Model 1900 guns had the same rifling: 60 grooves with tapering rifling from 1 in 50 to 1 in 25. The last four M1990s had 90 grooves and used rifling of 0 to 1 in 25.

Five different carriages were utilized to mount 10-inch seacoast guns. The types and even model similarities for these carriages closely resemble corresponding carriages for both the 8-inch and 12-inch guns of the period. First a non-disappearing barbette carriage was developed. Simple in design and the lowest in cost, it was perceived useful only on high bluffs on the Pacific Coast. Only 11 of these carriages were ever produced in the 1890s, and just nine were actually emplaced in service batteries in Puget Sound. Next a front-pintle disappearing carriage (the Model 1894) was produced. Shortly thereafter the center-pintle Model 1896 disappearing carriage was made in large numbers. Both a limited-fire model (74 produced) and an all-around fire model (just 3 produced) were made. Finally in the early 1900s sixteen Model 1901 disappearing carriages were made for both the Model 1900 gun and for Model 1895 guns. Despite limitations from the carriages to elevation and range, they were not deemed important enough to justify increasing their range in the 1910s. However, many emplacements were modernized for ammunition lifts, electrification, and expanded loading platforms before the First World War.

In 1917 thirty 10-inch guns, a combination of Models 1888s and 1895s were authorized for removal from inner lines to be used on railway carriages for Europe (along with ten extra tubes in storage). Others were removed in the early 1920s with the discontinuance of various harbor defenses. Still many others served until ultimately replaced by more modern armament of the 1940 Program from 1942-1944. In common with much of the rest of coast artillery material, they were not considered for memorial gun use, and often too large and bulky even for museum distribution. Fortunately, a few have survived in Canada and the Philippines and two 10-inch disappearing guns were brought back to the United States from Subic Bay in the late 1960s for public display.

10-inch DC M1901 at Fort Standish, Boston

10-inch Gun Model 1888
On 10-inch Barbette Carriage Model 1893

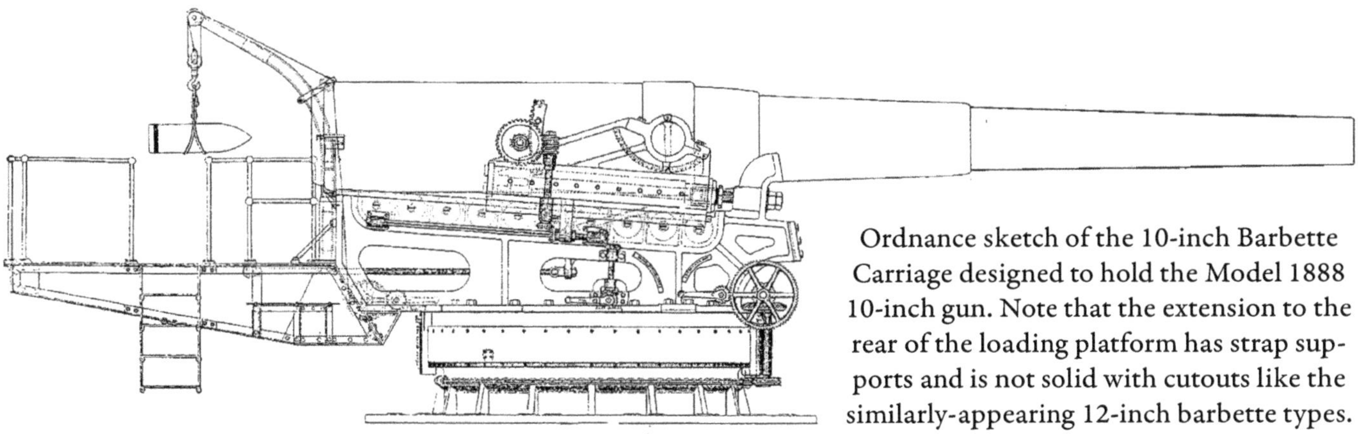

Ordnance sketch of the 10-inch Barbette Carriage designed to hold the Model 1888 10-inch gun. Note that the extension to the rear of the loading platform has strap supports and is not solid with cutouts like the similarly-appearing 12-inch barbette types.

The guns for this particular carriage were the standard 10-inch Model 1888 guns and modifications. Made were 24 Model 1888, 20 Model 1888M1 and 73 Model 1888MII. The only difference in these guns was the number of hoops used in their fabrication, otherwise their powder chambers and caliber of length were identical. The first Model 1888 was ordered in 1887, but others were built through 1901. It fired a 617-pound projectile using separate loaded powder of two sections. These guns were mounted on several different types of carriages, the barbette Model 1893 being one of the initial types.

The carriage was a center-pintle, gravity-return type. It had a base ring of cast iron bolted to a concrete emplacement platform. The gun traversed on 24 forged steel rollers on a racer. The cast steel chassis had two side members and transoms. The top of the chassis had rails inclined 4-degrees with nine rollers on each side to allow the top carriage to return by gravity to battery. The top carriage held the gun and two hydraulic recoil cylinders of 9.5-inch interior diameter. A capacious loading platform extended to the rear. Supporting this platform were triangular supports cut-out with an open framework, distinctly different from 12 and 8-inch barbette carriages and useful for photo identification. Like the 8-inch, but unlike the 12-inch, this size carriage used a chain and sprocket mechanism for achieving traverse. A prominent projectile hoist was attached to the chassis, as the breech height was such that it could be loaded only by lifting the shells rather than directly from carts. A rack and pinion elevation mechanism could provide -7 (useful on such carriages mounted on bluffs) to +15-degrees. By comparison, the contemporaneous disappearing

Morgan Engineering made five of the 10-inch Model 1893 barbette carriages; this is one their photos of a completed unit at their facility.

carriage for 10-inch could deliver just a -5 to +12-degree elevation range. A sight standard was on the right side of the chassis. All of the Model 1893 barbette carriages were intended to carry 10-inch guns Model 1888 or modification MI or MII gun tubes. There was a carriage nameplate on the right side of the chassis below the trunnion.

Eleven 10-inch Model 1893 barbette carriages were ordered, all from Watertown Arsenal. Five were ordered in January 1893, the others in 1898-99. Initially problems were encountered in obtaining the large steel castings Watertown required from contractors. Deliveries of finished carriages were made in 1895 and 1900. The 10-inch barbette was used almost exclusively in batteries in the Pacific Northwest. Five were mounted at Fort Worden, and four at Fort Flagler. The other two were used or stored at arsenals, though one being used in trial firings against experimental parapets at Fort Washington was temporarily moved to an active defense position during the Spanish-American War. The nine active carriages carried seven 10-inch Model 1888MII and two Model 1888 guns, all finally mounted in their emplacements in 1901-1902.

Four guns were dismounted in 1918 when their tubes were used to arm new railway mounts. The carriages for these guns were scrapped in place between 1920 (two) and 1932 (the last two). The other five served until early 1941 when they were transferred to new emplacements in Canada. The subsequent abandonment of these latter emplacements postwar in another national jurisdiction has meant a fortunate survival. One complete gun and carriage (originally from Battery Quarles, Fort Worden) and partial remains of another still exist on McNutt Island, Shelburne, Nova Scotia.

10-inch Gun Model 1888
On 10-inch Barbette Carriage Model 1893

Gun Characteristics
Bore size: 10-inch
Construction: Steel, built-up
Tube length overall: 367.25-inches
Nominal caliber: 34 caliber
Tube and breech weight: 67,183-pounds
Usual carriage: 10-inch Barbette Carriage Model 1893
Weight of gun and carriage: 144,637-pounds
Maximum elevation: 15-degrees
Loading: Separate
Shell weight: 617-pounds
Maximum muzzle velocity: 2,250-fps (with smokeless powder)
Maximum range: 14,062-yards

Builders: Guns by Watervliet Arsenal (No. 1-67) and Bethlehem Steel (No. 1-50)
Carriages by Watertown Arsenal (No. 1-11)

Number built: 2 Guns Model 1888 and 7 Model 1888MII used on 9 of the 11 carriages built

A mounted 10-inch barbette carriage Model 1893 at Fort Flagler, Puget Sound. It must be either Battery Rawlins or Revere.

One 10-inch gun on Model 1893 barbette was used during 1898 in trials for firing against various protection schemes at Fort Washington, Potomac River. It was relocated during the Spanish-American War for a defensive role, before being moved in 1900 for display at the Buffalo, New York Pan-Am Exposition.

Though partially damaged, one fairly intact Model 1893 barbette carriage survives. Two guns and carriages were supplied to Canada during WW2, and this one survives on McNutt Island, Shelburne, Nova Scotia.

10-inch Gun Model 1888
On 10-inch Disappearing Carriage LF Model 1894

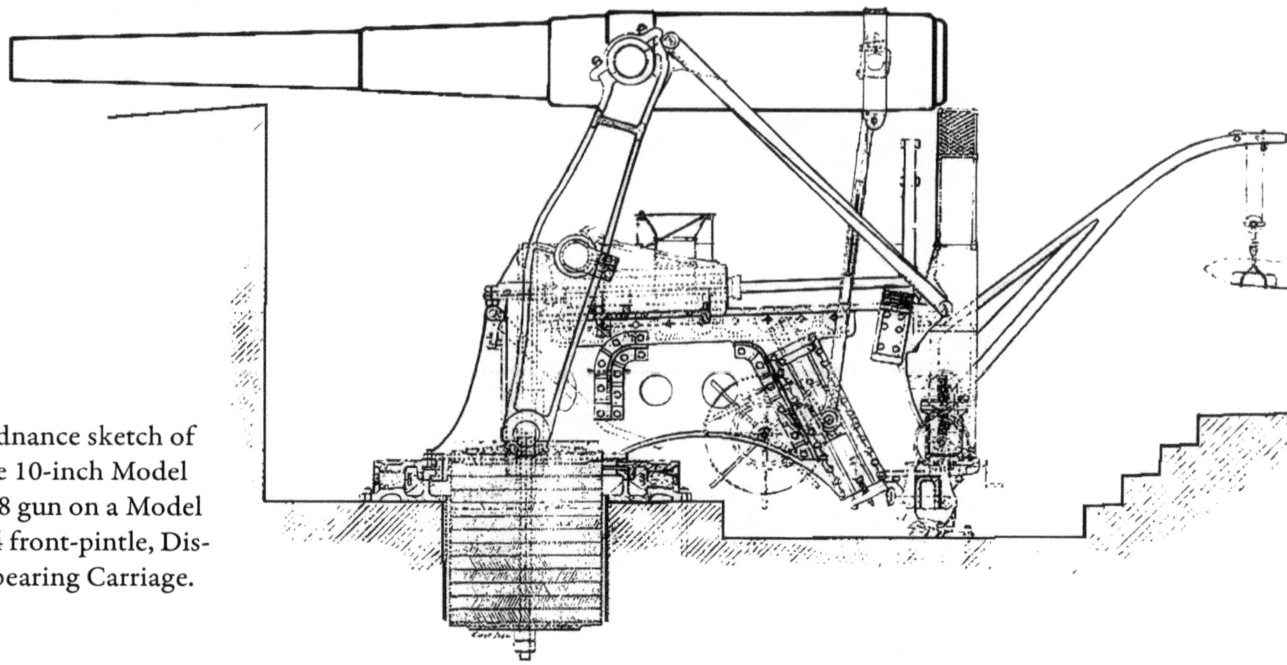

Ordnance sketch of the 10-inch Model 1888 gun on a Model 1894 front-pintle, Disappearing Carriage.

This carriage used the various Model 1888 types of 10-inch guns exclusively. This was a 34-caliber, steel built up gun with interrupted-screw breech mechanism developed in the late 1880s as one of the first modern heavy army guns. Through 1901 101 of the Model 1888 10-inch were produced and accepted for service.

The front-pintle limited-fire disappearing carriage was the early design type for this mechanism. Compared to the 8-inch version, it was more strongly built and had the gravity-return chassis rails placed at a 2-degree angle vs. being flat in the 8-inch type. The primary parts were the base ring and racer with 20 individual conical traversing rollers, a chassis of cast iron, top carriage of gun steel with recoil rollers and hydraulic recoil cylinders, gun levers, crosshead, lead counterweight, elevating arms, working platform and sighting platform. Recoil was absorbed by two hydraulic cylinders cast into the top carriage of 11-inches in diameter.

Upon firing the recoil action on the gun levers raised the counterweight. This counterweight was of lead and weighed about 79,000-pounds. The gun always returned to the same level to ease loading from hand trucks. The gun moved back into firing position through the action of the rollers on the inclined chassis top, just like a gravity-return carriage. The gun was raised again to firing position by releasing the pawls that dropped the counterweight. The counterweight was underneath the front part of the carriage, but the rear part was supported on wheels by a large racer ring. On this ring was a bronze azimuth circle marked in hundredths of degrees. It had a sighting platform on the right side of the chassis.

An experimental 10-inch (and the first 8-inch also of front-pintle type) was ordered from Southwark Foundry on June 22, 1891. It was delivered in early 1894 and evaluated and found satisfactory. Never given a serial number, this carriage was in fact rebuilt and emplaced itself in 1897. From late 1894 through 1896 34 disappearing carriages Model 1894 were ordered from Watertown Arsenal and four separate commercial contractors. Nineteen separate coast defense batteries in the continental U.S. had from one to three 10-inch carriages Model 1894. Generally, but not always, batteries were built to take just a single type of gun and carriage to simplify maintenance and training. All 35 carriages (including the original experimental carriage) were used in emplacements. In later years they were updated with a new tripping mechanism, electric retracting device and 3-inch telescopic sight. So modified they were designated as Model 1894M1 carriages.

10-inch Gun Model 1888 On 10-inch Disappearing Carriage LF Model 1894

Gun Characteristics
Bore size: 10-inch
Construction: Steel, built-up
Tube length overall: 367.25-inches
Nominal caliber: 34 caliber
Tube and breech weight: 67,183-pounds
Usual carriage: 10-inch Disappearing Carriage LF Model 1894
Weight of gun and carriage: 269,981-pounds
Maximum elevation: 12-degrees
Loading: Separate
Shell weight: 604-pound (short cap)
Maximum muzzle velocity: 2,250-fps
Maximum range: 12,259-yards

Builders: Guns by Watervliet Arsenal and Bethlehem Steel
Carriages by Southwark Foundry (No. Exp), Kilby Manufacturing (No. 1-10), Pond Machinery (No. 11-18), Bethlehem Steel (No. 19-22), Cramp Shipyard (No. 23-24) and Watertown Arsenal (No. 25-34)

Number built: Used 35 of 117 Model 1888 guns made 35 carriages Model 1894

The Corps of Engineers designed the 10-inch emplacement Type 1894 for use with this carriage. Usually paired, the guns were set at a distance of 124-feet apart. Each gun had its own separate set of magazines on the lower level of the flank of the emplacement, with an additional storage of 10-inch shot around the circular wall in front of the gun block. All were built in the 1890s, though modifications made over the next twenty years added wider loading platforms, electrification, and ammunition hoists.

They served in their coast defense role, though usually on the inner lines at most forts. A number were dismounted in 1917 to provide 10-inch guns for railway mounts, the carriages being scrapped soon thereafter. Others disappeared with the discontinuance of several smaller harbor projects in the early 1920s. Half (18) were scrapped and gone by 1930. However several lasted until the 1940s as an active part of the defenses, and were not eliminated until 1942-43. No 10-inch Model 1894 carriages still exist, though most of the emplacements still do.

Etching of the M1894 10 inch disappearing carriage (Scientific American)

Nice photographic view of a 10-inch Model 1894 Disappearing Carriage at Battery Sevier or Cullum, Fort Pickens, Pensacola Florida in the 1910s.

Battery Ward, Fort Strong in Boston was equipped with 10-inch Model 1894 Disappearing Carriages.

10-inch Gun Model 1888 and Model 1895 On 10-inch Disappearing Carriage LF Model 1896

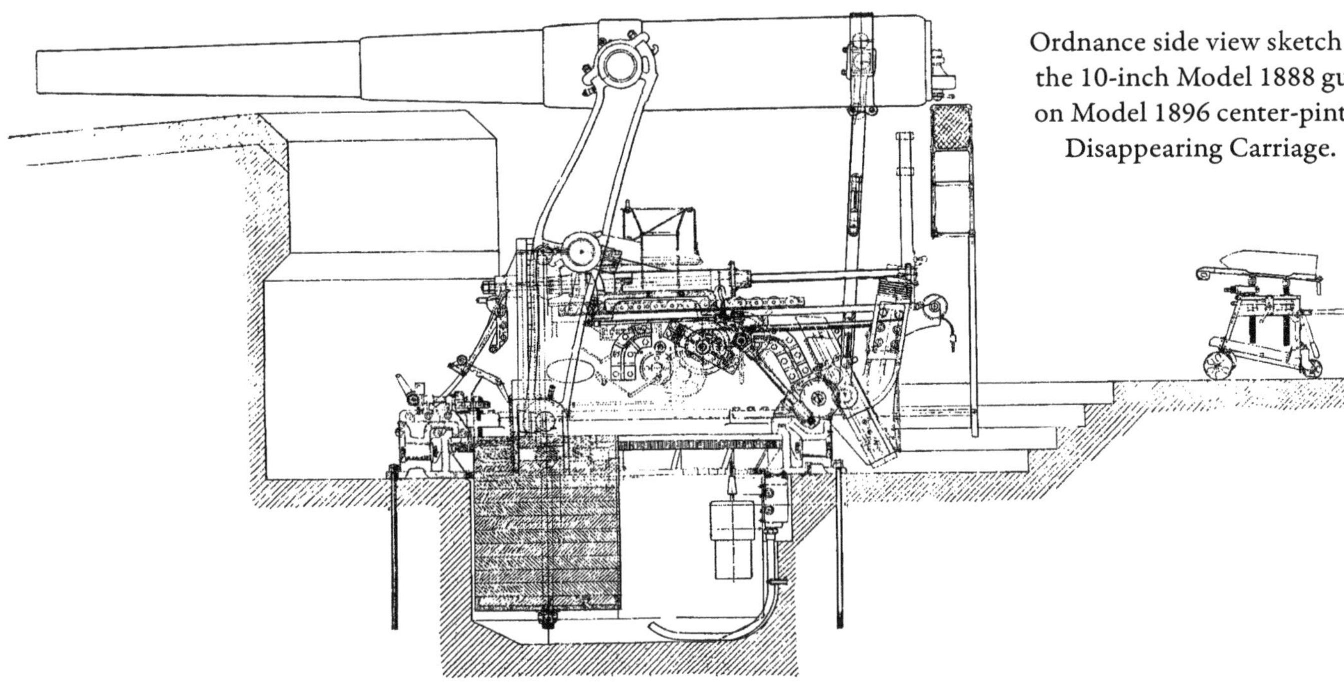

Ordnance side view sketch of the 10-inch Model 1888 gun on Model 1896 center-pintle Disappearing Carriage.

A new center-pintle disappearing carriage for the 10-inch gun was developed soon after the front-pintle Model 1894. While initially designed for the same Model 1888 gun, it could with minor adjustments also accommodate the new Model 1895 gun. The latter was an improved 10-inch bore gun of 35-caliber length. Its major innovation was the use of the Stockett-type breech mechanism that with a single handle was easier and safer to open and close. The first 48 Model 1896 carriages were built for the Model 1888 10-inch gun, the last 26 could carry both the Model 1888 and Model 1895 gun.

This limited-fire disappearing carriage was typical in design of the Buffington-Crozier type of mechanism. The primary parts were the base ring and racer with individual traversing rollers (30), a chassis, top carriage with recoil rollers and hydraulic recoil cylinders, gun levers, crosshead, lead counterweight, elevating arms, working platform and sighting platform. These parts were mostly made of cast iron or cast steel as necessary. The lead counterweight weighed 77,638-pounds. Upon firing the action by the carriage's gun levers raised the counterweight. The recoil was stopped by the action of the hydraulic pistons of the two recoil cylinders. The gun always returned to the same level to ease loading from hand trucks. It was moved back into firing position through the action of the rollers on the inclined chassis top, just like a gravity-return carriage. The gun was raised again to firing position by releasing the pawls that dropped the counterweight. A sighting standard and platform was usually attached to the left side of the chassis for the first 48 carriages, to the right on the last 26. The carriages were provided electrical firing and lighting from the beginning.

The carriage design was perfected in 1895-96, it was not thought necessary to build an experimental or type carriage. The first contracts for production orders were signed on August 17, 1896. A total of 74 Model 1896 LF Disappearing Carriages were ordered from 1896 through 1900 (Serial numbers 1-74). Five private contractors in addition to Watertown Arsenal were used as fabricators. All 74 were used in emplacements; initial gun armament was 56 armed with various Model 1888 guns, and 18 with Model 1895 guns, though that changed slightly over time as various batteries were disarmed and then rearmed.

The Corps of Engineers built 74 emplacements for this carriage as all or part of 34 separate batteries. Batteries had between one and four guns of this type.

The Type 1896, 1898, and 1900 battery designs all had guns separated by 124-feet, with separate powder and shell magazines protected on a lower level to the flank of each gun emplacement. At the carriage pit were four steps leading down from the loading platform to the working platform. After construction in 1896-1904 many were modified with widened loading platforms, revised ammunition lifts, new battery commander stations, improved access galleries and latrines.

Some of the guns were part of the 30 removed from batteries in 1917 for use on railway carriages, others were scrapped when smaller harbor defenses and redundant batteries were eliminated in the early 1920s. Still, many of the Model 1896 disappearing carriages lasted until replaced by more modern batteries in 1942-44. No examples of the carriage still exist, though a very few of each of the two gun barrel types exist with other carriage type survivors. Most of the emplacements still survive at state and national parks.

10-inch Gun Model 1895
On 10-inch Disappearing Carriage LF Model 1896

Gun Characteristics
Bore size: 10-inch
Construction: Steel, built-up
Tube length overall: 369.15
Nominal caliber: 35 caliber
Tube and breech weight: 66,700-pound
Usual carriage: 10-inch Disappearing Carriage LF Model 1896
Weight of gun and carriage: 193,000-pounds
Maximum elevation: 12-degrees
Loading: Separate
Shell weight: 604-pounds (short cap)
Maximum muzzle velocity: 2,250-fps
Maximum range: 12,259-yards

Builders: Guns by Watervliet Arsenal and Bethlehem Steel
Carriages by Niles Tool Works (No. 1-8 and 43-48), Southwark Foundry (9-18 and 62-64), Bethlehem Steel (19-30, 37-42, and 55-61), Watertown Arsenal (31-36 and 49-54), Walker Engineering (65-70) and Midvale Steel (71-74)

Number built: Used 56 10-inch M1888 and 18 10-inch M1895 74 Carriages

One of the M1896 10-inch Disappearing Carriages fabricated by Midvale Steel in 1900.

10 inch M1888 gun on a M1896 disappearing carriage at Fort Winfield Scott

A 10-inch gun on a Model 1896 DC in the raised position. Location or date unknown, but note the unusual camouflage paint scheme.

10-inch Gun Model 1888
On 10-inch Disappearing Carriage ARF Model 1896

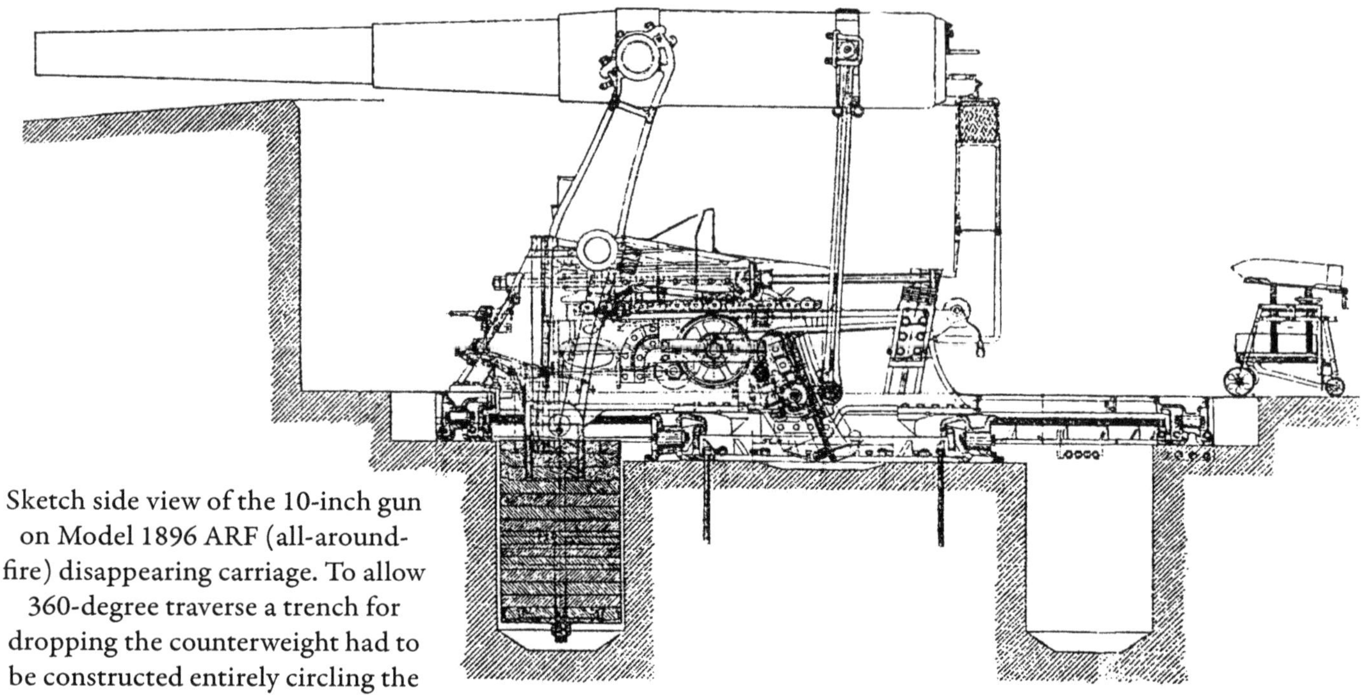

Sketch side view of the 10-inch gun on Model 1896 ARF (all-around-fire) disappearing carriage. To allow 360-degree traverse a trench for dropping the counterweight had to be constructed entirely circling the center pintle.

This was a unique gun-carriage model designed to provide 360-degree traverse. It had been anticipated that there would be some coast defense sites that required fields of fire to both the front and rear—and consequently guns might be needed to fire at all azimuth bearings. The Model 1896 ARF carriage was designed to satisfy this need. It was designed to carry either the Model 1888 or Model 1895 10-inch gun, but it turned out that only the Model 1888 was used in the relatively few such carriages actually fabricated. That type of gun was the first 10-inch type developed in the late 1880s. It was a serviceable weapon of 34-caliber length of built-up steel hoops using a common interrupted-screw breech mechanism. A total of 117 such guns were built, being used on several different types of coast defense carriages.

The all-round-fire disappearing carriage utilized the same basic actions as other Buffington-Crozier type carriages. The primary parts were the base ring and racer with individual traversing rollers, a chassis, top carriage with recoil rollers and hydraulic recoil cylinders, gun levers, crosshead, lead counterweight, elevating arms, working platform and sighting platform. These parts were mostly made of cast iron or cast steel as necessary. Upon firing the action gun levers raised the counterweight. The recoil of the gun was stopped by the piston rods in the hydraulic recoil cylinders. The gun always returned to the same level to ease loading from hand trucks. The gun moved back into firing position through the action of the rollers on the inclined chassis top, just like a gravity-return carriage. The gun was raised again to firing position by releasing the pawls that dropped the counterweight. There was a sight standard and platform on the left chassis accessible by attached ladder.

Unlike either front or center-pintle carriages, the center of the gun was further to the rear. The carriage rested on two concentric, 360-degree base rings. The smaller was in the center, in between it and the outer ring was a concentric, circular well 52-inches wide to receive the counterweight. Normally it was covered with a metal floor, and only open directly under the weight as it revolved around the well. It was a complicated and cumbersome mount; and the need for loading and protection at any bearing meant the same for the emplacement.

Three carriages were ordered at Watertown Arsenal, and completed by 1899. Only two batteries were actually built; a two-gun battery at Fort Stevens on the Columbia River, and one emplacement in a two-gun battery at Fort San Jacinto in Galveston. In both locations the forts were on a peninsula that could be approached from and fire on from the rear, thus the need for coverage in the reverse direction. All three were emplaced by mid-1900. However the Army quickly decided that the costs for the carriage and emplacement were disproportionately high, and no others were ever constructed. In 1910 it was decided that the three emplaced guns weren't really useful for reverse fire, and stops were put in to limit the coverage. That action greatly facilitated running electrical circuits to the guns and simplified their operation in several ways.

All three carriages received model 1888 10-inch guns. Despite the awkwardness of the emplacement even with modifications, they were not removed during or after the First World War. The Galveston emplacement was disarmed in 1943, the Columbia River battery was roofed over to provide for a Harbor Defense Command Post for Fort Stevens, but the guns were left inside—and were not removed and scrapped until after the end of the war. None of the carriages exist. The Fort San Jacinto emplacement is now gone, but the Fort Stevens battery remains intact, though roofed over with concrete.

10-inch Gun Model 1888
On 10-inch Disappearing Carriage ARF Model 1896

Gun Characteristics
Bore size: 10-inch
Construction: Steel, built-up
Tube length overall: 367.25-inches
Nominal caliber: 34 caliber
Tube and breech weight: 67,183-pounds
Usual carriage: 10-inch Disappearing Carriage ARF Model 1896
Weight of gun and carriage: 445,679-pounds
Maximum elevation: 12-degrees
Loading: Separate
Shell weight: 604-pounds (short cap)
Maximum muzzle velocity: 2,250-fps
Maximum range: 12,259-yards

Builders: Guns by Watervliet Arsenal and
Bethlehem Steel
Carriages by Watertown Arsenal

Number built: 3 guns Model 1888 used
of 117 built
3 carriages

The only battery was actually fully built to house the 10-inch ARF was Battery Mishler at Fort Stevens, Columbia River. This photograph shows that battery in service, with the two guns deep within its massive pits.

Gun Drill at Battery Mishler, Fort Stevens

One other 10-inch ARF was emplaced, at Battery Heileman, Fort San Jacinto, Galveston. In this case it was placed in a conventionally-configured emplacement. Still, the carriage appearance is distinctive.

10-inch Gun Model 1895 and 1900
On 10-inch Disappearing Carriage LF Model 1901

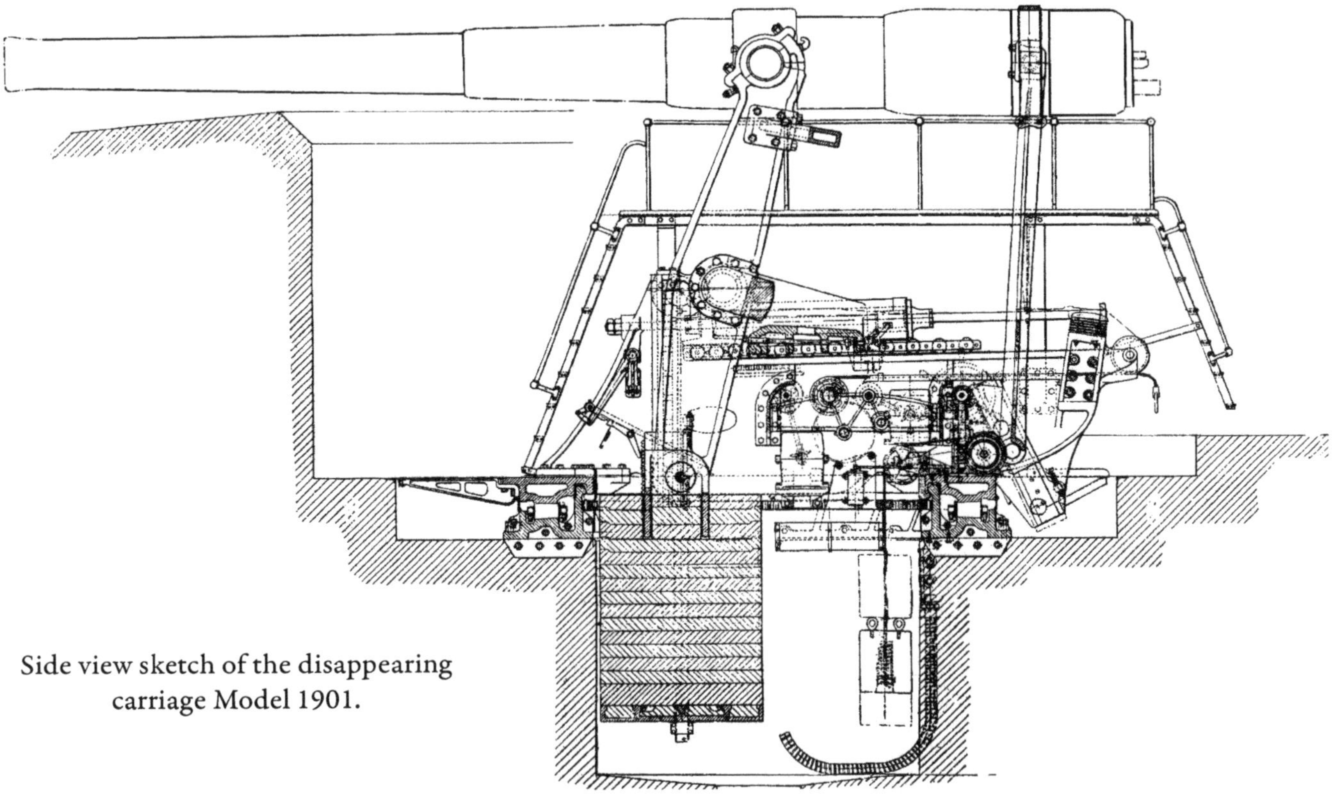

Side view sketch of the disappearing carriage Model 1901.

The final seacoast gun mount for guns of 10-inch size was for the more advanced 10-inch Model 1900. Designed to be a more powerful gun of greater length (40-caliber) and greater muzzle velocity (2500-fps), it was put into production in 1901. Unfortunately the larger powder charge and muzzle velocity led to much accelerated bore erosion. That was unacceptable for fixed guns, and in 1908 orders were issued to reduce the charge and in some guns to insert a liner to reduce the chamber size. This resulted in the guns only having the same velocity and general performance of the earlier Model 1888 and 1895 guns. In any event only 14 guns were fabricated at Watervliet Arsenal. The first 11 guns had 60 grooves with rifling of 1 in50 to 1 in 25, the last three had 90 grooves with rifling of 0 to 1 in 25.

This limited-fire disappearing carriage was typical in design of the Buffington-Crozier type of mechanism. The primary parts were the base ring and racer with individual traversing rollers, a chassis, top carriage with recoil rollers and hydraulic recoil cylinders, gun levers, crosshead, lead counterweight, elevating arms, working platform and sighting platform. These parts were mostly made of cast iron or cast steel as necessary. Upon firing the action of the carriage's gun levers raised the counterweight. The recoil was stopped by the action of the hydraulic pistons of the two recoil cylinders. The gun always returned to the same level to ease loading from hand trucks. It was moved back into firing position through the action of the rollers on the inclined chassis top, just like a gravity-return carriage. The gun was raised again to firing position by releasing the pawls that dropped the counterweight. Compared to the previous model of disappearing carriage it had a more robust construction. It also had two sighting platforms with ladders, one on each side of the chassis. For photo recognition, the rear of the lower carriage had a distinctive curved design. Sixteen carriages were all built at Watertown Arsenal. Eight were ordered in 1902, four in 1903 and four in 1905.

An emplacement design Type 1903 was specifically developed for this carriage. Distance between guns was extended to 137-feet. It had most of the features which later had to be added to the previous types of 10-inch emplacements. Only two steps led down from the loading level to the working platform. Five complete two-gun batteries were built, and two other single-gun emplacements being part of other batteries were built of this type and armed with the Model 1901 carriage. One of the latter (Battery Walbach, Fort Wetherill) actually was armed with an older Model 1888 gun; the others all got Model 1900 guns. Two additional dual batteries of this type were built in the Philippines, but their Model 1901 carriages received Model 1895 10-inch guns. Curiously the latter two emplacements had special, deep counterweight wells to accommodate cast iron rather than lead counterweights.

Despite the disappointment in the Model 1900 gun, there was no problem in using the modified weapon. The guns and carriages served until the Second World War. The U.S. batteries were taken out and scrapped in 1942-44. The two Philippine batteries were abandoned when American forces surrendered in 1942. The two guns and Model 1901 carriages still exist at Battery Grubbs on Corregidor, and the two from Battery Warwick at Subic Bay were relocated and are on display at Fort Casey State Park near Coupeville, Washington.

10-inch Gun Model 1900 On 10-inch Disappearing Carriage LF Model 1901

Gun Characteristics
Bore size: 10-inch
Construction: Steel, built-up
Tube length overall: 420.62-inches
Nominal caliber: 40 caliber
Tube and breech weight: 76,830-feet
Usual carriage: 10-inch Disappearing Carriage LF Model 1901
Weight of gun and carriage: 334,000-pounds
Maximum elevation: 12-degrees
Loading: Separate
Shell weight: 604-pounds (short cap)
Maximum muzzle velocity: 2,250-fps
Maximum range: 14,200-yards

Builders: Guns by Watervliet Arsenal
Carriages by Watertown Arsenal

Number built: 14 Model 1900 10-inch guns and 16 Model 1901 disappearing carriages

One of the 10-inch Model 1900s on M1901 carriage at Battery Russell, Fort Stevens, Columbia River. This shot clearly shows the noticeable hoop around the chamber area of the gun tube.

An emplaced 10-inch Model 1900 gun on Model 1901 Disappearing Carriage at Battery Morris, Fort Standish, Boston.

Four Model 1901 carriages survived the war in the Philippines, along with five Model 1895 10-inch tubes. The two guns and carriages from Fort Wint were relocated to Fort Casey in Washington state for display purposes. Even though residing in emplacement pits designed for an older carriage type (the Model 1896) they are impressive representatives.

Coast Defense 12-inch Guns

Of the Endicott and Taft Systems

Guns of 12-inch bore were adopted as the primary heavy coast defense gun of the early systems. This was estimated to be the largest practical size for the country's manufacturing capacity. Additionally, it was almost universally the gun size of choice for battleship main battery. With the theory that you needed to engage attacking ships with like-sized armament, it was taken on by the Army for developing, procuring, and emplacing. Seacoast gun batteries of 12-inch rifled ordnance were the nation's primary defender from 1890 right up until the 1940 Program just prior to the Second World War.

Three heavy guns were developed simultaneously by the Army—the 8, 10, and 12-inch breechloading, rifled guns Model 1888. The 12-inch was built-up from gun steel of numerous hoops, over a short period it was found possible to reduce the hoops in the M1888 from 7 A hoops, 8 B hoops, 10 C hoops and 5 D hoops to just 5, 4, 2, and 2 in the final Model 1888 MII. The gun was 34 caliber. All the 12-inch guns had 72 grooves and rifled from 1 in 50 to 1 in 25 at the muzzle. The breech was an interrupted screw, Welin type. A tray assisted loading. Like the similar 10 and 8-inch guns, the designs were not particularly innovative. However, they held their own in comparison to foreign designs, and many had long lives.

Obstacles were encountered with producing the first guns. The American steel industry was still discovering how to forge large pieces of steel. The first 12-inch type gun was assembled from forgings purchased from Schneider in France; all subsequent guns were of American manufacture from forgings coming from Bethlehem or Midvale Steel. Fifty-six Model 1888 12-inch guns were produced. Ten were fabricated by Bethlehem and 46 by Watervliet Arsenal. A slightly improved Model 1895 gun was then developed. It had different hoop arrangements, was slightly longer at 35-calibers, and had a simplified breech opening mechanism. Ninety of these were made, all by Watervliet except for 15 assembled by Bethlehem. However, the final Model 1900 an attempt was made to make a longer (40-caliber) gun of higher muzzle velocity to get more range and armor penetration. Eleven of these guns were made at Watervliet. Unfortunately, the design suffered from short bore life and eventually the gun could only be fired at a lower velocity that sacrificed all its other potential. All three models continued to serve well into the Second World War. The Model 1895 was altered to the M1895A1 for railway mountings and the M1895A2 for a new long-range barbette carriage.

A number of carriages were developed and produced for these guns. The first one was for M1888 guns on elevator-like "lifts" and called the Model 1891 Gun Lift Carriage. This was followed by the Model 1892 Barbette Carriage where the guns could traverse and elevate but always stayed exposed at a fixed height. Three unused gun lift carriages were converted to a similar type and emplaced in the Pacific Northwest. The Army's real preference was for the Buffington-Crozier type of disappearing carriage. Numerous carriages were built of the M1896 (mostly for the Model 1888 guns), M1897 (for the Model 1895 gun) and M1901 (originally for the Model 1900 gun but also used for the M1895 type). Coast defense guns in these days were mostly organized into named batteries containing 2-3 guns of the same type bore and carriage.

The 12-inch guns were gradually removed and either scrapped or redeployed to other uses in the First World War or with the reduction in the number of harbors defended. Finally in 1942-44 all the remaining batteries were removed as obsolete and replaced with more modern armament.

12-inch DC M1897 at Battery Palmer, Fort Michie

12-inch Gun Model 1888
On 12-inch Gun Lift Carriage Model 1891

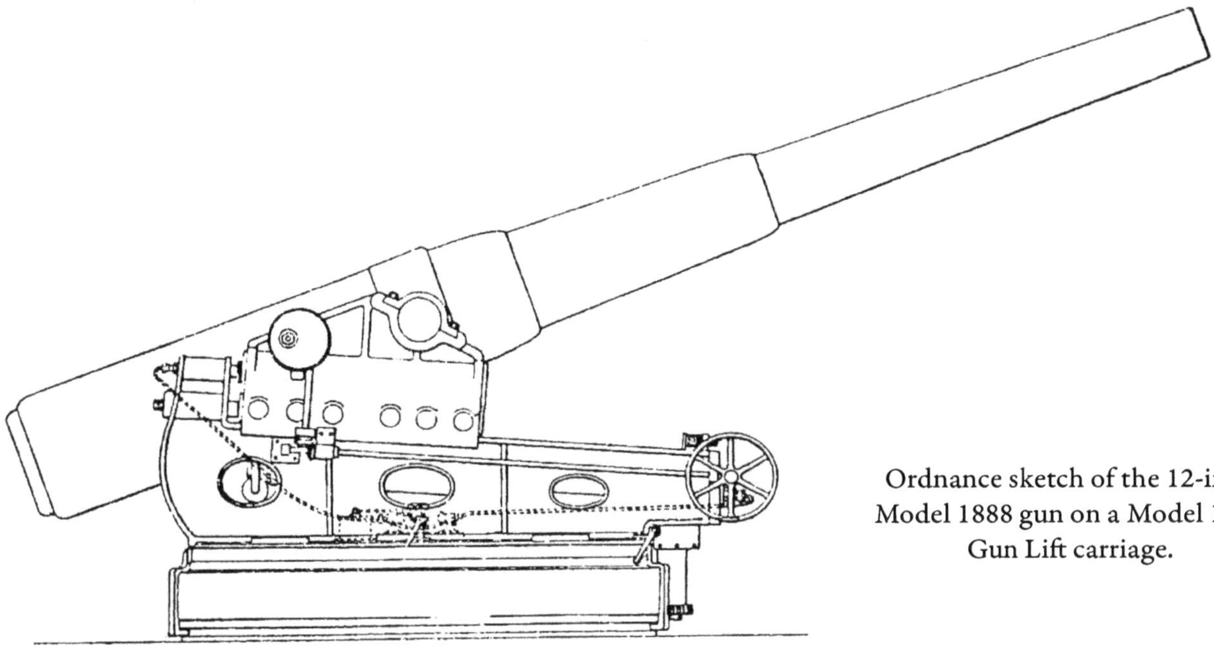

Ordnance sketch of the 12-inch Model 1888 gun on a Model 1891 Gun Lift carriage.

The gun lift concept was an original suggestion of the Endicott Board that recommended U.S. Coast defense policy in the late 1880s. Conceptually guns and carriages were to be lifted from within a large battery structure on elevator-like platforms into their firing positions. Once fired, they could be lowered for protection and reloading. The guns themselves were to be standard types. The barbette-like carriages, once lifted and locked into position, had the usual traversing and elevating capabilities. The lift mechanism and special carriages were begun as Army projects very early in its development schedule for modern seacoast armament.

The carriage was a gravity-return barbette carriage with dimensions to fit the projected elevator. A French type carriage was selected as a starting point, but with strengthened piston rods for the recoil cylinders. It was a fairly simple design – two large side frames connected with front and rear transoms. Trunnions held the gun on a moving top carriage. Movement in recoil was by rollers held in notches on the top carriage. Recoil was braked by the hydraulic cylinders, one beneath the gun, and two on the sides of the top carriage Return was by gravity on inclined bearing surfaces. Elevation on the carriage was -7 to +20-degrees. Both elevation and traversing were by hand-cranked gears.

As the emplacement and its mechanism were very large and expensive, it was logical to use this type of emplacement only for the most powerful weapons. Consequently, the gun selected for the emplacement was the 12-inch Model 1888. The first Model 1888 type gun (serial No. 1) had been assembled using French (Schneider) forged steel parts in May 1891. The gun was 34-calibers long and of built-up steel. A total of 56 guns of this type were built, the first two going into this gun lift emplacement. The first carriage was also adopted from the French. Henri Schneider & cie. received an order on July 12, 1889 for one of its large gun carriages with custom dimensions to fit the projected size of the intended elevator. Even though not as difficult to fabricate as a gun tube, it took almost three years before this carriage was completed. Difficulties with continuous design changes, labor conditions in France, and weather combined to delay the delivery to Sandy Hook until June 20, 1892.

The emplacement itself (eventually named Battery Potter) was a massive concrete box built at Fort Hancock on Sandy Hook, New Jersey. This fort held a major position in protecting the southern entry into New York Harbor. Inside the huge emplacement was a steam plant and hydrostatic water pumps for elevating the platforms. These were built by the Continental Iron Works under a contract of February 20, 1891. Physical work was done in 1891-1892. Eventually the emplacement was turned over the coast artillery troops in early 1898 at an estimated cost (with armament) of $455,028—a very high number for that time. The first gun was mounted on the Schneider carriage in September 1892. It did work as designed, taking only 20 seconds to raise or lower. The second gun and carriage were not emplaced until 1896.

While it worked as designed, the cost and effort of the design were way out of proportion to other options. Another battery in Fort Delaware was begun but changed to a disappearing battery during construction. The carriages built for it were later altered to barbettes. Battery Potter remained the sole example of this emplacement type and carriage. The guns served a few years, but about 1906 were removed. The old gun lifts were abandoned in place for a number of years, not being salvaged until 1917-1919. The emplacement was later used as an advantageous spot for fire control structures and still remains under National Park Service care.

12-inch Gun Model 1888
On 12-inch Gun Lift Carriage Model 1891

Gun Characteristics
Bore size: 12-inch
Construction: Steel, built-up
Tube length overall: 439.92-inches
Nominal caliber: 34 caliber
Tube and breech weight: 117,127-pounds
Usual carriage: 12-inch Gun Lift Carriage Mod 1891
Weight of gun and carriage: n.a.
Maximum elevation: 20-degrees
Loading: Separate
Shell weight: 1,046-pounds
Maximum muzzle velocity: 2,250-fps
Maximum range: n.a.

Builders: Guns: Watervliet Arsenal
Carriages: 1 by Schneider et. Cie.,
1 by Watertown Arsenal

Number built: Guns: 2 12-inch Model 1888 used in gun lift batteries
Carriages: 2

The only gun-lift battery built by the U.S. Army—Battery Potter, Fort Hancock

The first gun mounted at gun-lift battery Potter at Fort Hancock was fabricated by Schneider in France. Mounting was completed in late 1892.

The same gun of Battery Potter being lowered on its lift to the re-loading and servicing level.

12-inch Gun Model 1888
On 12-inch Altered Gun Lift Carriage Model 1891

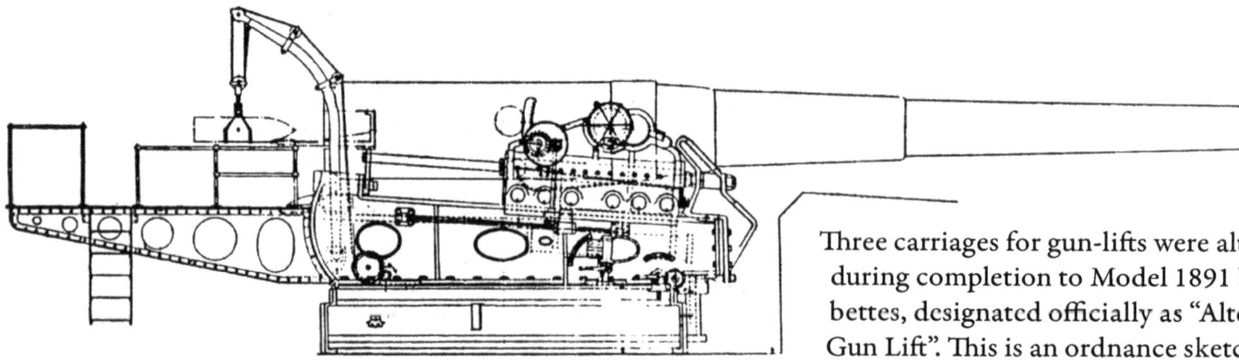

Three carriages for gun-lifts were altered during completion to Model 1891 barbettes, designated officially as "Altered Gun Lift". This is an ordnance sketch of their appearance, notice the new rear loading platform and shell hoist which had to be added.

After the relative success of the first gun lift battery, plans were cast for several others. A second such battery was begun inside old Fort Delaware on the Delaware River protecting the approaches to Philadelphia. This time the battery was to be for three gun lifts. Its total height calculated to fit nicely into the three-story old brick fort. Work began in late 1894, but by the end of 1895 (after $210,000 had been expended) work was suspended pending a review of the type of battery. It was decided to finish the battery as a disappearing gun emplacement; and eventually Battery Torbert was transferred to coast artillery troops on April 18, 1901. By the time the change was ordered, considerable work had been accomplished on the three gun lift carriages intended for this emplacement. Two new lift carriages (No. 3 and No. 4) had been ordered on January 9, 1893, and the fourth (No. 5) on September 7, 1894. The first pair were reported complete and the third unfinished but most of the major components fabricated. The intended gun tubes were the "standard" 12-inch Model 1888, and they were simply diverted to other barbette or disappearing carriage match-ups.

Orders were given to Watertown Arsenal to convert the carriages to barbettes on June 6, 1896 and that was reported accomplished in late 1897. In fact, it does not appear that much had to be done to accomplish this, though a new extended loading platform to the rear of the gun was added at this time and it was equipped with a new ammunition crane. The recoil system was modified by essentially reversing the piston to pull rather than a push through its hydraulic cylinder. The pattern of cut-outs in the support under the loading platform and shape of the front and rear transom lines of the chassis are distinguishing photo cues to differentiate photos of the altered gun lift from those of the standard Model 1892 barbette carriage. While the conversions served for several years, they were never entirely successful and caused numerous problems in service.

Once converted the carriages were considered for use as any other barbette carriage, but it did make sense to keep them together in the same geographic location. All three carriages were sent to Puget Sound on March 23, 1898 to be used in the armament program there. Because of the height of the bluff at Flagler, that site was considered suitable for barbette batteries versus disappearing ones. Battery Wilhelm, of two 12-inch guns, was constructed at Flagler in 1897-98, transferring in August 1902. Two completed altered carriages (renumbered as No. 1 and No. 2) were used for this battery, carrying 12-inch guns Model 1888. The third carriage was used in emplacement No. 5 of a seven-gun sequence at nearby Fort Worden. This was initially named Battery Quarles and carried AGL carriage No. 3. In 1905 one of the carriages at Wilhelm was seriously damaged in a service firing. Consequently (in 1907) the Worden carriage was moved to Flagler to be used in reconstructing the damaged unit and replaced at Fort Worden with a Model 1892 barbette carriage.

Despite the chronic problems with the mounts, the two remaining AGL carriages lasted at Battery Wilhelm at Fort Flagler through the interwar period, not being ordered for salvage until October 24, 1942. None survive today, though the two emplacements remain at Fort Flagler State Park.

12-inch Gun Model 1888 On 12-inch Altered Gun Lift Carriage Model 1891

Gun Characteristics
Bore size: 12-inch
Construction: Steel, built-up
Tube length overall: 439.92-inches
Nominal caliber: 34 caliber
Tube and breech weight: 117,127-pounds
Usual carriage: Altered Gun Lift Carriage Model 1891
Weight of gun and carriage: n.a.
Maximum elevation: 20-degrees
Loading: Separate
Shell weight: 1046-pounds
Maximum muzzle velocity: 2,250-fps
Maximum range: n.a.

Builders: Guns by Watervliet Arsenal and Bethlehem Steel
Carriages by Watertown Arsenal

Number built: 3 12-inch Model 1888 guns used
3 carriages built (No. 1 – 3)

One of the Model 1891 Altered Gun Lift carriages and its Model 1888 12-inch gun emplaced at Battery Wilhelm, Fort Flagler, Puget Sound. Even though initially being a somewhat convenient expedient, they lasted in service until the Second World War.

12-inch Gun Model 1888
On 12-inch Barbette Carriage Model 1892

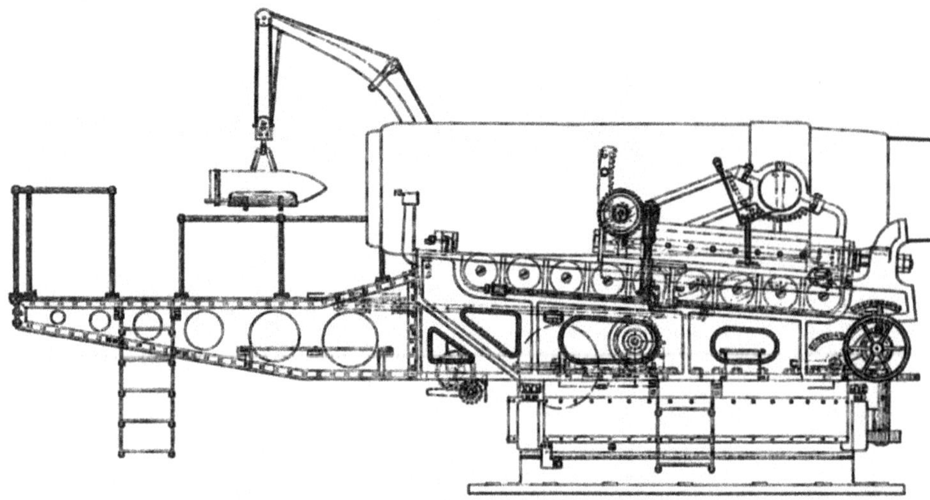

Ordnance sketch of the conventional Model 1892 barbette carriage for the Model 1888 12-inch gun.

Pending the successful design of new gun lift and disappearing carriages, the Army prepared designs for simpler gravity-return (Vavasseur pattern) barbette carriages for all of its new heavy guns. It developed the new Model 1892 barbette carriage without problem, testing it in 1894. As crew protection figured heavily into design consideration, barbettes were normally employed only at sites with considerable elevation; it being assumed that engaged ships would not be able to fire back against such emplacements. The first batch of such carriages ordered from Watertown Arsenal were specifically intended for emplacing on the bluffs on either side of the Golden Gate in San Francisco.

The carriage was a center-pintle, gravity-return type. It had a base ring of cast iron bolted to a concrete emplacement platform. The gun traversed on 24 forged steel rollers on a racer. The cast steel chassis had two side members and transoms. The top of the chassis had rails inclined 4-degrees with nine rollers on each side to allow the top carriage to return by gravity to battery. The top carriage held the gun and two hydraulic recoil cylinders. An attached loading platform extended to the rear. Supporting this platform were triangular supports cut-out with six circular perforations, useful for photo identification. A prominent projectile hoist was attached to the chassis, as the breech height was such that it could be loaded only by lifting the shells rather than directly from carts. A pinion elevation mechanism attached to a forged steel rack on the right side of the gun. It could provide -7 (useful on such carriages mounted on bluffs) to +15-degrees. All of the Model 1892 barbette carriages were intended to carry 12-inch guns Model 1888. There were 56 of these 34-caliber guns of various subtypes produced from 1891 through 1900. They were used on these, the gun lift, and first generation of disappearing carriages. The gun fired a 1046-pound common or AP shell with a 2250-fps muzzle velocity.

The first carriage was ordered from Watertown in late 1891 and delivered in January 1894. Eight more followed from Watertown until 1898. During the Spanish American War, a big order of 17 carriages was placed with Morgan Engineering. There was an excess of Model 1888 tubes available, and the relatively simple barbette carriage made at a private manufacturing establishment was the quickest way to get these gun mounted. Two more Watertown carriages were ordered in 1899 and 1900. Altogether 28 Model 1892 barbette carriages were made. Complete usage was made of this inventory, all 28 were mounted in batteries by 1902. Fifteen batteries utilized barbettes (to at least some extent, two batteries had mixed barbette and disappearing type carriages). Somewhere along the line the concept of using these only at elevated sites was lost, batteries in Charleston, Key West,

Wilmington and Baltimore were practically at the waterline. Altogether the fifteen batteries contained one to three each Model 1892 12-inch barbette carriages.

Several barbette batteries were salvaged when their harbor defenses were closed down in the 1920s. Generally, the removed gun tubes were saved as spares while the carriages were scrapped in place. The remaining guns and carriages were authorized for removal and scrapping in 1942-44 as their function was being replaced by more modern armament. Unfortunately, while most of the original emplacements still exist, none of these gun and carriage combinations survive.

12-inch Gun Model 1888
On 12-inch Barbette Carriage Model 1892

Gun Characteristics
Bore size: 12-inches
Construction: Steel, built-up
Tube length overall: 439.92-inches
Nominal caliber: 34 caliber
Tube and breech weight: 117,127-pounds
Usual carriage: 12-inch Barbette Carriage Model 1892
Weight of gun and carriage: 22,927-pounds
Maximum elevation: 15-degrees
Loading: Separate
Shell weight: 1,046-pounds
Maximum muzzle velocity: 2,250-fps
Maximum range: 17,342-yards

Builders: Guns by Watervliet Arsenal and Bethlehem Steel
Carriages by Watertown Arsenal (11), Morgan Engineering (17)

Number built: 28 Model 1888 guns used on 28 M1892 carriages

Morgan Engineering fabricated under contract seventeen Model 1892 barbette carriages for the Model 1888 gun. This is one of their photographs showing a completed example.

The 12-inch gun and barbette Model 1892 carriage at Battery Read, Fort Dupont, Delaware as completed in service in 1902.

Another 12-inch on Model 1892 barbette carriage at one of the two single-gun batteries at Fort Screven, Tybee Island, outside Savannah, Georgia in the early 1920s.

12-inch Gun Model 1888 or 1895 On 12-inch Disappearing Carriage LF Model 1896

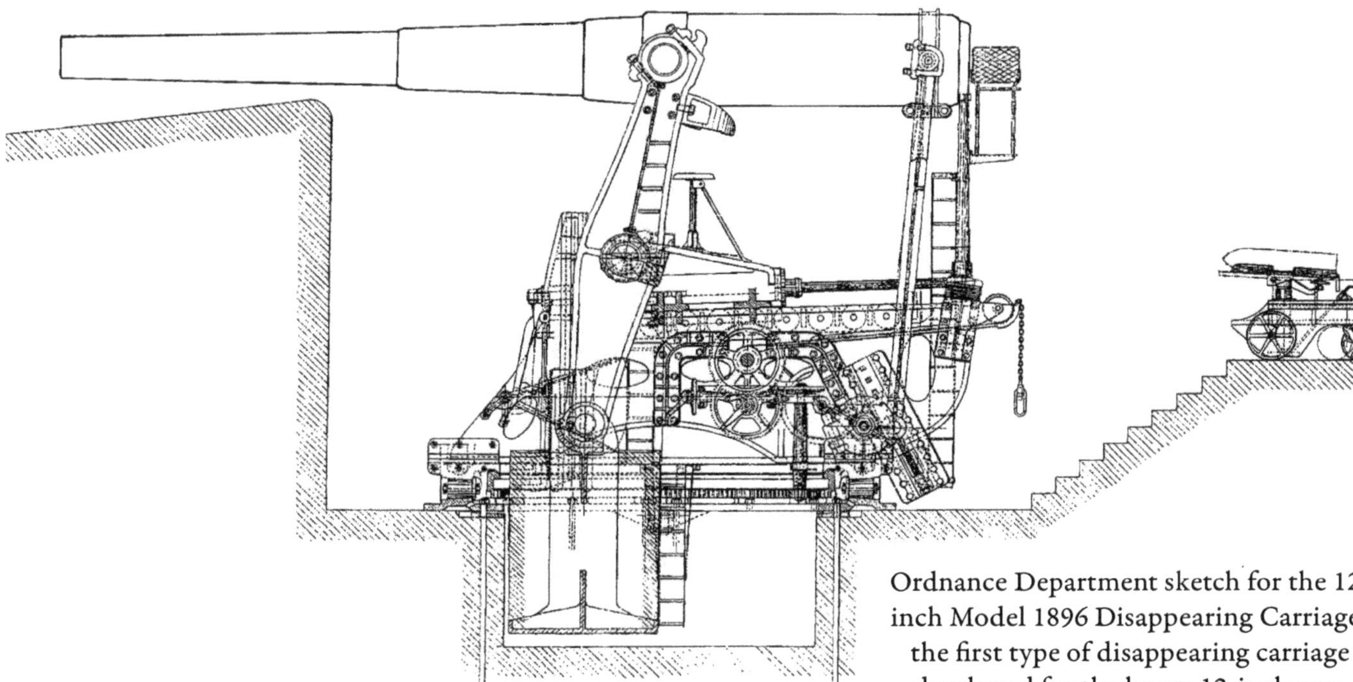

Ordnance Department sketch for the 12-inch Model 1896 Disappearing Carriage; the first type of disappearing carriage developed for the heavy 12-inch gun.

The limited-fire disappearing carriage for the 12-inch gun was similar in design of this type being developed for both the 8-inch and 10-inch guns. The primary parts were the base ring and racer with individual traversing rollers (24), a chassis, top carriage with recoil rollers and hydraulic recoil cylinders, gun levers, crosshead, lead counterweight, elevating arms, working platform and sighting platform. The action of the carriage caused by the recoil raised the counterweight, the recoiling gun was stopped by this and the action of the recoil cylinders. The gun always returned to the same level to ease loading from hand trucks. The gun moved back into firing position through the action of the rollers on the inclined chassis top, just like a gravity-return carriage. The gun was raised again to firing position by releasing the pawls that dropped the counterweight. The Model 1896 lower carriage cheeks had an arch or circular cut-out at its base on either side; this is different from later disappearing carriages and thus an aid in identification.

Watertown Arsenal built these carriages, beginning with an order for the first in February 1896 to the last batch in April 1898. Altogether 27 carriages were built and were soon emplaced. Fourteen batteries, all built during the last five years of the 19th century, were armed with either one, two, or three Model 1896 disappearing carriages. All except four carriages carried the 12-inch gun Model 1888 or its minor modifications. Battery Torbert at Fort Delaware was armed with the improved Model 1895 12-inch gun, and one Model 1888 gun at New York's Battery Hudson was replaced with a Model 1895 gun in 1904. For the M1895 guns the sighting standard was on the left side of the chassis rather than the right. The emplacements had most of the familiar features of disappearing gun batteries: individual platform for a single gun with ammunition service on either side, two story design with the magazines protected below, crow's nest or battery commander station between and behind the emplacement. The unusual height of the carriage meant an unusually deep pit below the crest. This model carriage of considerable height usually had 7-9 steps from the loading level to the base ring, which can often be used to help in battery identification in period photographs. The emplacement limited the traversing field to 170-degrees. The carriage could elevate from -3 to +10-degrees. A single sighting standard and platform was provided.

In service the gun, carriage, and emplacements proved adequate enough. There were modifications over time. The early emplacements were given widened loading platforms, communication galleries between emplacements on the loading level were provided, and often a more substantial battery commander's station added. Hoists were also installed and kept modernized to accelerate reloading. During 1907 the carriages were modernized with electrical power and firing circuits. In 1915-1917 25 carriages Model 1896 were given new parts to allow an increase in maximum elevation. These were new elevating arms and an elevating band with arm attachments at the top rather than sides. New elevation allowed was 15-degrees.

Some emplacements were disarmed in the 1920s when a few harbor defenses were discontinued and some individual batteries were eliminated when found to be in redundant locations. All remaining units were classified obsolete and guns and carriages scrapped in 1942-44. While most of the concrete emplacements still remain, there are no surviving 12-inch Model 1896 carriages or Model 1888 gun tubes.

12-inch Gun Model 1888 On 12-inch Disappearing Carriage LF Model 1896

Gun Characteristics
Bore size: 12-inch
Construction: Steel, built-up
Tube length overall: 439.92-inches
Nominal caliber: 34 caliber
Tube and breech weight: 117,127-pounds
Usual carriage: 12-inch Disappearing Carriage LF Model 1896
Weight of gun and carriage: 480,002-pounds
Maximum elevation: 10-degrees (some later modified to 15-degrees)
Loading: Separate
Shell weight: 1,046-pounds
Maximum muzzle velocity: 2,250-fps
Maximum range: 11,636-yards (later modified to 17,300-yards)

Builders: Guns by Watervliet Arsenal and
Bethlehem Steel
Carriages by Watertown Arsenal

Number built: 23 Model 1888 guns and 4 Model 1895 guns used on 27 Disappearing carriages LF Model 1896

Army photograph of an early, completed 12-inch Model 1896 Disappearing Carriage. The gun is in its retracted/loading position.

A Morgan Engineering photograph of one of its 12-inch Model 1896 Disappearing Carriage, with the gun in its fully elevated, firing position.

An emplaced 12-inch Model 1896 Disappearing Carriage, at either Battery Alexander or Bloomfield at Fort Hancock. This carriage was so "tall" that it required an abnormally deep mounting pit. It usually took nine steps to connect the level of the holding bolts to the level of the loading platform. That feature greatly assists identification of period photos of disappearing guns.

12-inch Gun Model 1895
On 12-inch Disappearing Carriage LF Model 1897

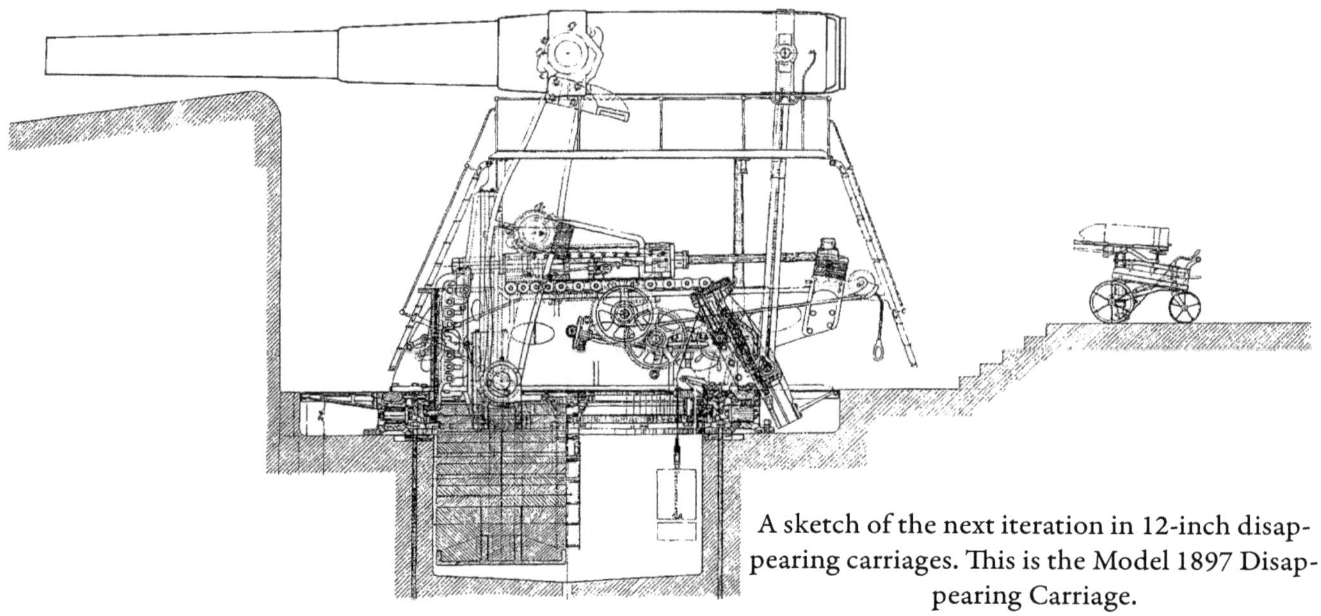

A sketch of the next iteration in 12-inch disappearing carriages. This is the Model 1897 Disappearing Carriage.

The 12-inch limited-fire, disappearing carriage Model 1897 was a progressive improvement over the Model 1896. The primary parts were the base ring and racer with individual traversing rollers (24), a chassis, top carriage with recoil rollers and hydraulic recoil cylinders, gun levers, crosshead, lead counterweight, elevating arms, working platform and sighting platform. The action of the recoiling carriage raised the counterweight, the recoiling gun stopped by this effort and the action of the recoil cylinders. The gun always returned to the same level to ease loading from trucks. The gun moved back into firing position through the action of the rollers on the inclined chassis top. The tube was raised again to firing position by releasing the pawls that dropped the counterweight. While designed with the Model 1895 gun in mind, with different gun levers it could accommodate the Model 1888 types also.

The carriage differed from its Model 1896 predecessor in having less overall height. As a consequence, only five steps were needed between the loading level and the lower working level. This difference helps in identifying photographs of battery emplacements. Total weight of lead for the counterweight was 138,635-pounds, which was suspended in this carriage on four vertical rods. Originally the carriage could elevate to 10-degrees. In 1906-07 the carriages were electrified with conduits and electric firing circuits. Around 1912 a systematic modernization program was made to add a 3-inch telescopic sight to the sighting standard, to provide electrical retracting and lights for night firing. Also a second sighting standard with ladder was added so it now carried one on each side of the chassis. Soon thereafter, in 1916-1917, a more major change came with modifications for increased elevation. This involved using new elevating arms and a new attached elevating band for these on the rear of the barrel. This increased the elevation to 15-degress and added 4000-yards to the maximum range. While 29 of the 35 carriages were modified, some batteries were not changed due to the need to retain depression.

Thirty-five carriages Model 1897 were ordered between April 1898 and July 1900. The last one was delivered in 1902. Five plants were used to build the carriages: Watertown Arsenal (5), Poole Engineering (12), Morgan Engineering (5), Bethlehem Steel (5) and Midvale Steel (8). All 35 carriages built were emplaced in 17 batteries with from one to three model 1897 disappearing carriages. Early in the 20th century changes were made to batteries to widen loading platforms, add electric hoists for ammunition, add latrines, improve communications between individual emplacements and provide new commander's stations

12-inch Gun Model 1895
On 12-inch Disappearing Carriage LF Model 1897

Gun Characteristics
Bore size: 12-inch
Construction: Steel, built-up
Tube length overall: 442.56-inches
Nominal caliber: 35 caliber
Tube and breech weight: 115,000-pounds
Usual carriage: 12-inch Disappearing Carriage LF Model 1897
Weight of gun and carriage: 477,959-pounds
Maximum elevation: 10-degrees (some later modified to 15-degrees)
Loading: Separate
Shell weight: 1,046-pounds
Maximum muzzle velocity: 2,250-fps
Maximum range: 11,636-yards (later modified to 17,300-yards)

Builders: Guns from Watervliet Arsenal and Bethlehem Steel
Carriages from Poole Engineering (No. 1-12),
Morgan Engineering (No. 13-17),
Bethlehem Steel (No. 18-22),
Watertown Arsenal (No. 23-27), and
Midvale Steel (No. 28-35)

Number built: 35 guns Model 1895 used on 35 Disappearing carriages

and walks. The standard emplacement limited a field of fire to 170-degrees, but cutting away some concrete in the forward wall as a breech recoiling space allowed some batteries to gain an additional amount of coverage on key flanks. In all cases they were initially armed with the 12-inch gun Model 1895. In 1917-18 a number of these batteries were relieved of their gun tubes for conversion to railway mounts for use in France. A few batteries were left without guns and their carriages scrapped, but nine of the carriages received older model 1888 guns in 1918 and remained in service.

Many of the batteries were retained right into the Second World War before being authorized for abandonment and scrapping in 1943-45. Although some have been heavily modified or filled-in, all the battery emplacements still at least partially exist, There are no surviving representatives of this important carriage type, but there are some Model 1895 gun tube survivors in the Philippine defenses.

A promotional photograph by the Morgan Engineering Company of one of its Model 1897 Disappearing Carriages

A mounted 12-inch gun Model 1895 on its Model 1897 Disappearing Carriage at Battery Mendell at Fort Barry, San Francisco before the First World War.

Closeup of the open breech of a Model 1895 12-inch gun mounted on M1897 Disappearing Carriage at Battery Pensacola, Fort Pickens, Pensacola Florida in 1912.

12-inch Gun Model 1895 and 1900 On 12-inch Disappearing Carriage LF Model 1901

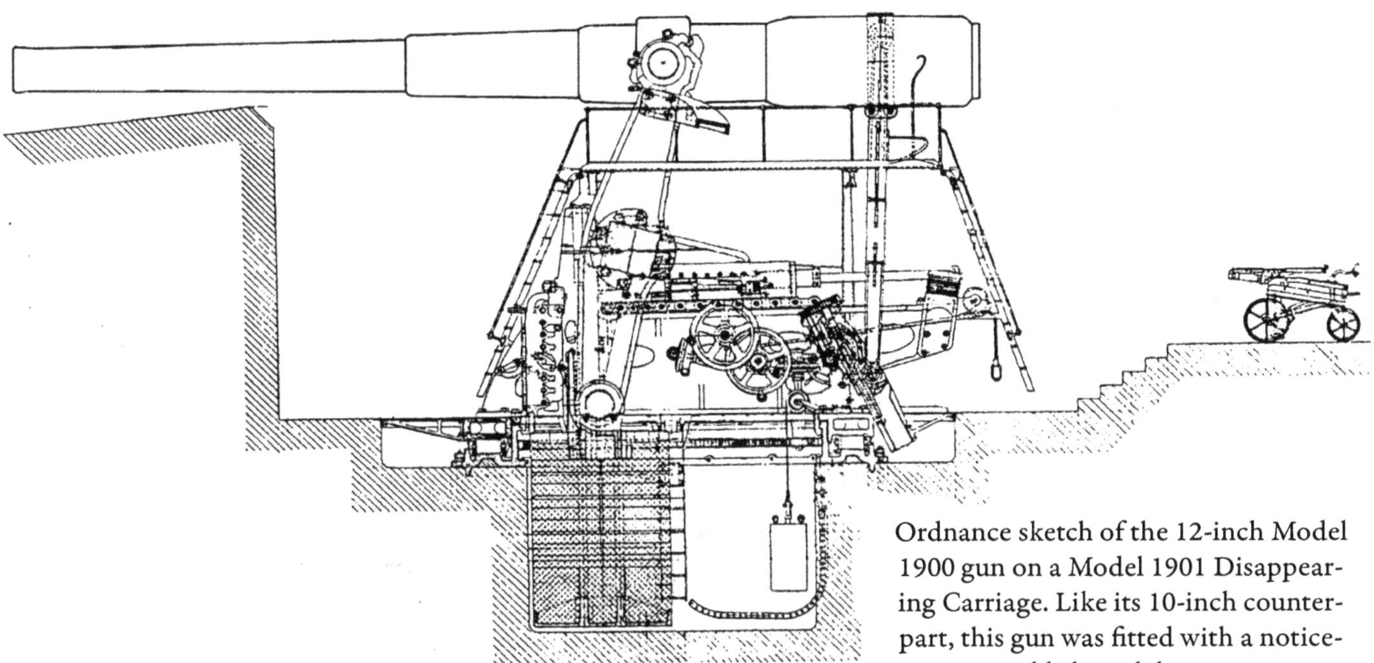

Ordnance sketch of the 12-inch Model 1900 gun on a Model 1901 Disappearing Carriage. Like its 10-inch counterpart, this gun was fitted with a noticeable breech hoop.

The Ordnance Department hoped to develop a more powerful generation of coast defense guns with their Model 1900 12-inch guns. Using a larger charge in a larger powder chamber and with improvements in alloy steel for the guns, the goal was to get a higher muzzle velocity and thus improved range and particularly penetration ability. The new tube was lengthened to 40-calibers. The powder chamber was enlarged to 16,514 cubic inches from 12,185. That should have delivered a muzzle velocity of 2550-fps, but that turned out to be impractical. A limited number of guns were designed and built, but it was found that the bore erosion from this increase was excessive. The guns wouldn't last for what was projected to be a normal engagement time. Ordered in 1901, only 11 tubes were actually produced at Watervliet Arsenal. The physical appearance of the tube is different from the Model 1888s and 1895s. The breech-end area around the powder chamber is enlarged and has a swollen appearance.

These tubes were emplaced in five new batteries but soon had to be modified to enhance their service life. Initially they were restricted to just using smaller powder charges, but then a project was undertaken to reduce the size of the powder chamber by inserting a steel sleeve, the resultant new charge gave the same 2250-fps muzzle velocity as the previous 12-inch models.

The disappearing carriage was fairly conventional in design. Like its predecessors, the parts were the base ring and racer with individual traversing rollers, a chassis, top carriage with recoil rollers and two hydraulic recoil cylinders, gun levers, crosshead, lead counterweight, elevating arms, working platform and sighting platform. The action of the recoil raised the counterweight, and the recoiling gun was stopped by this combined with the resistance of the recoil cylinders. The gun always returned to the same level for ease of reloading. The carriages placed overseas were sometimes given counterweights of iron rather than lead, requiring deeper pits to accommodate the larger size required. The 1901 carriage was electrified from the beginning and had sighting platforms on both sides of the chassis.

Twenty-one Model 1901 carriages were acquired. Four suppliers were used, orders stretched from 1901 to 1908. The first 11 used the Model 1900 gun, the subsequent ten used Model 1895 tubes. These latter were emplaced either overseas (six guns at Fort Mills, two at Fort Kamehameha) or at the last 12-inch disappearing battery built, at Fort Worden, Puget Sound. The emplacement design for this final generation of 12-inch disappearing batteries was for enlarged, spacious positions. There were still just five steps from loading to working level, but the spacing between guns was increased. By their relatively late construction date they were designed with features like enlarged loading platforms, shot truck closets, covered hoist delivery galleries, large battery commander stations, internal speaking tubes, and adequate storerooms. All batteries were eventually equipped with mechanical range transmitters.

Even with measures to mitigate bore wear, two batteries had to have their Model 1900s replaced with Model 1895s. Battery Richardson lost her tubes in 1910, and Battery Parrott followed in about 1925. All the Model 1901 carriages were all given modifications for higher elevation in 1916-1918. These were important batteries that survived until the Second World War. At that time (1942-44) they were declared obsolete and the armament scrapped. The Japanese attack of 1942 captured the three batteries' worth of guns at Fort Mills, Manila Bay. In an ironic way that saved them from scrapping. All six guns (in fact more as there are also three spare barrels) and six Model 1901 carriages, though of course damaged from both war and subsequent illegal scrapping attempts, exist in place.

12-inch Gun Model 1900
On 12-inch Disappearing Carriage LF Model 1901

Gun Characteristics
Bore size: 12-inch
Construction: Nickel steel, built-up
Tube length overall: 504.35-inches
Nominal caliber: 40 caliber
Tube and breech weight: 132,380-pounds
Usual carriage: 12-inch Disappearing Carriage LF Model 1901
Weight of gun and carriage: 642,730-pounds
Maximum elevation: 10-degrees (later modified to 15-degrees)
Loading: Separate
Shell weight: 1,046-pounds
Maximum muzzle velocity: 2,550-fps (2,250-fps after modification)
Maximum range: 13,400-yards (10-degrees), modified to 17,300-yards (15-degrees)

Builders: Guns: Watervliet Arsenal
Carriages: Watertown Arsenal (10), Midvale Steel (7),Bethlehem Steel (2), Alliance Machinery (2)

Number built: 11 guns Model 1900 and 10 guns Model 1895 used on 21 disappearing carriages Model 1901

Army instructional diagram showing a 12-inch Model 1900 on Model 1901. The actual unit used in the photograph is Battery Parrott at Fort Monroe, Virginia.

A Model 1895 12-inch tube mounted on a Model 1901 Disappearing Carriage of Battery Wheeler, Fort Mills, Manila Bay. The photo is from 1933. Notice that the elevating arms have trunnions high on the breech side—indicating that the battery has been modified for greater elevation that occurred in the 1916-1919 time period.

Five Model 1901 12-inch Disappearing Carriages survive, though showing clear damage from the war and subsequent scrappers, all at Fort Mills, Corregidor Island, the Philippines. This is a recent photograph of Battery Crockett.

Coast Defense 14 and 16-inch Guns

Of the Taft Period

The increasing size and power of naval armament in the first years of the 20th century was one of the prompts for creating the Taft Board. A project for a new 14-inch gun was initiated. The Department still wanted performance that had initially been promised by the high-powered Model 1900 12-inch gun. When that model failed, they turned to a larger bore size but with a relatively low muzzle velocity. That gave rise to the 14-inch Model 1907 gun. This gun had a shorter caliber than the 12-inch, 40-caliber it was intended to replace. Technically it might fit into the older style emplacements, in practice this new gun was only ever utilized in entirely new emplacements. Five of these guns were made before a much improved 14-inch Model 1910 was developed. This new design was a full 40-calibers and delivered excellent performance.

The Army still preferred the disappearing carriage for its large seacoast armament. Long since virtually every other nation in the world had abandoned these expensive, complicated devices prone to mechanical failure, the Ordnance Department continued to stress their relative protection for operating crew and rapidity of fire. While the Army did equip a unique fort in Manila Harbor with 14-inch turrets, the other 14-inch emplacements built in the Philippines, Hawaii, the Panama Canal Zone and Los Angeles retained the same disappearing carriage technology.

Made larger and stronger with numerous small improvements, the Model 1907 DC (for the Model 1907 gun) and its sister the 1907M1 DC (for the Model 1910 gun) were not innovative designs—but they worked without any significant problems. During manufacture or early installation, they were modified to provide an increase to 20-degree of elevation. The original Endicott Board report had projected turrets mounting heavy guns. Despite successful application of turret technology in Europe, the United States only once utilized turrets prior to the Second World War. Plans were made for fortifying El Fraile Island in Manila Bay by covering the small island with an encasing concrete structure. The primary armament was to be two twin superfiring turrets with 14-inch guns. The long period of time necessary to design the structure and replacing the rocky island with the concrete fort allowed for the substantial technical challenge of designing the armament. While thought was given to adopting naval-type turrets, in the end the Army's Ordnance Department developed a new model of 14-inch gun and turret for use just in this location. The Model 1909 14-inch guns were wire-wound in construction, but performance was not dissimilar to the built-up Model 1910 types.

The original Endicott plan had recommended the development and deployment of 16-inch coast defense guns. Initially they were beyond country's manufacturing capability. Eventually a single 16-inch tube was designed and completed as the Model 1895 gun. This example of a 16-inch gun, made more to prove the manufacturing technology than anything else, had lain around Army arsenals for years, being occasionally displayed at various fairs and exhibits. Finally, it was given a new Model 1912 disappearing carriage and emplaced as one of the final pieces of the defenses of the Panama Canal. This carriage was basically a scaled up version of the recent 14-inch design. In service it never really proved to be a reliable weapon.

All of these larger 14 and 16-inch guns would have been scrapped with no existing examples if the Philippine forts hadn't been captured in 1942. Two 14-inch guns on disappearing carriages (and the 14-inch gunned turrets of Fort Drum) exist today because of this event. Heavily damaged, partially covered with cement, and rusting badly the two 14-inch Model 1910 guns on Model 1907M1 carriages at Fort Hughes on Caballo Island, Manila Bay still exist.

14-inch M1910 in Panama

14-inch Gun Model 1907 and 1907M1 On 14-inch Disappearing Carriage LF Model 1907

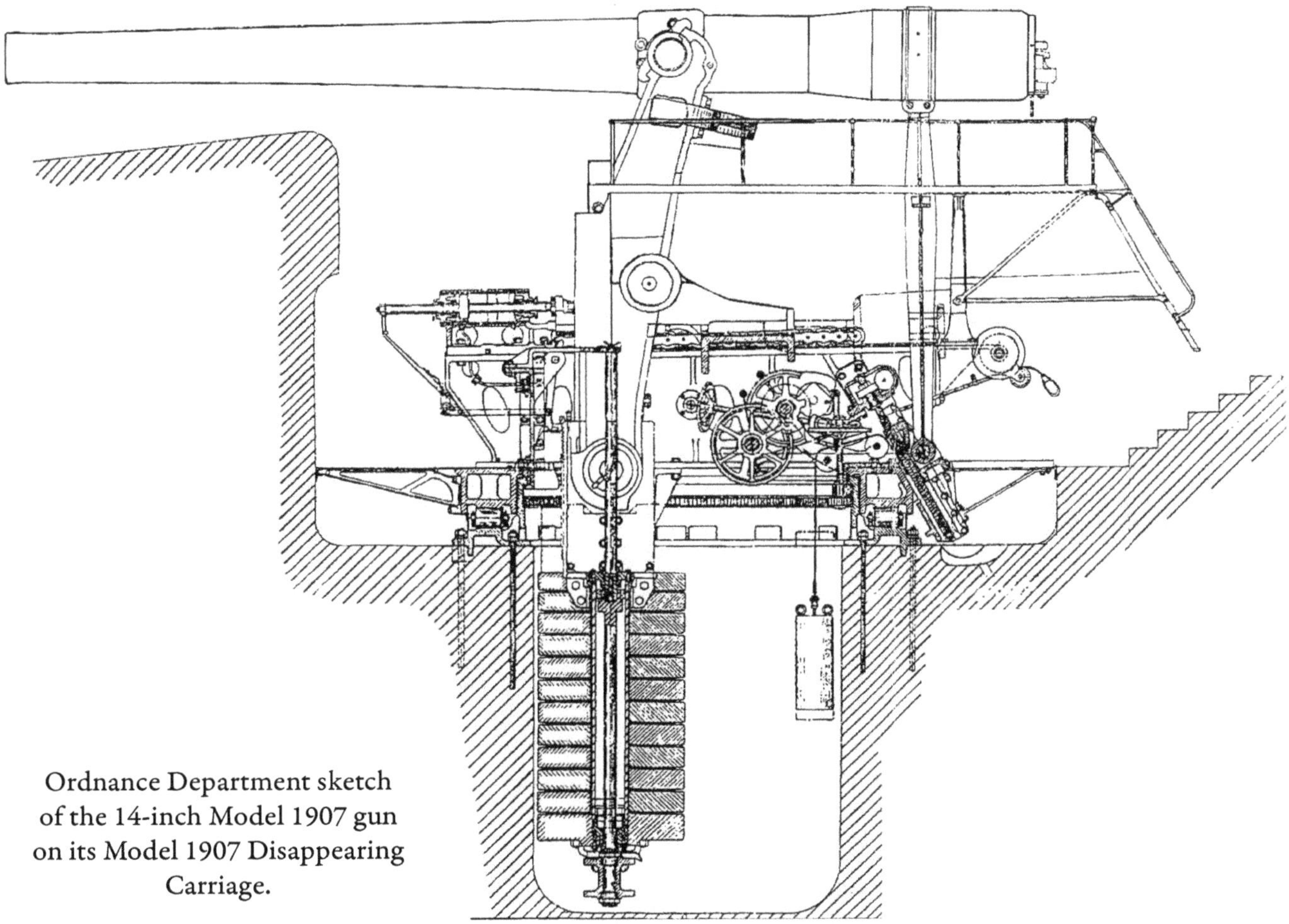

Ordnance Department sketch of the 14-inch Model 1907 gun on its Model 1907 Disappearing Carriage.

The Army's first 14-inch gun was developed to replace the flawed, high-velocity Model 1900 12-inch gun. The latter was found to have too much bore wear at its designed velocity, and performance was reduced when the powder charge had to be diminished. A larger bore but lower velocity could achieve the same performance. In fact, a length (34 caliber) was purposely chosen to allow the new 14-inch to fit the dimensions of new 12-inch batteries under construction or projected. The type gun (the sole M1907) was of wire-wrapped type, but the four production guns (M1907M1) were of steel, built-up construction. The gun was of conventional design with a Weilin step-type breech of 12 segments (9 thread, 3 blank). It had 126 grooves and rifling 1 in 50 increasing to 1 in 25 near the muzzle.

The limited-fire disappearing carriage was typical in design for this type of mechanism. The primary parts were the base ring and racer with individual traversing rollers, a chassis, top carriage with recoil rollers and hydraulic recoil cylinders, gun levers, crosshead, lead counterweight, elevating arms, working platform and sighting platform. These parts were mostly made of cast iron or steel as necessary. The carriage used the action of the recoil to raise the counterweight; the recoiling gun was stopped by this drag and the recoil cylinders. The gun moved back into firing position through the action of the rollers on the inclined chassis top, just like a gravity-return carriage. The gun was raised again to firing position by releasing the pawls that dropped the counterweight. The carriage was quite similar to the previous 12-inch types, except for obvious changes in dimensions and a different arrangement of the recoil cylinders.

Production of guns and carriages proved a trouble-prone, protracted process. The five guns were ordered in 1907 and delivered between 1909 and 1912. In 1911 the powder chambers were expanded, and a severe accident at the proofing ground in 1912 forced a breech design change. Similarly, the carriages (all ordered in 1908) experienced weak lever arms and parts had to be redesigned and strengthened. Shipment and emplacement were not finally completed until 1915.

Three batteries were built and equipped with this combination of gun and carriage. Battery Randolph at Fort DeRussy in Honolulu, Hawaii was a conventional 2-gun battery in a shared emplacement, somewhat similar to recent 12-inch battery designs. Ammunition was stored in a lower magazine and raised to the loading level by hoists. It was built between December 1908 and 1910 but not transferred until late 1913. It served until eliminated in 1944. The guns and carriages were scrapped in place. The other two service carriages went into separate single emplacements also built in 1908-10 at Fort Frank, Manila Bay. Because of the local topography there was not room for a normal dual emplacement. Each single gun had an adjacent, but lower level magazine with once again ammunition service by hoists.

Like the more modern 12-inch carriages, efforts were taken in 1913-16 to increase the elevation of the carriage. This involved new elevating arms and bands to give an increase of 5-degrees elevation. Also to maintain balance and equilibrium in the emplacement, the entire carriage had to be shifted. Concrete had to be cut-out of the forward face of the parapet wall to allow an extension of the metal platform to fit. The Fort Frank pair saw heavy usage during the 1941-42 campaign. All the guns and carriages are now gone, the abandoned guns in their emplacements in the Philippines were removed by scrap metal thieves in the 1980s. The Honolulu emplacement, though somewhat damaged in an attempt to destroy it, is the current home of the U.S. Army Museum of Hawaii.

14-inch Gun Model 1907M1 On 14-inch Disappearing Carriage LF Model 1907

Gun Characteristics
Bore size: 14-inch
Construction: Steel, built-up
Tube length overall: 495-inches
Nominal caliber: 34 caliber
Tube and breech weight: 118,700-pounds
Usual carriage: 14-inch Disappearing Carriage LF Model 1907
Weight of gun and carriage: 636,073-pounds
Maximum elevation: 20-degrees (after modification)
Loading: Separate
Shell weight: 1,660-pounds
Maximum muzzle velocity: 2,150-fps
Maximum range: 20,000-yards

Builders: Guns: Watervliet Arsenal
Carriages: Watertown Arsenal

Number built: Guns: 5, Carriages: 4

14-inch gun on a M1907 disappearing carriage

The initial Model 1907 14-inch carriage was extensively trialed at the Sandy Hook Proofing Ground.

While not of good resolution, this photo shows the initial installation, before modification, of one of the two 14-inch Model 1907 guns on Model 1907 Disappearing Carriage at Battery Randolph, Fort DeRussy, Oahu Hawaii in 1916.

14-inch Gun Model 1910 and 1910M1
On 14-inch Disappearing Carriage Model LF Model 1907M1

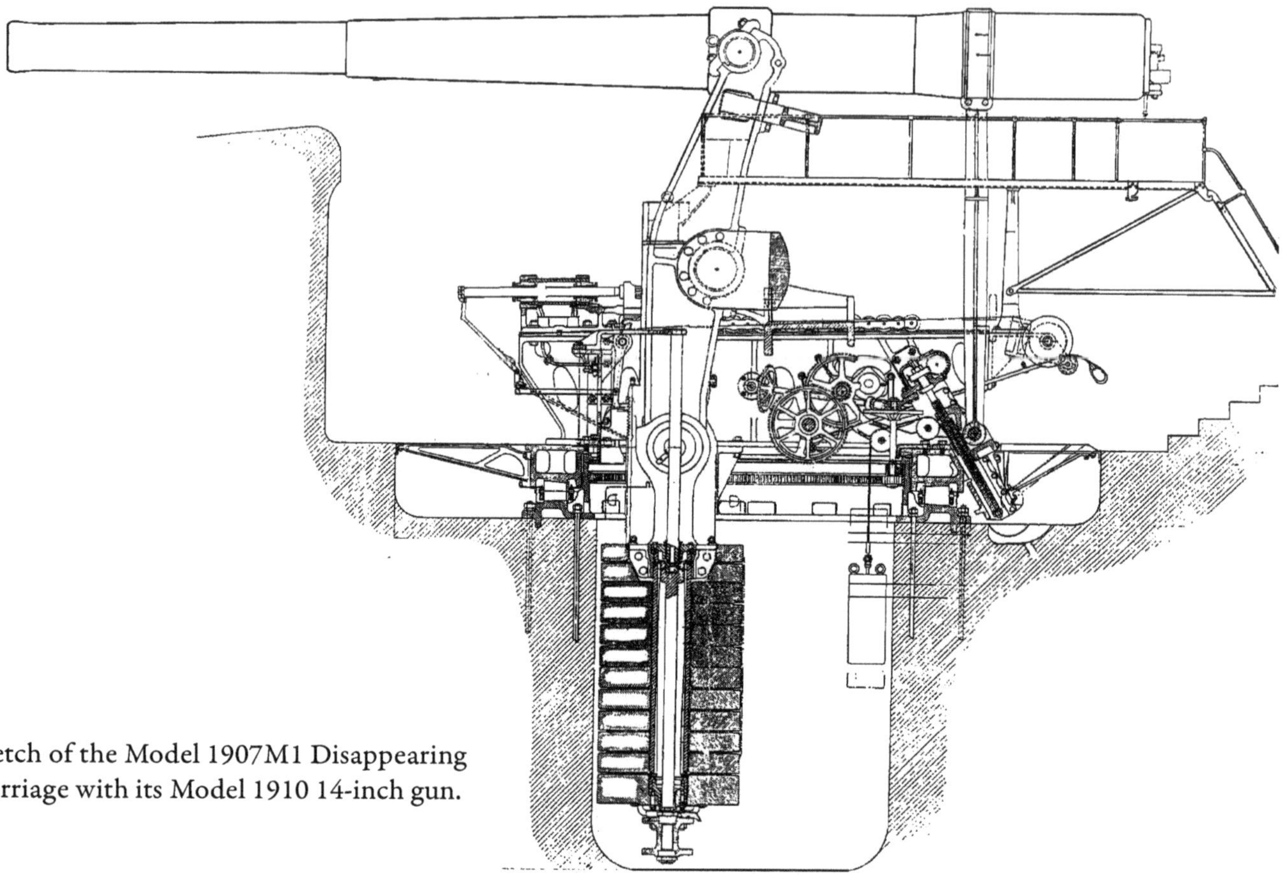

Sketch of the Model 1907M1 Disappearing Carriage with its Model 1910 14-inch gun.

A follow-up design for a new 14-inch gun resulted in the Model 1910 type. It chiefly differed from its predecessor by being a longer, 40-caliber size. Similar in many respects to the Model 1909 used for turrets, it was wire-wrapped. The gun had 98 grooves and a right-hand uniform rifling twist of 1 in 25. It used the now "standard" Weilen stepped-breech of twelve segments (nine threaded and three blank). The tube had a trunnion hoop as they were intended for disappearing carriages. After production of 13 guns, the following five (18 total) had stiffer chase hoops, a thickened tube, and jacket and B tube of nickel steel. These were designated Model 1910M1. The guns initially had 15,000-cubic inch powder chambers, but they were expanded to 19,323 either during production or right after installation from 1915 to 1917.

A slightly modified Model 1907 disappearing carriage (used with the previous Model 1907M1 14-inch gun) was produced for this gun. The limited-fire disappearing carriage was typical in design for this type of mechanism. The primary parts were the base ring and racer with individual traversing rollers, chassis, top carriage with recoil rollers and hydraulic recoil cylinders, gun levers, crosshead, lead counterweight, elevating arms, working platform and sighting platform. The carriage used the action of the recoil to raise the counterweight; the recoiling gun being stopped by the action of recoil cylinders. The gun was raised again to firing position by releasing the pawls that dropped the counterweight. In its emplacement the gun was centered 15-foot from the parapet wall.

Eighteen guns were produced at Watervliet Arsenal, starting with an initial order in 1910. Sixteen carriages M1907M1 were fabricated at Watertown Arsenal up until 1918. All 16 of these were emplaced with their guns, two guns remained as spares or on proving ground assignment. In service they proved successful and lasted until late in the Second World War.

This weapon was an important element of the Taft-era defenses. Twelve named batteries were constructed from 1912-1917. Most of these new emplacements featured expansive emplacement pits and loading platforms at ground level. Ammunition was stored in an adjacent magazine, sometimes at a depressed level with ramps in order to provide maximum overhead protective cover. Shells were delivered from magazines on overhead trolleys to a shell table which in turn fed a ramped shell corridor directly to the gun. In some cases, local terrain dictated retention of the two-story plan. Battery commander stations, plotting rooms, and power stations along with other support facilities completed the structure. Considerable efforts were made to fit and blend the structures into local topography, so each battery wound up being fairly unique rather than standardized.

These batteries were just being armed when the changes were adopted to increase the gun elevation. The usual changes in the elevating band and arms were made and strengthening of the carriage made by addition of a kick-down mechanism. Maximum elevation increased from 15 to 20-degrees. By 1920 all these batteries were in service and modified for long-range. They served actively during the war. The Manila Bay batteries saw action early in the war and were surrendered at Fort Hughes. Batteries in Los Angeles (two connected structures of four total guns but organized as four separate batteries) and Panama (four dual batteries and two single-gun batteries) served until deactivated and scrapped right at the end of the war.

14-inch Gun Model 1910
On 14-inch Disappearing Carriage LF Model 1907M1

Gun Characteristics
Bore size: 14-inch
Construction: Steel, built-up
Tube length overall: 580.96-inches
Nominal caliber: 40 caliber
Tube and breech weight: 138,000-pounds
Usual carriage: 14-inch Disappearing Carriage LF Model 1907M1
Weight of gun and carriage: 682,234-pounds
Maximum elevation: 20-degrees (after modification)
Loading: Separate
Shell weight: 1,660-pounds
Maximum muzzle velocity: 2,350-fps
Maximum range: 22,800-yards

Builders: Guns: Watervliet Arsenal
Carriages: Watertown Arsenal

Number built: Guns: 18, Carriages: 16

Fortunately, though in poor condition, the two guns and carriages at Fort Hughes still exist on a Philippine Naval Station.

Proving ground photograph of the 14-inch Model 1910 on Model 1907M1 Disappearing Carriage.

Panama received ten of the 40-caliber 14-inch Model 1910 guns on disappearing carriages. This photograph is of one in its emplacement.

There are two surviving combinations of the 14-inch Model 1910 gun on Model 1907M1 Disappearing Carriage. Both the single-gun batteries at Fort Hughes in Manila Bay, though somewhat damaged, survive. This is Battery Gillespie at that location.

14-inch Gun Model 1909
On 14-inch Turret Mount Model 1909

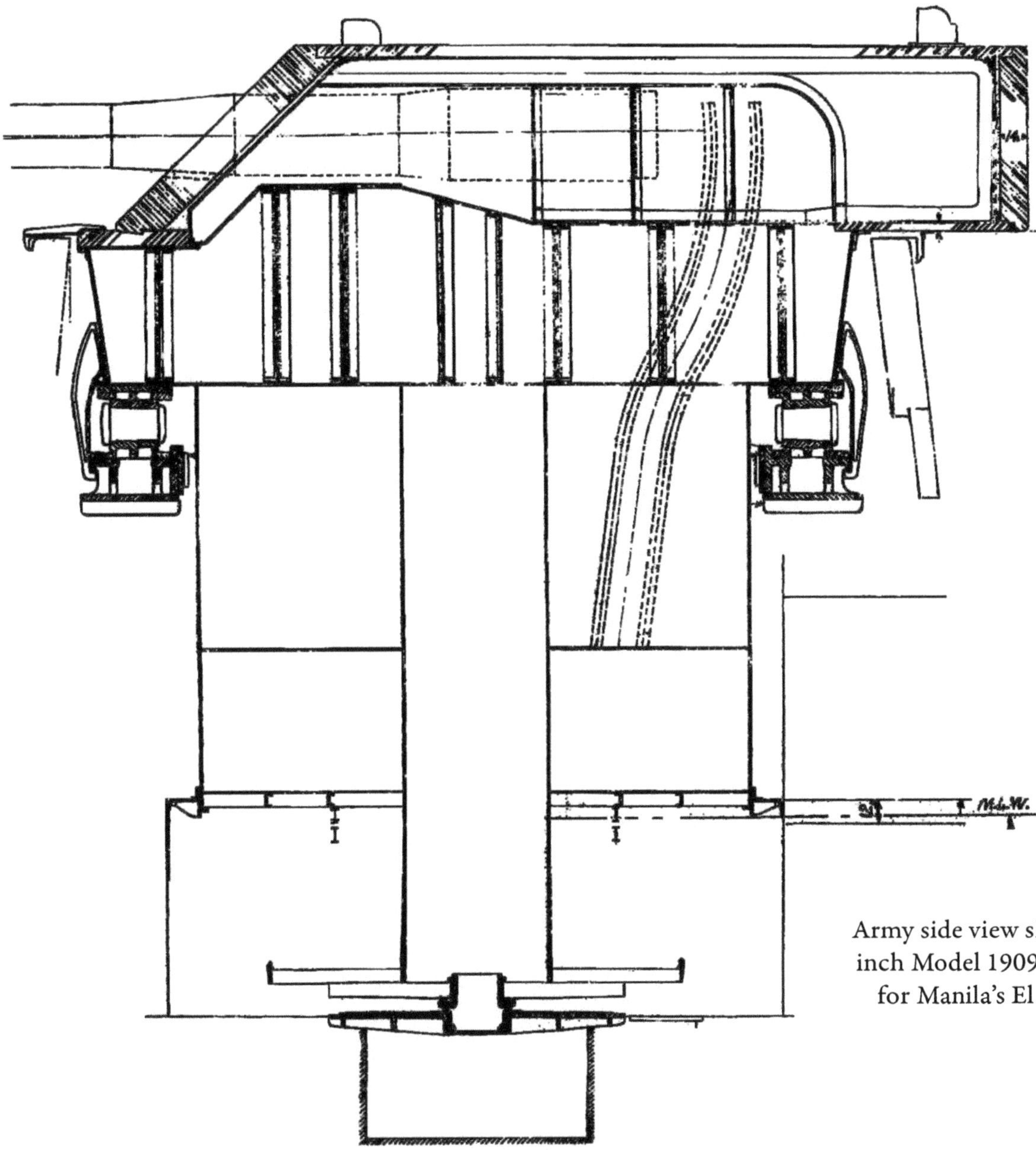

Army side view sketch of the 14-inch Model 1909 turret planned for Manila's El Fraile Island.

The design to completely rebuilt El Fraile Island in Manila Bay to a fortified island was approved in 1908. The constricted size of the reservation argued against disappearing emplacements, the concept of mounting turrets here—particularly with the distinct possibility of exposure to land-based artillery was highly favored. By 1909 work began on what would become Fort Drum and the Ordnance Department began developing new 14-inch guns, mounts, and heavy army turrets for the project. Ultimately two dual turrets of original design were fabricated and mounted, though the complicated engineering and production problems delayed completion of the fort until June of 1918.

A new 14-inch gun was designed. It differed primarily from the previous M1907 by being longer (40 caliber) and being wire-wound. Two inner gun steel tubes were wrapped by wire and a full jacket. The jacket was threaded for the breech housing; the Welin stepped-breech had twelve sectors. Four guns were ordered, two with right-hand and two with left-hand breech mechanisms as befit the turret arrangement.

There was no trunnion band, as the guns were keyed to a sleeve carried on the carriage. Four guns were ordered from Watervliet in December 1909 using Midvale forgings. There were development problems with American seacoast 14-inch guns. In 1913 the powder chambers were expanded to alleviate loading problems, then again in 1915 they were again enlarged and other breech modifications made before any guns were installed. The four tubes were finally finished and shipped to Fort Drum in late 1916.

The mounts for the guns were provided by Watertown Arsenal and consisted of a cast steel cradle in a yoke to hold each gun. A single recoil cylinder and four recuperator cylinders were mounted with each cradle. Much more complicated was the actual turret structure. An upper and lower turret were fabricated by Newport News Shipbuilding and Dry Dock Company. The armor was very thick in comparison to naval types—18-inches in front, 14 on sides, and for the roof 4.5-6 inches. The two guns were partitioned internally. The Army developed its own type of trolley hoists to raise powder and shells from the lower level magazines. Gas ejectors with compressed air equipped each gun. Delays were experienced in delivery of the turrets, the parts (for assembly locally) were not sent to Manila until late 1916. On the deck of the "concrete battleship" the turrets sat atop concrete-encased steel barbettes 18-inches thick. As mounted the upper turret (Battery Wilson) had 360-degree traverse while the forward, lower turret (Battery Marshall) just 230-degrees.

Once installed, the armament of Fort Drum served successfully. Fire control was provided with the use of coincidence range finders mounted in a cage mast behind the battery. During the siege of 1941-1942 the guns were extensively used for counter battery fire against both the Ternate and Bataan coasts and their Japanese siege batteries. Despite intense bombardment, the turrets proved immune to enemy fire, and they were still in service when the fort was surrendered in May, 1942. Damaged by the surrendering Americans and subsequent air bombardment, they were incapable of being restored by the Japanese. In subsequent years scrappers have removed most of the internal equipment and parts of the turrets, leaving just the bulk of the barrels and the outer plates of the turret armor. In that condition, they still exist on El Fraile Island.

14-inch Gun Model 1909
On 14-inch Turret Mount Model 1909

Gun Characteristics
Bore size: 14-inch
Construction: Gun steel, wire-wound
Tube length overall: 579-inches
Nominal caliber: 40 caliber
Tube and breech weight: 139,240-pounds
Usual carriage: 14-inch Model 1909 gun mount and turret
Weight of gun and carriage: 2,137,000-pounds turning weight
Maximum elevation: 15-degrees
Loading: Separate
Shell weight: 1,660-pounds
Maximum muzzle velocity: 2,350-fps
Maximum range: 20,400-yards

Builders: Guns: Waterlivet Arsenal
Carriages: Watertown Arsenal
Turrets: Newport News Shipbuilding & Dry Docks
Number built: 4 guns & carriages & two twin turrets

The two turrets for Model 1909 14-inch guns installed and practice firing at Fort Drum, El Fraile Island, Defenses of Manila Bay. Photograph from the mid-1930s.

One of the 14-inch Model 1909 gun dual turrets being assembled and proofed at the Sandy Hook Proofing Grounds.

The unique structure of Fort Drum with at least the outer shell of its Model 1909 14-inch turrets, survives to this day. Heavily damaged by its war experience, unauthorized scrapping, and the toll of time, it is difficult to access.

16-inch Gun Model 1895
On 16-inch Disappearing Carriage LF Model 1912

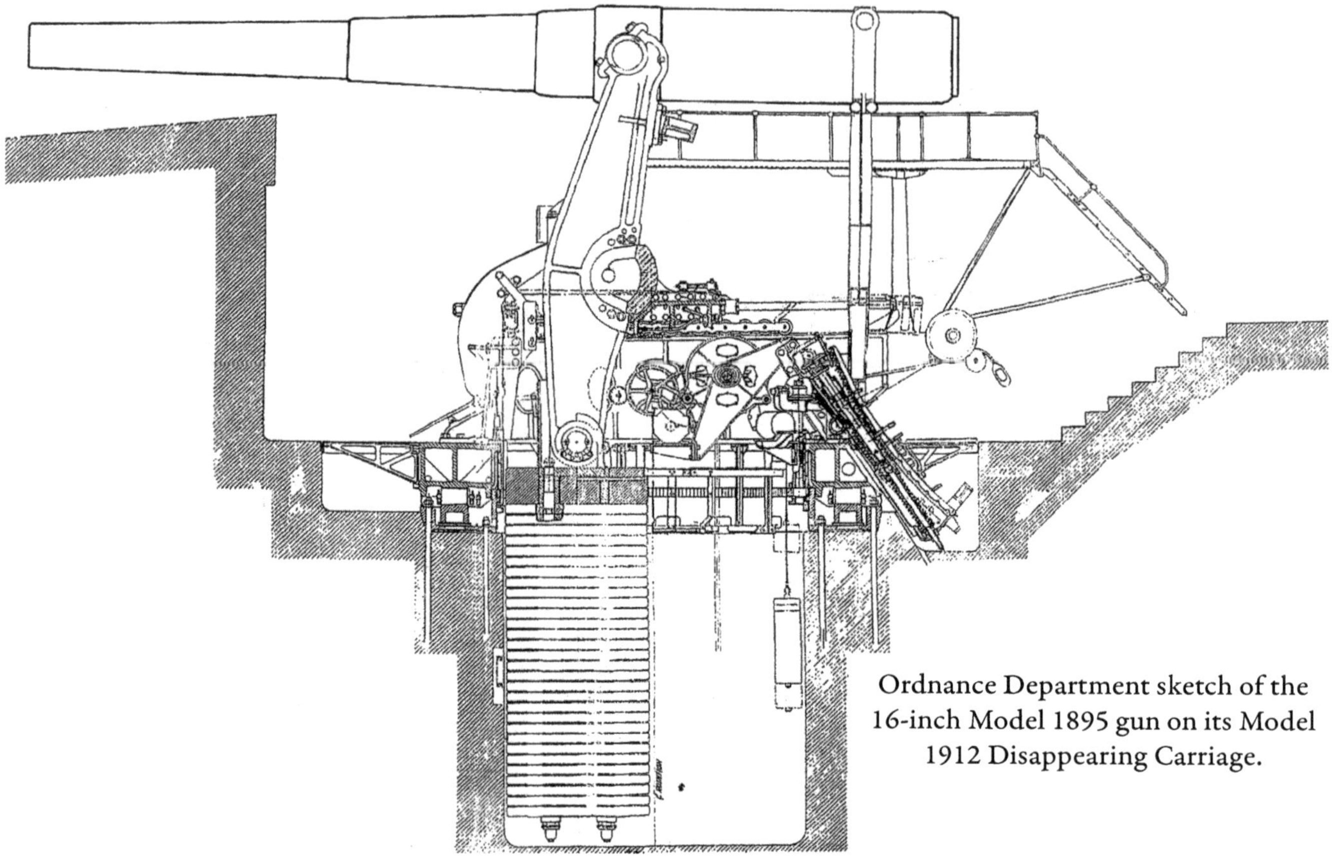

Ordnance Department sketch of the 16-inch Model 1895 gun on its Model 1912 Disappearing Carriage.

The Endicott report had strongly recommended 16-inch guns as part of American defenses. Such a large technical undertaking was initially beyond both government and private manufacturers, so the first generation of guns developed were of the 8, 10 and 12-inch size. Soon it was found that these were adequate for the efficient defense of protected harbors, and the 16-inch would no longer be necessary. Still, there was a certain inertia to actually design and produce at least an example of such a super weapon. Many years and resources were consumed in this project, and finally a single Model 1895 16-inch gun was produced at Watervliet Arsenal. It was a gun steel, built-up tube of 35-calibers length. It had the usual type of Ordnance Department interrupted-screw breech. In fact except for its size, it was of fairly conventional construction and performance. The gun fired a 2340-pound projective with separate loading at 2250-fps muzzle velocity. The rifling was of 96 grooves and progressed from 1 in 50 to 1 in 25 near the muzzle. It was estimated to have a service life of 200 rounds.

At the end of 1896 Bethlehem Steel was awarded the contract for the gun's forgings. It took just about two years to finish these forgings at Watervliet. A certain notoriety was obtained with the gun; it appeared in numerous articles and at some of the international fairs of the time—representing the achievements of American military science.

After several years the design for a disappearing carriage began in 1911. The limited-fire disappearing carriage was typical in design for this type of mechanism. The primary parts were the base ring and racer with individual traversing rollers, chassis, top carriage with recoil rollers and hydraulic recoil cylinders, gun levers, crosshead, lead counterweight, elevating arms, working platform and sighting platform. These parts were mostly made of cast iron or cast steel. The action of the carriage used the recoil to raise the counterweight; the recoiling gun was stopped by this and the action of the recoil cylinders. The gun always returned to the same level to ease loading from hand trucks. The gun was raised again to firing position by releasing the pawls that dropped the counterweight. This counterweight was of 33 pieces of cast iron and weighed 444,808-pounds. Four vertical rods held the counterweight pieces together. Two sighting stands were on both sides of the chassis. It was equipped with a 3-inch M1912 telescopic sight. Elevation and traversing could be achieved both by electrical and hand controls.

Initially designed for a maximum 15-degree elevation, during fabrication the carriage was changed to allow 20-degrees primarily by affixing a new 8-inch hydraulic kick-down cylinder at the ends of the elevating arms and adjusting the configuration of the elevating band. The maximum range was thus increased from 19,000 to 23,000 yards.

It was eventually decided to use the gun as part of the defenses of the Panama Canal. Battery Newton, for just the single 16-inch disappearing gun, was emplaced on the top of Perico Island. Typical of late Taft batteries, the protected magazines were to the flank of the gun platform and on the same level—avoiding the need for heavy lifts to carry the ammunition. A plotting room and emergency power generation room and other general support rooms completed the rather substantial emplacement. The emplacement limited traverse to 170-degrees.

In service the gun proved awkward on its massive, complicated carriage. While it was accepted into service, the emplacement was never supplied with more than a fraction of its intended ammunition supply. As soon as May of 1929 the battery was reduced to just emergency reserve status. Eventually it was authorized for removal in June 1943. No traces remain of the gun or carriage, but the battery emplacement, now in Panamanian hands, still survives.

16-inch Gun Model 1895
On 16-inch Disappearing Carriage LF Model 1912

Gun Characteristics
Bore: 16-inch
Construction: Gun Steel, built-up
Tube length overall: 610-inches
Nominal caliber: 35 caliber
Tube and breech weight: 284,000-pounds
Usual carriage: 16-inch Disappearing carriage LF Model 1912
Weight of gun and carriage: 1,273,992-pounds
Maximum elevation: 20-degrees (after modification)
Loading: Separate
Shell weight: 2,400-pounds
Maximum muzzle velocity: 2,250-fps
Maximum range: 23,000-yards

Builders:,Gun Watervliet Arsenal
Carriage Watertown Arsenal

Number built: 1 gun and carriage

The installed Model 1895 16-inch gun on Model 1912 16-inch Disappearing Carriage at Battery Newton, Fort Grant, Panama.

The only 16-inch Model 1895 gun ever built during its trials at Sandy Hook.

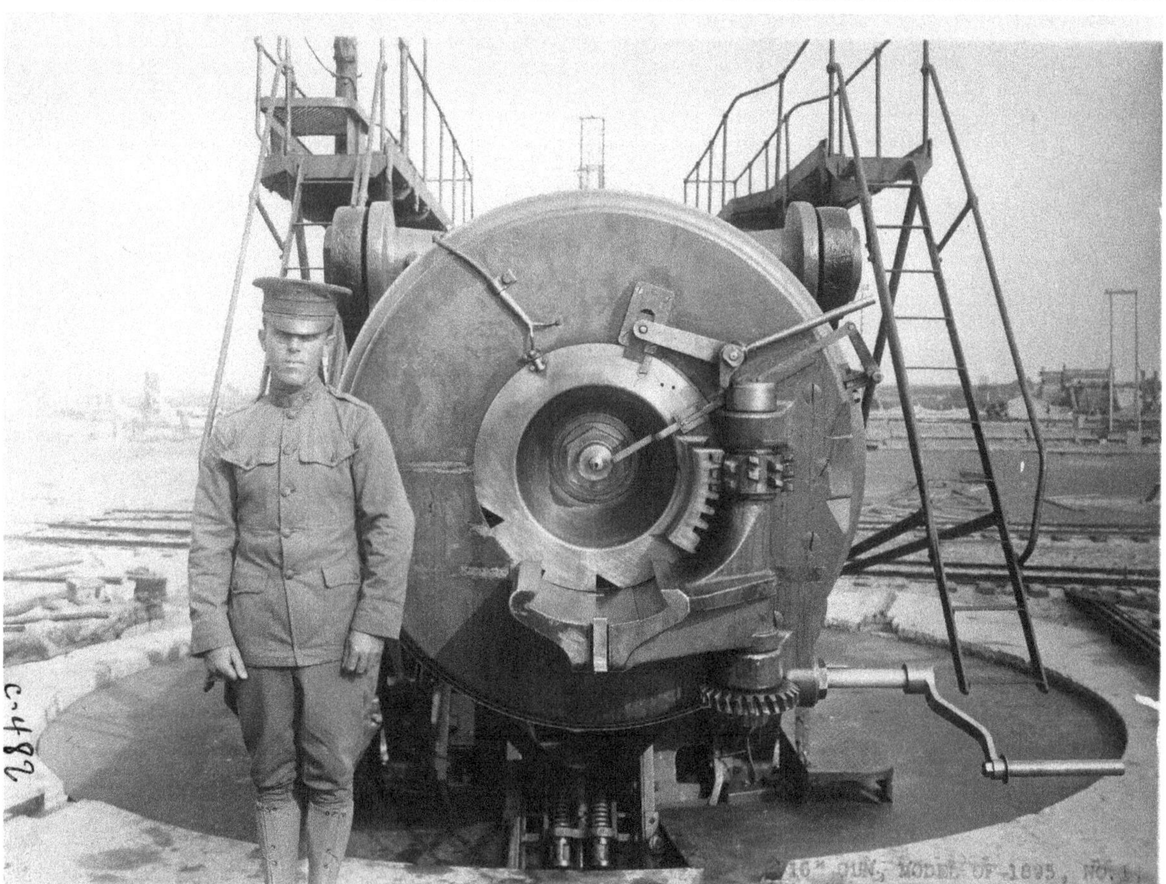

Breech view of the 16-inch Model 1895 gun on its trials. Note the gun was already equipped with its "high-trunnion elevating bands" to allow increased angle of fire and thus range.

Coast Defense Guns of the 1915 Board of Review

In 1915 the Army conducted a review of the status of American coast artillery. With a view to cope with recent improvements in naval artillery (both an increase in the bore size of main battery guns and greater range due to the increase in gun elevation) it recommended some very specific steps. While some harbors needed specific enhancement (like the entrance to Chesapeake Bay), most of the emphasis was on providing a new generation of heavy, long range weapons that could counter naval artillery developments. Most of the Endicott and Taft generation defenses were to stay in place, but new heavy batteries were to be selectively added to the most important harbors to enhance the defense deemed necessary.

Three new gun carriages were recommended. First a new long-range barbette carriage was to be deployed with the existing stock of 12-inch gun barrels. Generally mounted two per battery, these new emplacements would considerably increase the range of the defense. For the most worrisome or highest priority locations the Army would develop and deploy a new, modern 16-inch gun. Also on a very high elevation carriage, these guns would be the most powerful possible and capable of dealing with any current or reasonably foreseeable naval threat. Finally, a new 16-inch bore mortar, this time more accurately termed a howitzer would provide powerful plunging fire for the most important channels. While not totaling more than 4-6 guns or howitzers per harbor, the implementation of this program was meant to bring American defenses once more into the tactical supremacy congress demanded.

One other unique coast defense gun was emplaced. In addition to the barbette types, the Board had called for the use of either disappearing or turreted 16-inch guns in low-lying positions. An entirely new Model 1917 16-inch disappearing carriage was developed and one of the new Model 1919 16-inch guns coupled to it and emplaced at Fort Michie in Long Island Sound. But the day of the disappearing carriage was over, and this carriage experienced problems throughout its service life. After this unit, no new disappearing carriages were produced again in the U.S.

Of course, another world event happened between 1914 and 1918 considerably altered the program. Still, the new 12-inch long-range guns were for the most part built and emplaced. With the gun tubes already existing, and the fort reservations already in government hands, the protective magazine emplacement was about all that was needed. The development of the new 16-inch tube took longer. However, the resultant model 1919 16-inch, 50-caliber gun was a magnificent piece. It is rightly called the most powerful service piece ever produced by the United States. The new carriage to carry this gun was also a technical challenge but ultimately served well once adopted. This gun and carriage entered production right when America cut back immensely on its military spending, and only a few were made. Likewise, only five of the new howitzers were produced and they went to the defense of a sole location. Thus, the actual numbers of new seacoast guns and emplacements that came about from the 1915 Board was small. However, the new 16-inch carriage was basically used throughout the Second World War, as were all but a very few of the guns and emplacements themselves.

While the emplacements for these virtually all still exist, and except for a pair of barbette 12-inch guns on Corregidor, there are no survivors from a most interesting generation of powerful weapons.

16-inch barbette carriage at Watertown in 1922

12-inch Gun Model 1895A1 On 12-inch Barbette Carriage Long Range Model 1917

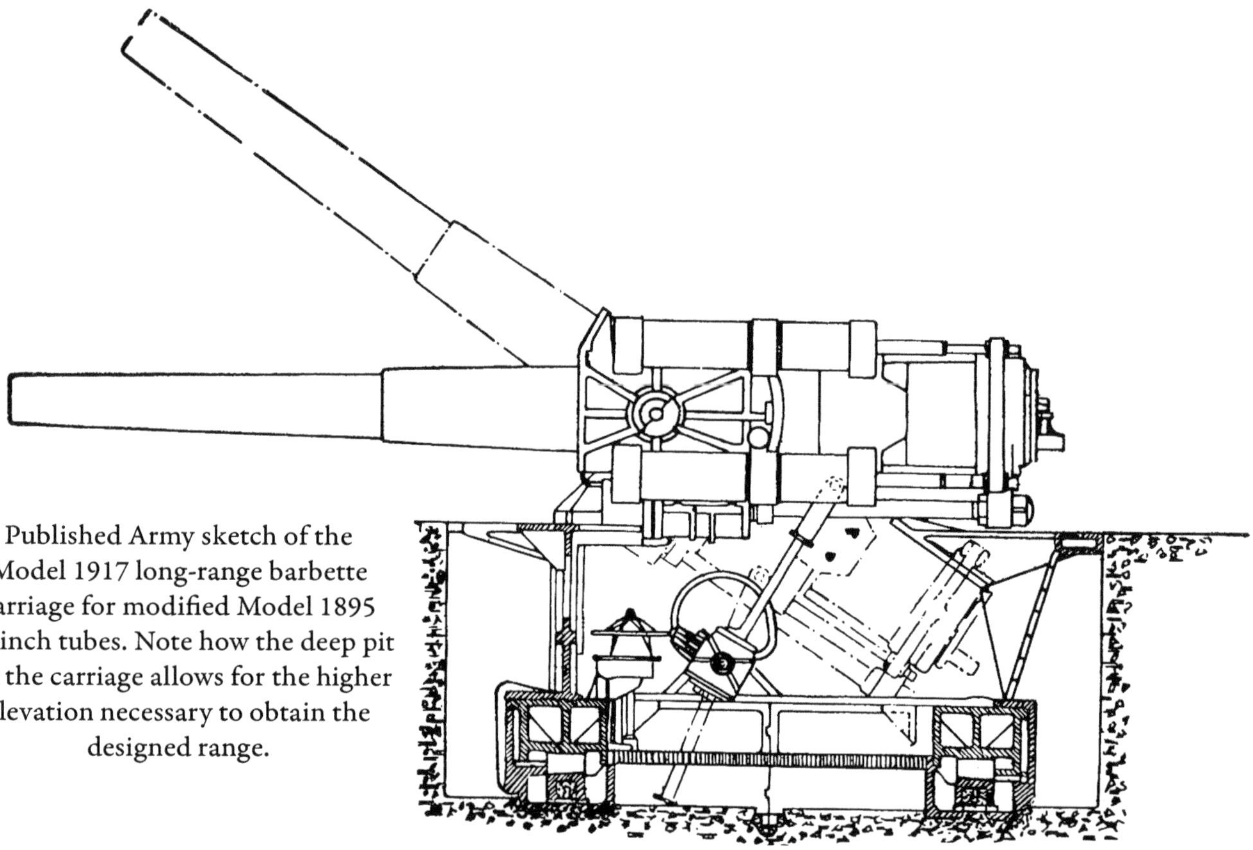

Published Army sketch of the Model 1917 long-range barbette carriage for modified Model 1895 12-inch tubes. Note how the deep pit for the carriage allows for the higher elevation necessary to obtain the designed range.

This was a fairly straightforward scheme to get longer range from an existing gun barrel. A considerable number of spare 12-inch M1895 barrels existed, and eventually more would become available when a project to mount them as railway mounts was curtailed. The 1915 Board of Review initially called for 36 new carriages to be made to arm 18 dual batteries, but that was soon reduced to 32 for 16 batteries. The adopted Model 1917 carriages were built in 1918-1920, being produced both by Watertown Arsenal (21) and Robert Poole (11).

The carriage had two cast steel side frames connected by transoms. The lower part of the carriage, was emplaced below ground level in order to provide adequate space for the recoil of the gun at maximum elevation. A new collar was fitted to the gun which, with old trunnion band removed then became the Model 1895A1. Attached to the collar were the four spring counter recoil cylinders. The hydraulic recoil cylinder was mounted below the tube. It allowed for a 30-inch recoil. An elevation of 0 to 35-degrees was obtained using bevel gears. Power elevation was used, but hand power could also be used if necessary. Observation platforms were on each side of the carriage, and an attached 3-inch telescopic sight Model 1912 was used. The gun was electrically fired using the Seacoast M1903 firing mechanism. Emplacing the gun on an open, unobstructed platform allowed for 360-degree traverse. Initially the gun used an ex-caliber 75mm M1916 gun but later was fitted with a subcaliber firing gun within the tube. Shells supplied were the Mk VI AP shell and a Mk X HE round.

Sixteen batteries were begun and 32 carriages were produced. However, one battery was modified upon completion to take 16-inch guns (Battery Frank Long at Fort Duvall), so two carriages remained as spares. A new type of emplacement was utilized. The two guns sat in open, unprotected platforms some 420-feet apart. An intervening cut-and-cover protected structure contained the powder and shell magazines, power plant and other support rooms. Two observing stations were located on the roof at each side supporting the corresponding gun. Mechanical range indicators were built into the battery with display boards near each gun. Both a closed-back and open-backed emplacement variety were built. Usually just a single battery of two guns was allocated to a fort, but in a couple of instances two batteries were built in close proximity to provide four such guns.

In the 1940 Program most of the batteries were modernized. The guns in their original positions were given casemates with heavy overhead protection and connected by tunnel to the magazines. Casemating restricted the guns to a 145-degree traverse. Open-back variety emplacements were completely covered. Most of this work was done in 1940-1942. Three new batteries were built in the program: one in Charleston, Delaware, and the San Juan, Puerto Rico defenses. The latter used the two spare carriages, the other two had guns and carriages relocated from two batteries that weren't otherwise casemated. Also, late in the war a project started to modify the guns and provide an electric rammer and gas evacuation system. Some were modified in 1945-1946. Virtually all the casemated type of emplacements continued to serve until the end of the Coast Artillery service in 1947-48. Almost all the guns and carriages were then scrapped, the only emplacements surviving with armament are those at Fort Mills, Manila Bay.

12-in Gun Model 1895M1A4
On 12-inch Barbette Carriage Long Range Model 1917

Gun Characteristics
Bore size: 12-inch
Construction: Steel, built-up
Tube length overall: 454.6-inches
Nominal caliber: 35 caliber
Tube and breech weight: 114,700-pounds
Usual carriage: 12-in Barbette Carriage Model 1917
Weight of gun and carriage: 416,700-pounds
Maximum elevation: 35-degrees
Loading: Separate
Shell weight: 1,046-pounds
Maximum muzzle velocity: 2,250-fps
Maximum range: 26,800-yards

Builders: Guns: Watervliet Arsenal (No. 1-75)
and Bethlehem Steel (No. 1-15)
Carriages: Watertown Arsenal (No. 1-11, 21-29, & 32)
and Robert Poole & Son (No. 12-20, and 30-31)

Number built: 90 Guns, 32 Carriages

Testing the newly-installed 12-inch Model 1895A1 gun on its 12-inch Model 1917 Barbette Carriage of Battery Kingman, Fort Hancock, New Jersey in June of 1919.

This view of a 12-inch Model 1895A1 on Model 1917 barbette clearly shows the increased barrel elevation the new carriage could provide. This is Battery Closson at Fort Kamehameha, Oahu in the 1920s.

It seems during the interwar years the 12-inch on BCLR Model 1917 carriage was particularly attractive to serviceman amateur photographers. This is a good example.

The new 1940 Program BCLR emplacements, along with most of those built in the 1920s and 30s, were modernized with overhead concrete protection. This photo is from 1943 inside the gunroom of Battery Reed, Fort Amezquita, San Juan Puerto Rico.

While all of the Model 1917 BCLR emplacements still exist, just the two wartime abandoned guns and carriages on Corregidor at Fort Mills still survive. This is a recent photo of Battery Smith.

16-inch Gun Model 1919
On 16-inch Disappearing Carriage LF Model 1917

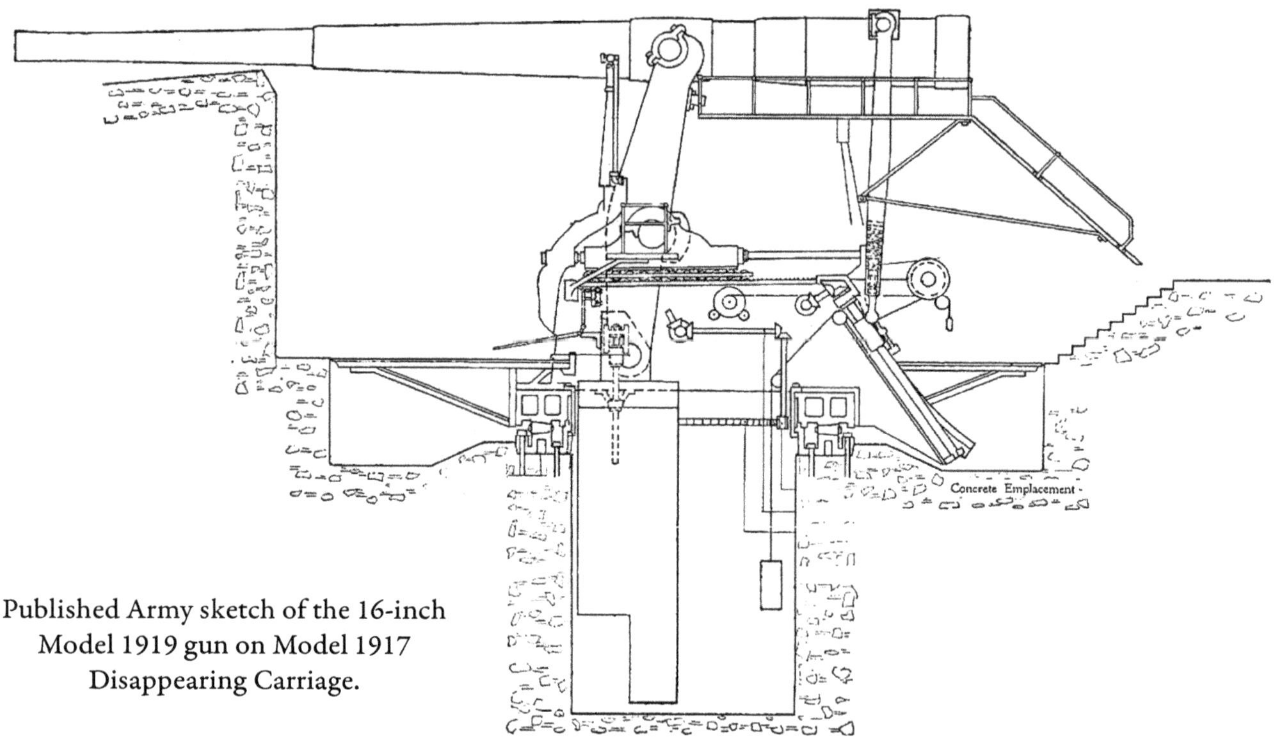

Published Army sketch of the 16-inch Model 1919 gun on Model 1917 Disappearing Carriage.

The army's new 16-inch gun began development right before America's entry into the First World War. In early 1917 two prototype Model E guns were ordered, but the war suspended work and when it was restarted it was with an improved design. The first production order of 12 tubes was issued to Watervliet on July 24, 1921, but by April of the following year it was cut to just nine tubes.

The gun was a heavy wire-wound tube of 50-caliber length. The gun had 144 grooves and was rifled from 1 in 50 to 1 in 25 near the muzzle. The M1919 gun fired a new 2400-pound AP projectile with a substantial muzzle velocity of 2700-fps. Like other disappearing carriages, it was loaded by truck and rammed manually. For this model carriage the gun did not use a gas ejector and had a hinge on the breech face carrying a loading tray.

The carriage understandably borrowed much from previous 14-inch and the 16-inch Model 1912 types. The limited-fire disappearing carriage was typical in design for this type of mechanism. The primary parts were the base ring and racer with individual traversing rollers, a chassis, top carriage with recoil rollers and hydraulic recoil cylinders, gun levers, crosshead, lead counterweight (weighing 315-tons), elevating arms, working platform and sighting platform. The carriage used the action of the recoil to raise the counterweight, being caught by pawls. The recoiling gun was stopped by this effort and the action of the recoil cylinders. The gun always returned to the same level to ease loading from hand trucks. After loading the gun was raised again to firing position by releasing the pawls that dropped the counterweight.

There were two features that were new. The gun was modified right after installation to provide higher angle of fire. The increase was from 20 to 30 degrees. That required a new arrangement for the elevating band, and a substantial kick-down cylinder in the front of the carriage to absorb some of the shock from firing at such an angle. In fact, this modification never really worked well, and the gun was actually fired very rarely and even then with due caution. Also, a large, angled shield for protection from shell fragments was placed around and in front of the crew servicing area, although of course the barrel was still free to elevate and fire.

The emplacement for this single disappearing carriage built was Battery J.M.K. Davis at Fort Michie, defenses of Long Island Sound. The mission here was to prevent a "run-by" of enemy ships intent on approaching New York through the sound. Consequently, the range was short (just 5000-yards), the elevation low (barely above sea level) and the need for protection was high. At first a turret mount was recommended for the selected site, but the funding available eventually drove the selection to a disappearing carriage battery instead. The battery was built in 1919-1922. In fact, the current technology had moved past the era of disappearing guns, and it would have better had it been constructed as a barbette battery.

As mentioned above the gun was not considered suitable for combat, the emplacement of one of the finest and most powerful gun tubes on an anachronistic carriage proving somewhat an embarrassment to the Army. The 1940 Program called for the battery's elimination. Still it remained in active service until late in the Second World War, being authorized for removal and scrap only on March 2, 1944. The carriage and gun were scrapped, but the emplacement still exists.

16-inch Gun Model 1919
On 16-inch Disappearing Carriage LF Model 1917

Gun Characteristics
Bore size: 16-inch
Construction: Steel, wire wound
Tube length overall: 861.3-inches
Nominal caliber: 50 caliber
Tube and breech weight: 340,601-pounds
Usual carriage: 16-inch Disappearing Carriage LF Model 1917
Weight of gun and carriage: 1,671,876-pounds
Maximum elevation: 30-degrees (after modification)
Loading: Separate
Shell weight: 2,400-pounds
Maximum muzzle velocity: 2,700-fps
Maximum range: 41,600-yards

Builders: Gun: Watervliet (1 of 9 Model 1919)
Carriage: Watertown Arsenal

Number built: 9 guns built, 1 disappearing carriage

Watertown Arsenal photograph of a new 16-inch Model 1919 gun on the only Model 1917 Disappearing Carriage in December, 1920.

The mounted 16-inch Model 1919 gun on its Model 1917 disappearing carriage with protective side shields at Battery J. M. K. Davis, Fort Michie, Long Island Sound defenses.

Another photograph of the unique installation of Battery Davis and its 16-inch disappearing gun.

16-inch Gun Model 1919MII and 1919MIII On 16-inch Barbette Carriage Model 1919

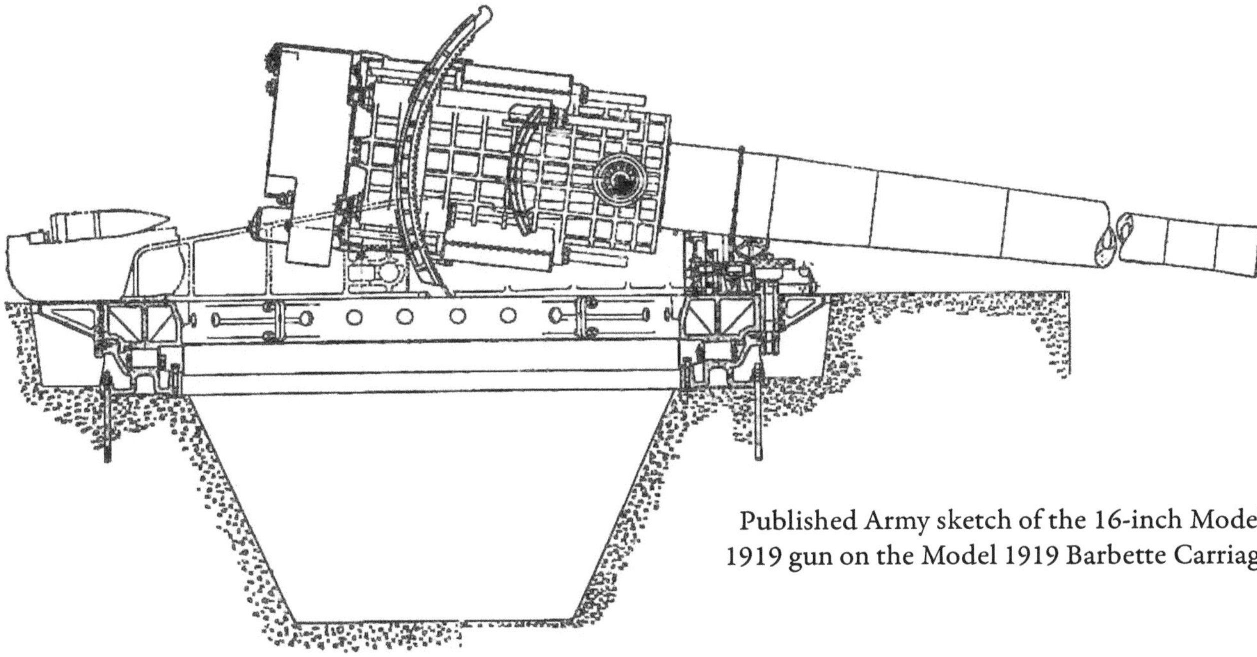

Published Army sketch of the 16-inch Model 1919 gun on the Model 1919 Barbette Carriage.

The new 16-inch gun was developed right before America's entry into the First World War. In early 1917 two prototype Model E guns were ordered, but the war suspended work and when it was restarted it was with an improved design. The first production order of 12 tubes was issued to Watervliet on July 24, 1921, but by April of the following year it was cut to just nine tubes (and three partial tubes stored, but ultimately never completed). This was due to the unanticipated availability of ex-navy 16-inch guns following the cancellation of capital ships by the Washington Navy Treaty. Minor changes were introduced, with recoil bands for barbette carriages guns No. 2-8 were Model 1919MII and No. 9 was the sole Model 1919MIII.

The gun was a heavy wire wound tube and jacket of 50-caliber length. It had 144 grooves and was rifled from 1 in 50 to 1 in 25 near the muzzle. The M1919 gun fired a new 2400-pound AP projectile with a substantial muzzle velocity of 2700-fps. It had a Welin, step-threaded breech. There were 16 sections on the breech, allowing it to be opened or closed with a relatively small turn.

The new M1919 barbette carriage designed for this gun was simple in concept, but the great weight of the gun carried imposed many technical challenges in engineering and fabrication. A 4-piece base ring was bolted together and emplaced to a concrete foundation. Atop the ring were the racer ring and traversing rollers. The top carriage was basically two heavy side cheeks with trunnion beds to hold a massive, complicated cradle. The cradle had trunnion extensions and completely encircled the gun tube. Four recoil cylinders (two long and two short) and two recuperators managed recoil. Electrically powered traverse could rotate the gun 360-degrees and powered elevation by rack could deliver elevation of 65-degrees. The great weight of the shells meant power loading and ramming—generally at a 4-degree elevation.

Only six barbette carriages Model 1919 were produced. After that run subsequent carriages (Models 1919M1-M5) were used for the Navy Mk II guns. The six original carriages were all made at Watertown Arsenal, being ordered in 1920 or 1921. They were emplaced at three, two-gun batteries built from 1921-24. The batteries featured entirely open, circular emplacements with the guns served by ammunition carts on rails connecting to enclosed, protected magazines. The emplacements were Battery Harris at Fort Tilden, Battery Williston at Fort Weaver, and Battery Long, Fort Duvall. The latter in fact was intended to be a battery for two 12-inch long-range barbettes, but the armament was changed to 16-inch before completion. The guns were about 1000-feet apart (except for the modified Battery Long). These three batteries were not actually completed until the late 1920s. The nine 16-inch M1919 guns produced were used in these barbette batteries (6 guns), the 16-inch disappearing battery at Fort Michie, with one gun staying at the proving ground and the final gun considered as a spare.

During the early 1940s the batteries were still considered important elements of the seacoast defenses. Two of the emplacements (at Tilden and Duvall) were given overhead, casemated protection and full fire control baselines. The third (at Oahu's Fort Weaver) was left uncasemated as the local command was unwilling to sacrifice the battery's ability to fire completely around the island without the restriction of traverse incurred by enclosing. There was a project to fabricate steel tunnel shields for the guns. These were apparently actually built and delivered but never installed late in the war. The three batteries with their armament were declared obsolete in 1948. Both the guns and carriages were subsequently scrapped; there are no surviving Model 1919 16-inch guns or Model 1919 barbette carriages.

16-inch Gun Model 1919MII and 1919MIII On 16-inch Barbette Carriage Model 1919

Gun Characteristics
Bore size: 16-inch
Construction: Alloy steel, wire-wound
Tube length overall: 852.938-inches
Nominal caliber: 50 caliber
Tube and breech weight: 385,847-pounds
Usual carriage: 16-inch Barbette Carriage Model 1919
Weight of gun and carriage: 1,000,000-pounds
Maximum elevation: 65-degrees
Loading: Separate
Shell weight: 2,400-pounds (C.I.)
Maximum muzzle velocity: 2,700-fps
Maximum range: 49,140-yards

Builders: Guns: Watervliet Arsenal
Carriages: Watertown Arsenal

Number built: 9 Guns
6 Barbette carriages Model 1919

Watertown Arsenal photograph of May 1921 showing a 16-inch Model 1919 gun being fitted to its Model 1919 Barbette Carriage.

The very high elevation obtainable by the 16-inch Model 1919 barbette carriage.
This was the gun at Battery Williston, Fort Weaver, Hawaii.

There were only three 2-gun batteries for the original Model 1919 gun on barbette carriage, this is the one at Battery Frank, Fort Duvall, Boston defenses.

16-inch Howitzer Model 1920 On 16-inch Howitzer Carriage Model 1920

Ordnance Department sketch of the 16-inch Model 1920 howitzer on its Model 1920 barbette carriage.

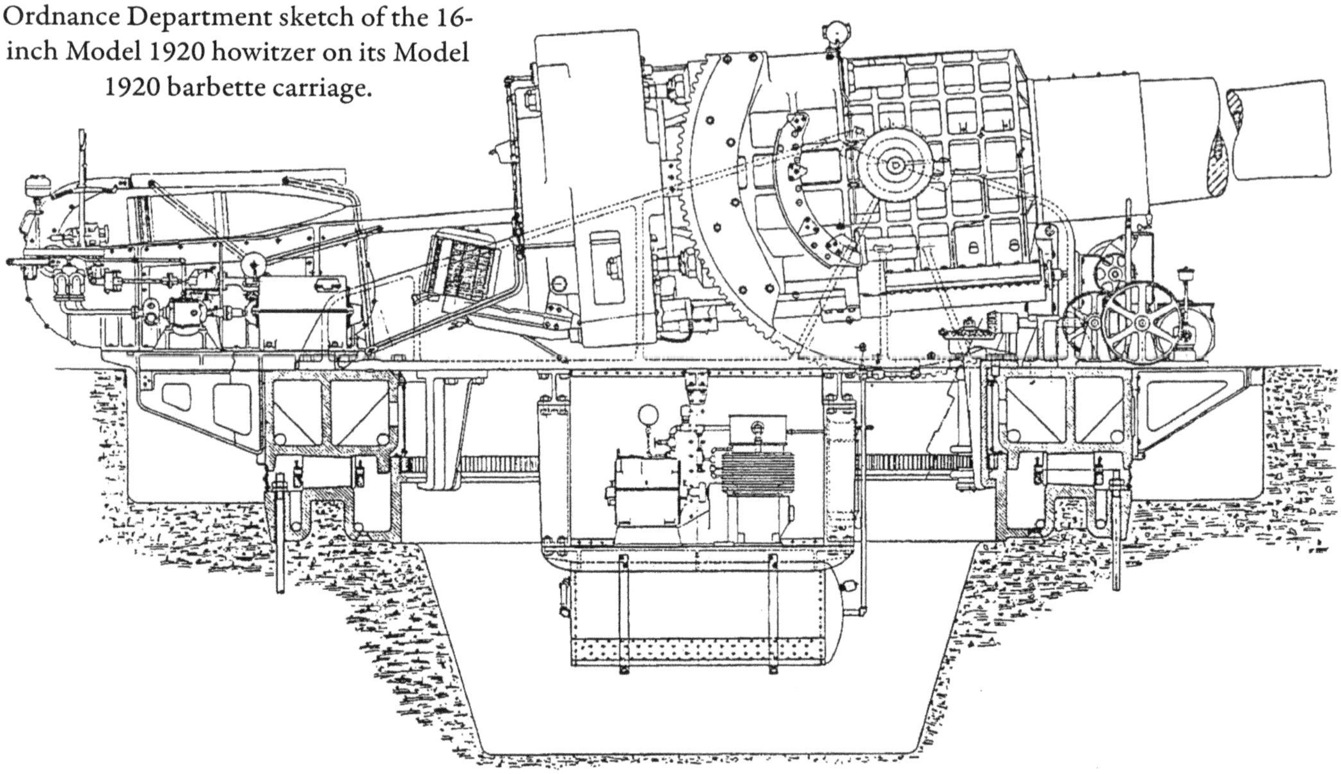

The 1919 revision to the Board of Review called for the emplacement of sixty 16-inch mortars. The weapon was based off of a technical staff proposal for a high-angle, 25-caliber gun of the largest possible bore. This new weapon was basically a reduced-length 16-inch gun. It was wire-wrapped, and of 25 calibers. There were 144 grooves and rifling of 1 in 50 increasing to 1 in 25. The reduced muzzle velocity of 1950-fps meant a greater expected tube life of about 800 rounds. Maximum range was 24,500-yards. The howitzer used the same breech as the M1919 gun. A large counterweight was affixed to the breech end of the gun to balance it and provide connectors for recoil pistons. A spline on the howitzer keyed to the cradle and prevented any rotation of the tube during firing.

The carriage closely resembled that for the 16-inch gun Model 1919, though slightly lighter and simpler due to the lighter weight of the howitzer tube. The base ring was bolted together and emplaced on a concrete foundation. Atop the ring were the racer ring and traversing rollers. The top carriage had two heavy side members with trunnion beds to hold a massive, complicated cradle. The cradle had trunnion extensions and completely encircled the gun tube. Four recoil cylinders of the dashpot type were carried, and a single recuperator cylinder was mounted on the bottom of the cradle. Electrically powered traverse could rotate the howitzer 360-degrees and powered elevation by rack could deliver elevation of -7 to +65-degrees. It was equipped with power loading and ramming.

While technically successful, the need for the howitzer was quickly called into question. Specifically, it was found that the newest 16-inch, 50-caliber gun on its Model 1919 barbette carriage could have an elevation of 65-degrees and basically achieve everything the shorter howitzer could. Even though much more expensive, the full gun could deliver both long range and plunging fire. In particular the relatively short 25,000-yard range of the howitzer was its downfall. Orders were changed in early 1922, and just four howitzers Model 1920 were produced by Watervliet Arsenal in addition to the one pilot gun already on hand. Only four Model 1920 barbette carriages were ordered and produced by Watertown Arsenal.

A four emplacement battery for all four production howitzers and carriages was constructed at Fort Story on the Virginia Capes in 1921-22. Initially named Battery Pennington, it was later organized into two batteries of two 16-inch howitzers each named Battery Pennington and Battery Walke. Similar to a 16-inch gun barbette battery, the gun platforms were entirely open, circular blocks allowing a full 360-degree traverse. They were joined by a rail track to separate powder and shell magazines—also of light, unprotected construction. The expectation was that the guns and magazines would be protected just by their dispersion and camouflage.

This battery served through the Second World War. In the 1940s light splinter shields were added around each gun, but they were not otherwise casemated or protected. However, the batteries were supplied with a bomb-proof plotting and switchboard room. They served until declared obsolete early postwar. Subsequently the guns, shields, and carriages were scrapped. While portions of the open concrete platforms and some of the rail and magazine system still exist, none of the guns or carriages survived.

16-inch Howitzer Model 1920
On 16-inch Howitzer Carriage Model 1920

Gun Characteristics
Bore size: 16-inch
Construction: Alloy steel, wire-wound
Tube length overall: 437.44-inches
Nominal caliber: 25 caliber
Tube and breech weight: 195,300-pounds
Usual carriage: 16-inch Howitzer Carriage Model 1920
Weight of gun and carriage: 900,300-pounds
Maximum elevation: 65-degrees
Loading: Separate
Shell weight: 2,400-pound (C.I.)
Maximum muzzle velocity: 1,950-fps
Maximum range: 24,500-yards

Builders: Guns: Watervliet Arsenal
Carriages: Watertown Arsenal

Number built: 5 howitzers, 4 carriages

Watertown Arsenal photograph of the pilot 16-inch Model 1920 howitzer and its barbette carriage.

1942 service photograph of one of the four 16-inch Model 1920 howitzers on like model barbette carriage at Fort Story, Chesapeake Bay defenses.

During the Second World War the Model 1920 howitzers at Fort Story were provided with new protective shields. This is a photograph of one of those shields on Gun No. 4 in 1947. (Special credit to Markland Collection, Casemate Museum via Shawn Welch).

Coast Defense 12-inch Mortars

The heavy seacoast mortar had emerged from the American Civil War with a very positive reputation. A new system of coast defense in the early 1870s attempted to integrate batteries of numerous mortars with heavy guns. The theory was that plunging fire from mortars would be devastating to lightly protected ship decks, while pits with mortars (howitzers) firing at high angles would be almost untouchable by return fire. Howitzers might be inherently less accurate, but this could be compensated for by placing and firing them in large numbers close together. Engineer General Henry L. Abbot designed a mortar emplacement consisting of four pits each of four mortars arranged in a 2x2 quadrilateral pattern to fire simultaneously. While only one prototype emplacement of this type was built in the early 1880s, there were no modern mortars available yet to arm it.

In 1883, even before the Endicott Board Report, a project was started to develop a new rifled, breech-loading howitzer. In America these guns were still defined as "mortars"—although that designation later was assumed by smaller, most frequently smoothbore infantry weapons. The Model 1886 mortar was only nine calibers in length and still had a cast iron body but was hooped with steel and had a modern breech. The follow-up Model 1890 Mortar was all steel, 10 calibers in length, and was built in very large numbers. The 310 such mortars built between 1892 and 1907 were by far the most numerous type of seacoast gun emplaced in the modern era. Two later wire-wound mortar types, the Model 1908 and Model 1912 were built in much smaller numbers primarily for insular possessions. All of these mortar used simple, but still modern and effective carriages that allowed for high-angle fire and had complete recoil mechanisms.

The emplacements used for these mortars also evolved over time. Initially the earliest mortars went into quadrilateral batteries of sixteen mortars of General Abbot's type. Shortly thereafter (circa 1896) it was found that the accuracy of the mortars was greater than anticipated, and they began to be fired by individual pit rather than sets of pits. Consequently four, and then later just two pits of four were arranged inline. Over time it was found that fewer and fewer mortars were needed to effectively deliver plunging fire on relatively close targets. Batteries went from sixteen to eight and then just four mortars between 1890 and 1916.

This reduction in requirements was applied to both new construction (primarily only in insular possessions and new West Coast harbors after about 1905) and to the armament of existing batteries. Several older batteries were rebuilt in 1910-1915. A major reduction of taking fully half of the mortars from each pit began in 1917. The mortars released were available for both railway mounts and for a few new temporary batteries emplaced during the First World War. Additional batteries were abandoned and their mortars and carriages scrapped when several harbor defenses were discontinued in the 1920s. Still, well into the 1940 Program and the Second World War most American harbors and insular stations had mortar batteries armed and actively in service. These were generally removed in 1942-1944 as further cutbacks hit the Coast Artillery Corps and more modern defenses were completed.

The mortars and carriages were never widely distributed as memorial or museum pieces. Fortunately, those (of three models) left in the Philippines and one battery overlooked for removal in St. Petersburg, Florida still exist, along with a single practice mortar in New York. Most of the battery emplacements themselves still survive in national, state, and local parks.

12-inch Model 1890M1 mortar on M896MI carriage of Battery Worth

12-inch Mortar Model 1886
On 12-inch Mortar Carriage Model 1891

Ordnance sketch of the early, cast-iron Model 1886 mortar on a Model 1891 mortar carriage. Note the stubby appearance of this short, just 9-caliber, weapon.

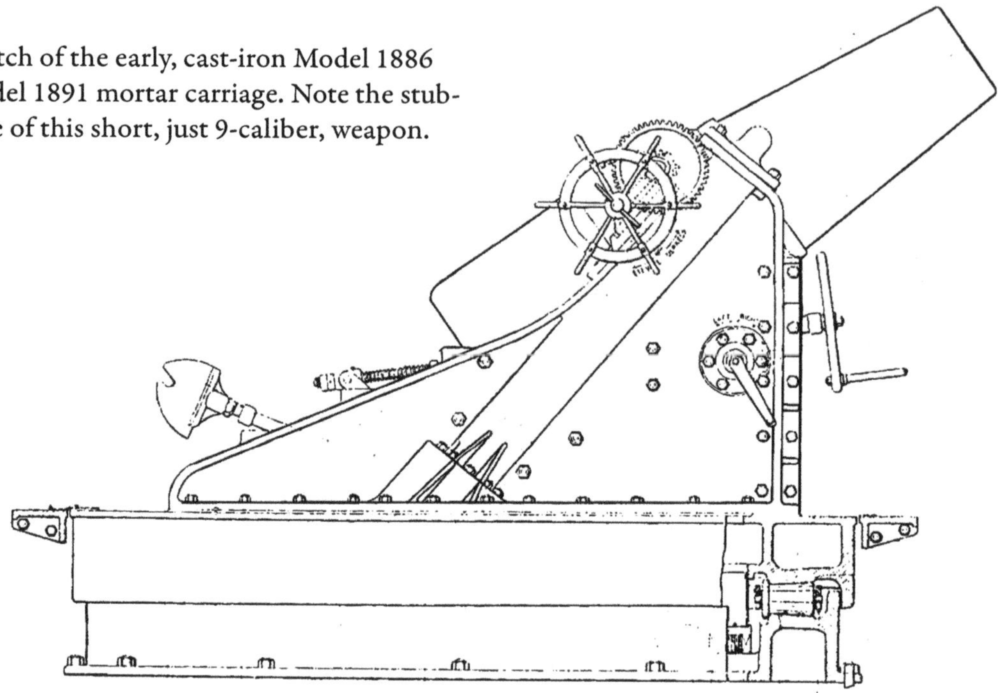

This was the first modern Army seacoast weapon developed for the prospective batteries of the Endicott Period. Frankly that was because the body was made of cast iron and it was relatively short and of low muzzle velocity. In other words, it could be designed and more importantly built by the nascent American arms industry. Design work began in 1883; the prototype mortar was tested at the proving grounds in August 1887. The mortar had a cast iron body with two rows of short steel hoops. A conventional breech block had an interrupted screw of six sections. It was operated by two cranks held parallel to the breech face. Later a safety bar catch was added. The Model 1891 carriage was developed after studying several purchased foreign designs. It had a base ring anchored to the concrete platform, 24 traversing rollers and a racer. On the racer were mounted a pair of gun-iron side pieces 7-foot high joined by a transom. The mortar tube sat in a slide trunnioned to the arms, and recoiling down a 50-degree slide. Cast in the side frame were two hydraulic recoil cylinders and spring counter recoil cylinders. Elevating gear giving an 80-degree elevation was a bronze rack and hand wheel.

Mortars were ordered beginning in August 1889. Builders Iron Foundry made all 73 mortars using their own and South Boston Iron Works bodies and Midvale Steel fabricated steel reinforcing hoops and steel breech blocks. A total of 85 carriages were ordered between 1891 and 1894, being delivered in 1896-97. The last 13 carriages were specifically intended to take all-steel mortars. Watertown Arsenal and three private contractors (Builders Iron Foundry, Robert Poole & Sons, and West Point Foundry) fulfilled the carriage orders.

This first mortar and carriage were almost exclusively designed to go with the first 1892 Type of mortar emplacement. Personally designed by General Henry L. Abbot, four new batteries were begun in 1891-1892. Two in New York, one each in Boston, and San Francisco were built for sixteen mortars in four pits. Slightly later a fifth battery was built at Fort Moultrie in Charleston (1896-98). Each battery was armed with 16 Model 1886 mortars on Model 1891 carriages. But as there were only 73 mortars available, to complete the Charleston battery the Ordnance Department had to order additional Model 1886-90 mortars in 1897-1904. These were all-steel versions of the Model 1886 but had the same length and ballistic characteristics so as to be compatible with the fire control of the other half of the battery. Nine of these mortars were built, eight for this battery and one to replace a Model 1886 destroyed in Boston in an accident. The five extra Model 1891 carriages were matched to Model 1890 mortars—four in one pit at Fort Monroe, and one for instruction at the West Point Military Academy.

With the advent of the all-steel mortars and improvements in the carriages, this first generation of guns was soon obsolete. They were seldom subject to the progressive modifications of the later types, and one of the original batteries at Fort Banks, Boston, was modernized and armed with more modern mortars in 1910-1915. Nonetheless the others served into the First World War. While the old Model 1886s were not selected for use on railway carriages, some were re-located to temporary batteries during the war. After the war, in June 1920, all the old Model 1886 and Model 1886-90 with their Model 1891 carriages were declared obsolete and scrapped. No examples of this pioneer piece of modern armament still exist.

12-inch Mortar Model 1886
On 12-inch Mortar Carriage Model 1891

Gun Characteristics
Bore size: 12-inch
Construction: Cast iron body, built-up with steel hoops
Tube length overall: 129-inches
Nominal caliber: 9 caliber
Tube and breech weight: 31,920-pounds
Usual carriage: Mortar Carriage Model 1891
Weight of gun and carriage: 111,634-pounds
Maximum elevation: 65-degrees
Loading: Separate
Shell weight: 824 - 1,046 pounds
Maximum muzzle velocity: 910 - 1,140 fps
Maximum range: 10,500-yards

Builders: Mortar: Builders Iron Foundry (M 1886)
Watervliet Arsenal (M1886-90)
Carriage: Builders Iron Foundry, Robert Poole, West Point Foundry, Watertown Arsenal

Number built: 73 Mortars Model 1886
(30 Model 1886, 43 Model 1886M)
9 Mortars Model 1886-90
85 carriages Model 1891

Demonstration unit of a 12-inch Model 1886 mortar on Model 1891 carriage.

One of the 12-inch M1886 mortars on its M1891 carriage as mounted at Battery McCook, Fort Hancock, New Jersey.

Mortars were emplaced in groupings or pits of four during the Endicott defense period. This is an entire pit of 12-inch M1886s at Battery McCook, Fort Hancock, New Jersey in July, 1894.

MODEL 1886. CAST IRON BODY-STEEL HOOPED – 14 TONS

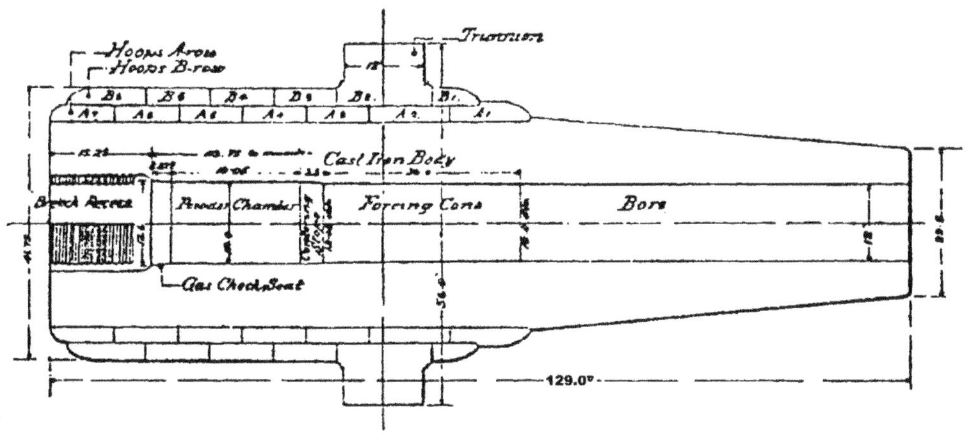

To complete the armament for the mortar battery building at Fort Moultrie, Charleston, the ordnance department had to place an order for additional 9-caliber mortars. It ordered a specially-designed, steel version of the Model 1886, known as the Model 1886-90. This diagram compares the design of the original and later variant.

MODEL 1886-90 MI. STEEL– 12 TONS

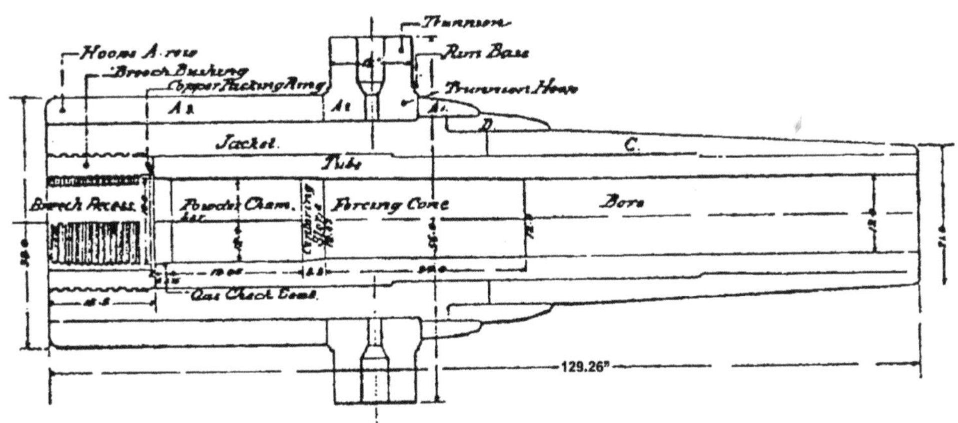

Period photograph of the Model 1886-90 12-inch mortar tubes on Model 1891 carriages at Battery Capron in Charleston.

12-inch Mortar Model 1890 and 1890M1 On 12-inch Mortar Carriage Model 1896M1 and 1896MII

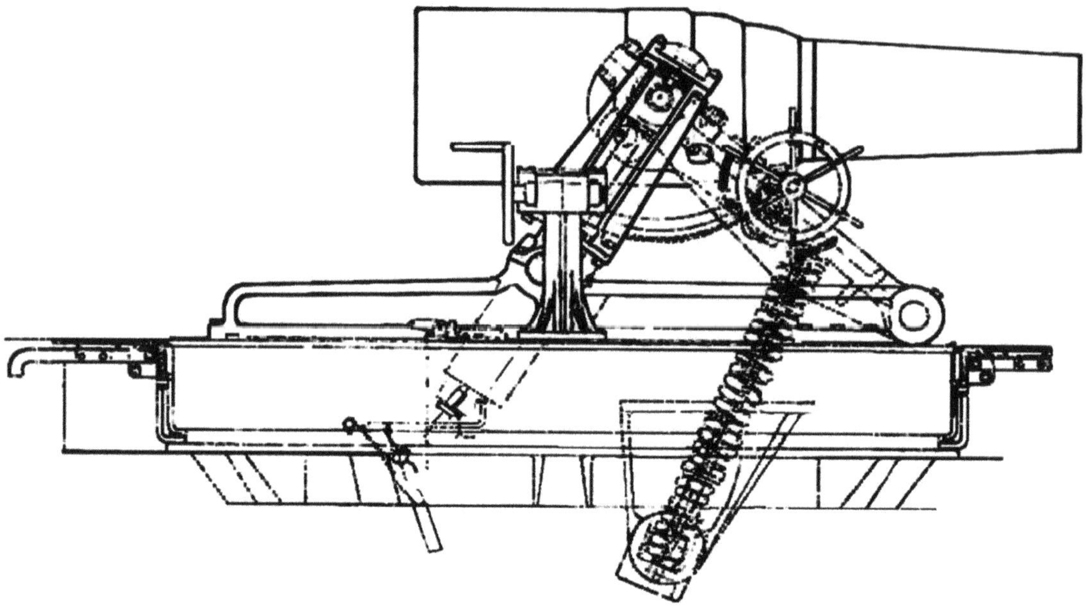

Ordnance Department sketch of the 12-inch Model 1890 mortar on original type of Model 1896 mortar carriage. Note how the carriage is attached to the ground platform with two independent support arms.

Soon after the acceptance of the cast-iron Model 1886 mortar, work commenced on a rifled, all-steel type. It was of built-up construction with numerous hoops or bands. It used a familiar interrupted-screw breech mechanism (though now supported by the jacket) with the usual breech rotating crank on the breech. It had 72 grooves and rifling of 1 in 40 to 1 in 20. The No. 1 type gun was ordered in August 1890. This first and the subsequent next seven ordered in 1893 were all Model 1890 12-inch mortars. Subsequent improvements in reducing the number of required bands led to the Model 1890M1. In fact, only eight principal parts were used in this model. An amazing total of 316 of this type were produced starting with contracts in August 1896. They were made (each with their own serial numbering series starting at No. 1) at Watervliet Arsenal, Niles Tool Co., Builders Iron Foundry and Bethlehem Steel.

A few of these mortars were mounted on older Model 1891 carriages, but the vast majority went onto the new spring-return Model 1896 12-inch Mortar Carriage. It was a significant improvement on its immediate predecessor. It had the usual base ring, traversing rollers and racer ring. Two large arms with trunnion beds held the tube. Extending from the trunnion to the rear were two large recoil pistons, extending into the recoil cylinders which were mostly in the lower working pit. Four heavy springs for counter recoil returning of the mortar to position were between the trunnion arms. A separate pedestal holding the hand wheel for traversing rose from the racer. This carriage was designated the Model 1896. After several years of development, test firings of the new prototype carriage was conducted in mid-1895, and contracts were issued beginning in 1896. Production was widely spread; Watertown Arsenal and nine separate private contractors produced 310 Model 1896 carriages through 1908.

The first carriages had structural problems and frequently broke parts. A reinforced, improved Model 1896M1 was developed. One of its primary changes was the placing of a large "I" beam between the trunnion arm and the racer. Existing M1896 carriages were modified in place with these beams and a new reinforcing plate inside the recoil well. Hundreds of kits were made and distributed to forts that modified their carriages in 1905-1910. In 1907 four final model 1896MII carriages were built for a battery pit in Hawaii.

A total of 46 permanent mortar batteries were, at least at some point of their service, armed with the Model 1890 types of mortar on Model 1896 carriages. Most of these were of the 1894 and 1898 emplacement types. However, one older 1892 type (at Fort Dupont) and several later Taft generation batteries in Hawaii and the Philippines also received this armament. Starting in 1910 some mortars were transferred between batteries. Most mortar pits were reduced to just two guns starting in 1918. Some of the released mortar tubes were used on railway carriages, while some complete guns and carriages were moved to temporary batteries. Many carriages were scrapped in the early 1920s when the removals were made permanent. Many mortar batteries continued to serve in the interwar years with just two mortars per pit. These were removed and the mortars and carriages scrapped in 1942-44 with the completion of the new system of armament.

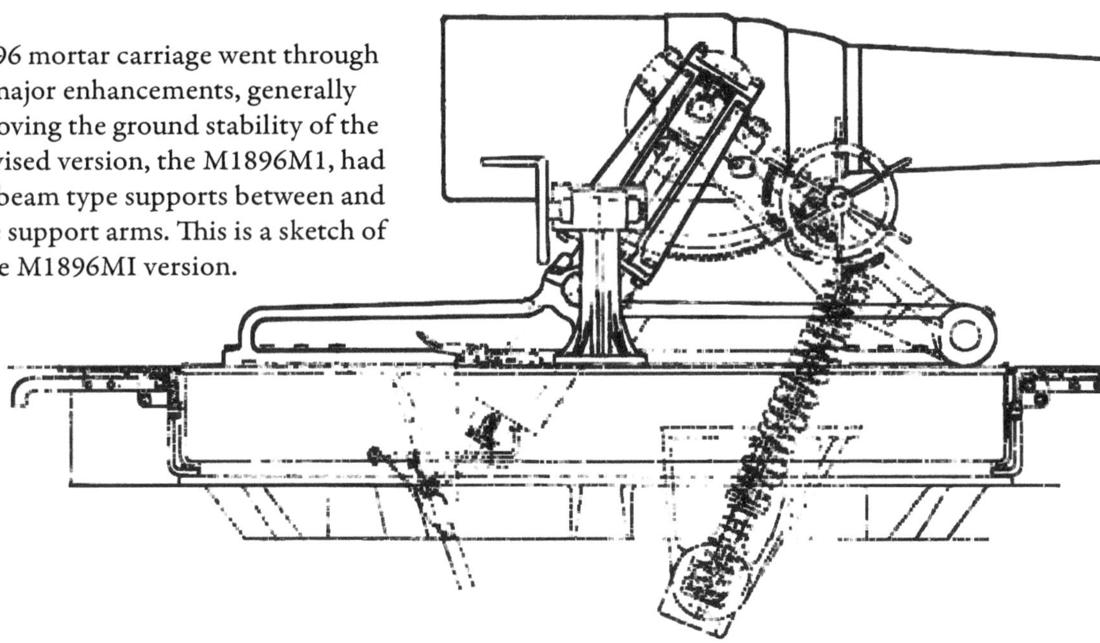

The Model 1896 mortar carriage went through a couple of major enhancements, generally aimed at improving the ground stability of the mount. The revised version, the M1896M1, had added lower I-beam type supports between and connecting the support arms. This is a sketch of the M1896MI version.

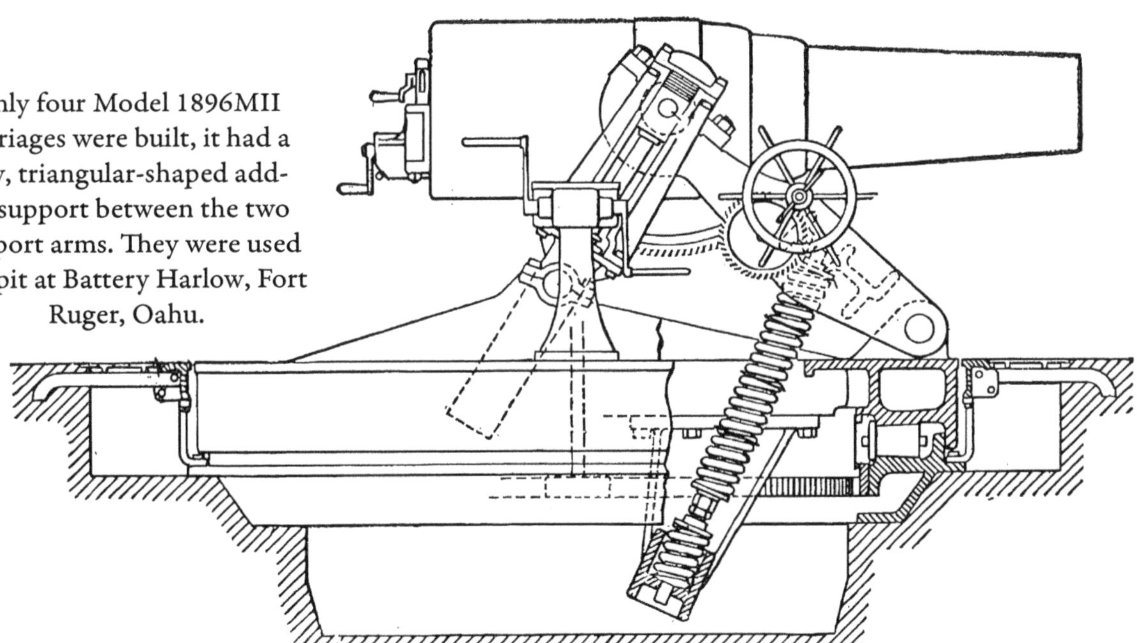

Only four Model 1896MII carriages were built, it had a low, triangular-shaped added support between the two support arms. They were used in a pit at Battery Harlow, Fort Ruger, Oahu.

12-inch Mortar Model 1890 and 1890M1 On 12-inch Mortar Carriage Model 1896, 1896M1 and 1896MII

Gun Characteristics
Bore size: 12-inch
Construction: Steel, built-up
Tube length overall: 141.12-inches
Nominal caliber: 10 caliber
Tube and breech weight: 29,120-pounds
Usual carriage: 12-inch Mortar Carriage M1896, M1896M1
Weight of gun and carriage: 105,631-pounds
Maximum elevation: 65-degrees
Loading: Separate

Shell weight: 824 and 1,046-pounds
Maximum muzzle velocity: 1,050 - 1,520 fps
Maximum range: 15,000-yards

Builders: Mortars: Watervliet Arsenal, Niles Tool, Builders Iron Foundry, Bethlehem Steel
Carriages: Watertown Arsenal, Robert Poole. Southwark Foundry, Providence Steam Engine, American Hoist and Derrick, Lake Erie Engineering Works, Builders Iron Foundry, Columbus Machine Co., Maryland Steel Co., Rarig Engineering Co.

Number built: 316 Mortars, 310 Carriages

Very few mortars still exist. Six Model 1890M1 on Model 1896M1 carriages are still mounted in their batteries at Fort Mills, Corregidor, Philippines. Four others, apparently overlooked during disarmament, are at a county park at Battery Laidley at Fort Desoto, St. Petersburg Florida.

Good photograph of the original 12-inch steel mortar M1890M1 on M1896 carriage, as emplaced at Battery Rodney, Fort Dupont, Delaware defenses.

View of at pit with four M1890MI mortars on their carriages just before the First World War. The modified side supports attached to each side of each mortar making these carriages all M1896MI.

The modified M1896MI mortar carriages.

A 12-inch mortar Model 1890M1 on a Model 1896M1 carriage with its noticeable side supports.

A 12-inch mortar Model 1890M1 on Model 1896M1 carriage with the side supports and M1896MII top carriage parts.

12-inch mortars Model 1890M1 on Model 1896MII carriages in Pit B at Battery Harlow, Fort Ruger Hawaii in the early 1920s.

Ten Model 1890M1 Mortars and Model 1896M1 carriages survive—six in the Philippines and four seemingly forgotten for removal in St. Petersburg, Florida. This is one of those now in Fort DeSoto Park, St. Petersburg, Florida.

12-inch Mortar Model 1908
On 12-inch Mortar Carriage Model 1908

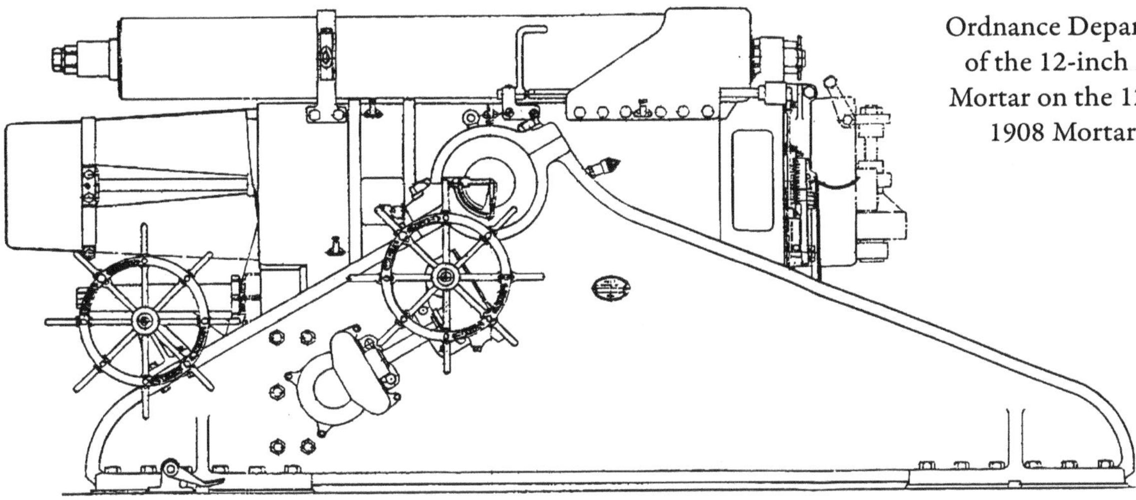

Ordnance Department sketch of the 12-inch Model 1908 Mortar on the 12-inch Model 1908 Mortar Carriage.

The next advancement in mortar development occurred after the start of the new century. A lighter, stronger tube was made possible through the use of wire-wrap construction. This was a technique where steel wire, one-eighth inch in cross section, was tightly wrapped around an inner steel tube and then covered with steel hoops. The resulting Model 1908 mortar tube was of the same length and had the same ballistic performance of the Model 1890 tubes. It utilized an improved breech design but still had the proven interrupted-screw mechanism. Rifling was changed to 108 grooves and lands with a uniform twist of 1 in 20.

An improved mortar carriage was also adopted at the same time. Still designed for high-angle fire up to 65-degrees, the carriage has a distinctly different appearance from other mortar carriages. It had the normal base ring, traversing rollers (30) and racer. Attached to the racer was the chassis with transom and bed to hold the trunnions of the cradle and attached sleigh. In this design the two large recoil cylinders were mounted below the barrel on either side between the cradle and sleigh. Above them were similarly arranged the two counter recoil spring cylinders. In appearance the mortar tube is equally surrounded by the distinctive four protruding cylinders. An elevating hand wheel connected to the elevating rack was on the left side of the carriage, the traversing hand wheel was on a separate pedestal attached to the racer on the right side.

All of the new mortars were fabricated at Watervliet Arsenal. Twenty-five mortars were built. A pilot Model 1908 carriage was built at Watertown Arsenal in 1908. It was followed by orders for 17 carriages at the same facility in October of that year, and eight others at Southwark Foundry in early 1909. All were from appropriations for armament at insular possessions. It does not appear that there were any major subsequent modifications of either gun or carriage once completed and emplaced.

The emplacements for these weapons followed the engineer Type 1906 plan. These were the last mortar designs intended for two adjacent pits of four mortars each. Battery Hasbrouck at Fort Kamehameha, Oahu, Battery Koehler at Fort Frank and Battery Geary at Fort Mills in Manila Bay were each built for eight of these mortars. However, during construction, in order to keep at least some of this type of mortar in the continental U.S., a change was made. Four mortars and carriages were withdrawn from Battery Geary (being replaced with older Model 1890M1 mortars on Model 1896 carriages) and emplaced instead at the recently rebuilt Battery Whitman at Fort Andrews in Boston Harbor. Once this change was made, the batteries served intact until the Second World War.

The mortars at Fort Andrews were authorized for scrapping in December 1942, those in Hawaii on May 1, 1945. The two batteries in the Philippines were abandoned with the surrender of the harbor defenses in May 1942. Because of that particular circumstance, there are survivors. In fact all 12 mortars in the Philippines survived the war. However, by 1980 the eight at Fort Frank had been salvaged by unauthorized scrappers. Still existing are all four tubes and parts of the carriages from Battery Geary Pit B, though heavily damaged and tossed about from the explosion of the battery's magazine in April 1942 during the siege of Corregidor.

12-inch Mortar Model 1908
On 12-inch Mortar Carriage Model 1908

Gun Characteristics
Bore size: 12-inch
Construction: Steel, wire wound
Tube length overall: 136.2-inches
Nominal caliber: 10
Tube and breech weight: 18,200-pounds
Usual carriage: 12-inch Mortar Carriage Model 1908
Weight of gun and carriage: 129,873-pounds
Maximum elevation: 65-degrees
Loading: Separate
Shell weight: 824 -1,046 pounds
Maximum muzzle velocity: 1,050 -1,500 fps
Maximum range: 15,000-yards

Builders: Mortar: Watervliet Arsenal
Carriage: Watertown Arsenal, Southwark Foundry

Number built: 25 mortars, 26 carriages
(1 pilot + 25 production)

Watertown photograph from 1913 of a 12-inch Model 1908 mortar on its Model 1908 carriage. The shot clearly illustrates how the mortar could be lowered for reloading while leaving the recoil mechanism in position.

A pit of four 12-inch Model 1908 mortars and carriages at Battery Whitman, Fort Andrews, Boston around 1917. This was, in fact, the only position armed with M1908 mortars in the continental United States.

The barrels of four 12-inch Model 1908 mortars still exist at the damaged remains of Battery Geary, Fort Mills, Philippines. This battery was torn apart when its magazine exploded during the siege of 1942, the mortars were tossed about and heavily damaged, but are still identifiable.

12-inch Mortar Model 1912
On 12-inch Mortar Carriage Model 1896MIII

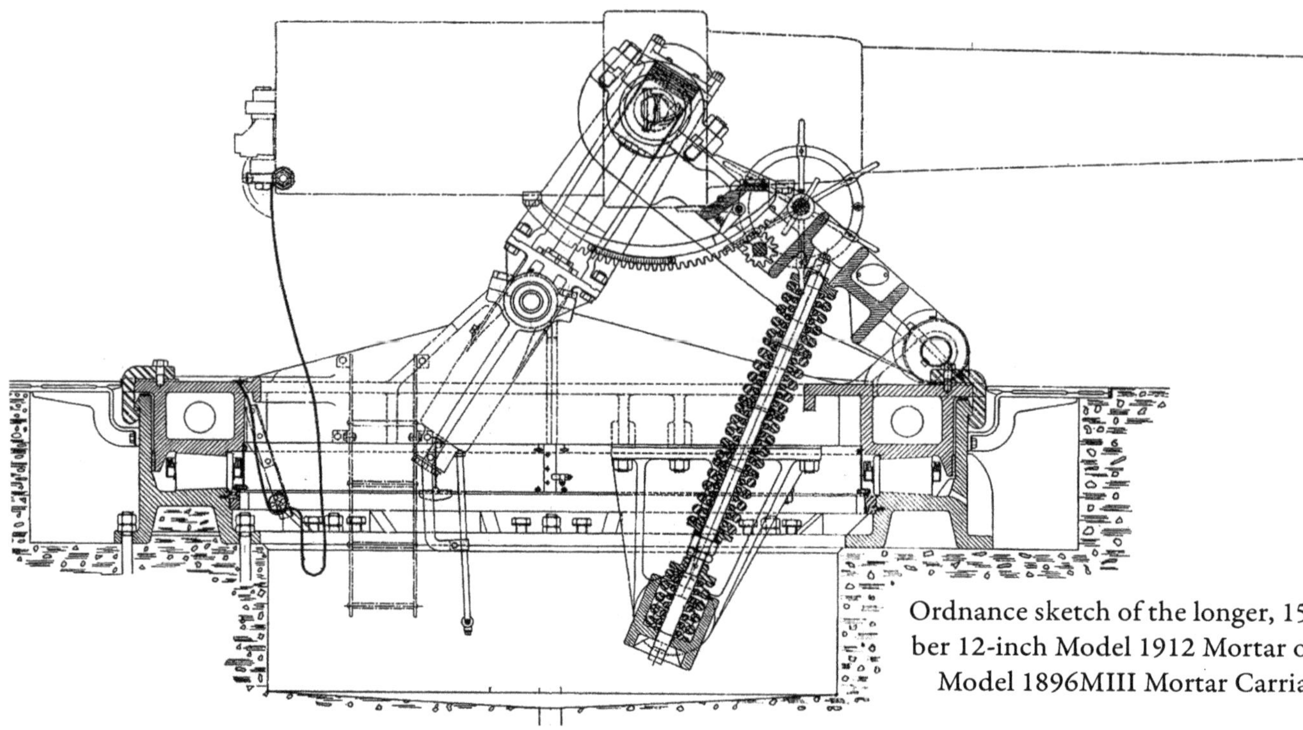

Ordnance sketch of the longer, 15-caliber 12-inch Model 1912 Mortar on the Model 1896MIII Mortar Carriage.

The final 12-inch mortar developed by the Ordnance Department was the Model 1912. The opportunity was taken to finally lengthen the tube to fifteen calibers vs. nine or ten in the earlier models. This mortar also would develop a higher muzzle velocity of 1850-fps with a 700-pound shell. The primary advantage of this was an increase in range, in this case adding five thousand yards. Otherwise, the tube closely resembled the previous Model 1908. It was also wire-wound in construction and used a very similar breech mechanism with tray and operating crank. Rifling was again changed. For the Model 1912 it was 108 grooves and lands with a twist of 1 in 40 increasing to 1 in 20 and then maintaining that for 34 inches. The first order for 12 mortars was placed on April 2, 1912. Altogether Watervliet Arsenal fabricated 41 Model 1912 mortars, finishing deliver in early 1914. The first six mortars had their powder chambers enlarged, the subsequent mortars had the expanded chamber from original manufacture.

For the carriage, the Ordnance Department went back to the successful Model 1896 type. Modifications were needed in strengthening certain parts and balancing weight to allow it to carry the longer, more powerfull 15-caliber mortar. With these changes it was designated the Model 1896MIII. It had the usual base ring, 25 rollers, and racer. The cast steel top carriage basically had two strong arms joined together by a heavy web. The arms ended in the trunnion beds. Recoil was braked by two hydraulic cylinders, which oscillated on longitudinal ribs connected to the racer. Counter recoil was provided by a hydraulic buffer, and the mortar was returned to position by four columns of heavy springs supporting the top carriage. Carriages were built simultaneously with the mortars, 41 being built. They were fabricated either by Watertown Arsenal or Builders Iron Foundry.

By the time this new mortar was developed the army was convinced that the accuracy of these weapons justified just the emplacement of two guns per pit. New emplacement designs had been developed in the early 1910s. All of the new batteries intended for the Model 1912 mortar were armed from the start with two guns per pit. Twenty-eight Model 1912 mortars of Model 1896MIII carriages went to Panama Canal Defenses: eight at Fort Sherman, eight at Fort Randolph, and twelve at Fort Grant. Eight were mounted at Batteries Barlow and Saxton at Los Angeles' Fort

MacArthur. The final four were mounted at Battery Craighill at Fort Hughes in Manila Bay.

The single battery of four guns at Fort Hughes in the Philippines was surrendered with the American forces in May 1942 and recaptured in 1945. The Model 1912 mortars in the Panama Canal were authorized for elimination in 1943, followed by the two batteries in Los Angeles in 1944. Subsequently these were all scrapped. However, the Philippine mortars and carriages, all still mounted in their emplacement, after the war were turned over to Philippine forces. As Fort Hughes on Caballo Island is still used for ammunition storage by the Philippine Navy, access to the battery and the four surviving mortars is quite difficult.

12-inch Mortar Model 1912
On 12-inch Mortar Carriage Model 1896MIII

Gun Characteristics
Bore size: 12-inch
Construction: Steel, wire wound
Tube length overall: 208-inches
Nominal caliber: 15 caliber
Tube and breech weight: 33,854-pounds
Usual carriage: 12-inch Mortar Carriage Model 1896MIII
Weight of gun and carriage: 132,000-pounds
Maximum elevation: 65-degrees
Loading: Separate
Shell weight: 824 -1,046 pounds
Maximum muzzle velocity: 1,850-fps
Maximum range: 20,000 yards

Builders: Mortars: Watervliet Arsenal
Carriages: Watertown Arsenal,
Builders Iron Foundry

Number built: 41

One of the 12-inch Model 1912 mortars in the defenses of Manila Bay, during the 1930s.

A 12-inch Model 1912 Mortar on its Model 1896MIII carriage at the Sandy Hook Proving Grounds about 1914.

The Model 1912 mortars were utilized in the late-Taft generation defenses. This is one of the mortars at Fort Grant, Panama

One of the four surviving Model 1912 12-inch mortars and the M1896MIII carriages which remain at Battery Craighill, Fort Hughes, Manila Bay.

Coast Defense Guns of the 1940 System

The final 1940 system standardized just a single type of new, primary armament. The fortuitous availability of large numbers of ex-navy 16-inch Mk II guns could be utilized on a modified Model 1919 16-inch barbette. These guns and carriages, emplaced in new, heavily protected casemate emplacements would be the mainstay of the new system. A limited number of 16-inch gun batteries built after the First World War with similar type of carriage were also provided with new casemated, overhead protection.

Just one gun and battery type was initially selected to be the secondary type of armament for the 1940 Program. A well-protected 6-inch gun capable of good elevation and rapid traverse fit the requirements. In early 1940 a new, shielded barbette carriage was developed. It utilized the substantial stock of 6-inch Model 1903 and 1905 gun tubes that had been held in storage since their removal from the expedient field artillery mounts Model 1917. However, to provide for the increasing number of projected sites and for spare tubes (there were practical limitations preventing the relining of the M1903 tubes), a new 6-inch T-2, 50-caliber gun also entered development and then limited production. The new 6-inch barbette batteries were widely emplaced in the continental U.S. and overseas territories and even foreign bases in 1940-45.

A small number of new 8-inch gun batteries were added to the program in 1941-42. Two prototype batteries had been built for this size gun in the late 1930s. In many ways they were scaled-up 6-inch battery designs. The army had received 74 ex-navy 8-inch, 45-caliber gun tubes in the 1930s, and while some had been used for a new railway mount, enough extras were on hand to consider for fixed emplacements. Viewed as a heavier intermediate gun to counter enemy cruisers, they were intended for defenses that lacked 16-inch guns (Alaskan harbors and Roosevelt Roads) or required coverage over landing beaches (Oahu). Twenty new M1 carriages were authorized in 1941 for use with a like number of guns. While not all were completed, a number of these batteries were armed late in the war.

One final element was added to the original guns of the 1940 System. The European war demonstrated a new, dangerous threat from very fast motor torpedo boats. The Italians, Germans, and British all used and had to defend anchorages from the assault of these unconventional attackers. In 1941 the Americans began a project to use the standard 90mm M1 Anti-aircraft gun for seacoast, anti-torpedo boat usage. They could also be useful in modernizing the AA defense and in replacing older 3-inch guns for minefield and boom protection. Basically only requiring a new carriage modification with depression coverage and a splinter proof shield, the new T3 mount was standardized and produced as the M3 carriage. Several hundred of these were produced and emplaced on simple gun blocks in 1943-45 throughout the United States and at selected overseas bases.

While most of the emplacements of this final generation still exist in parks or still on government reservations, precious little of the armament survived scrapping postwar. A few navy 16-inch tubes are still at Dahlgren, though not generally on public display. Six 6-inch guns in shields survived because they were in out of the way places, and remains of a couple 6-inch and 8-inch guns and carriages are still in place in Alaska after attempted demolition. Similarly, a few 90mm guns in their M3 shielded carriages fortunately missed destruction.

Battery Townsley, Fort Cronkhite

90mm Gun M1A1
On 90mm Gun Mount M3

Ordnance publication drawing of the M3 anti-motor torpedo boat mounting for the M1 90mm gun.

Development of new 90mm anti-aircraft gun for a coast defense, anti-ship role started in mid-1941, with design specs issued on December 18, 1941. The tactical need was for a gun of rapid-fire and rapid traverse to supplant or replace the old 3-inch pedestal mounts. They would be mounted at existing seacoast forts to provide coverage of minefields, to act as examination batteries, and to prevent fast motor torpedo boats and submarines from slipping into protected harbors. Technically the modifications of the 90mm for anti-tank service (improved automatic loading and full depression for surface targets) could assist the development of a new carriage. A full project began in February 1942 for development of this mount.

As standardized, the new AMTB gun used the same M1A1 90mm gun tube. The mount underwent considerable change. The gun slide, recoil system, and autoloader were mounted on a simple pedestal mount replacing the bottom carriage and outriggers. This mount had side cheeks with trunnion beds to hold the cradle and in turn was placed on a racer with traversing rollers. Both manual and power drives could elevate and traverse the gun. It used the full -8 to +80° elevation of the mount (limited by the new shield) and would be able to traverse 360-degrees. A large, box-like shield encompassed the mount. Flat-sided armor plates were held in a hexagonal pattern around the gun, with only the rear portion missing to allow full recoil and ammunition servicing of the gun. A sliding plate on top would move back to allow firing in higher, anti-aircraft elevation. Normally it could range to 8,000-yards, but equipment could track further to 12,000-yards. The mount was equipped with the M6 8-power telescope. Normally it was tied to both an M8 (surface) or M9 (AA) director and the AN/MPG-1 radar set, all through the T13 type gun data computer.

The M3 gun was usually emplaced as part of a fort AMTB battery consisting of two fixed emplacements, along with two mobile M1A1 guns and two smaller 37 or 40mm light AA guns. The emplacement consisted of two concrete platforms separated by 120-feet. The platform had hold-down bolts for the M3 mount. A combined traverse magazine held 400 rounds/gun and included a power plant and fire control equipment room. An adjacent battery commander station was usually also built, often on a tower or height between and behind the gun emplacements. Often the two mobile guns occupied prepared positions on the flank of either fixed gun. Usually supplied with the high explosive M71 round with point-detonating fuse, ready ammunition pits contained 25 rounds near each gun. Gun crew was 15 men—nine in the gun squad and six with the ammunition squad.

On September 18, 1943 a project was approved to construct 123 batteries for 246 guns and mounts M3 (although seven of these were non-emplaced reserve batteries). While there were some small adjustments, virtually all of these were actually emplaced. The M1A1 guns came from the pool of the over 7000 90mm that were constructed. Fisher Body was contracted to produce 254 of the M3 mount through September of 1943. Many of these M3 mounted 90mm guns lasted at coast artillery posts until the very end of that service in 1949. None were supplied to any other nations or services. They were subsequently removed and scrapped in the late 1940s and early 1950s. There are three known survivors: a gun with shield still in its emplacement in Battery Parrott at Fort Monroe; a carriage and shield (but no gun tube) awaiting restoration at Fort MacArthur in California; and a complete mount and parts of several others remain on Shemya Island, Alaska .

90mm Gun M1A1
On 90mm Gun Mount M3

Gun Characteristics
Bore size: 90mm
Construction: Alloy steel monobloc
Tube length overall: 186.15-inches
Nominal caliber: 50 caliber
Tube and breech weight: 1,465-pounds
Usual carriage: 90mm Gun Mount M3
Weight of gun and carriage: 5,564-pounds (w/o shield)
Maximum elevation: 80-degrees
Loading: Fixed
Shell weight: 23.4-pounds
Maximum muzzle velocity: 2,700 fps
Maximum range: 8,000-yards

Builders: Guns: Various (see 90mm AA Gun M1)
Carriage: Fisher Body Co.

Number built: 254 Carriages

Surviving 90mm AMTB M1 gun on an M3 carriage recently restored at Tactical Battery No. 23 at Battery Parrott, Fort Monroe, Virginia.

The two 90mm AMTB guns on M3 carriages of the prototype Tactical Battery No. 21 at Fort Story, Virginia in 1943.

6-inch Guns M1903, M1905, and T2 On 6-inch Barbette Carriage M1, M2, M3, and M4

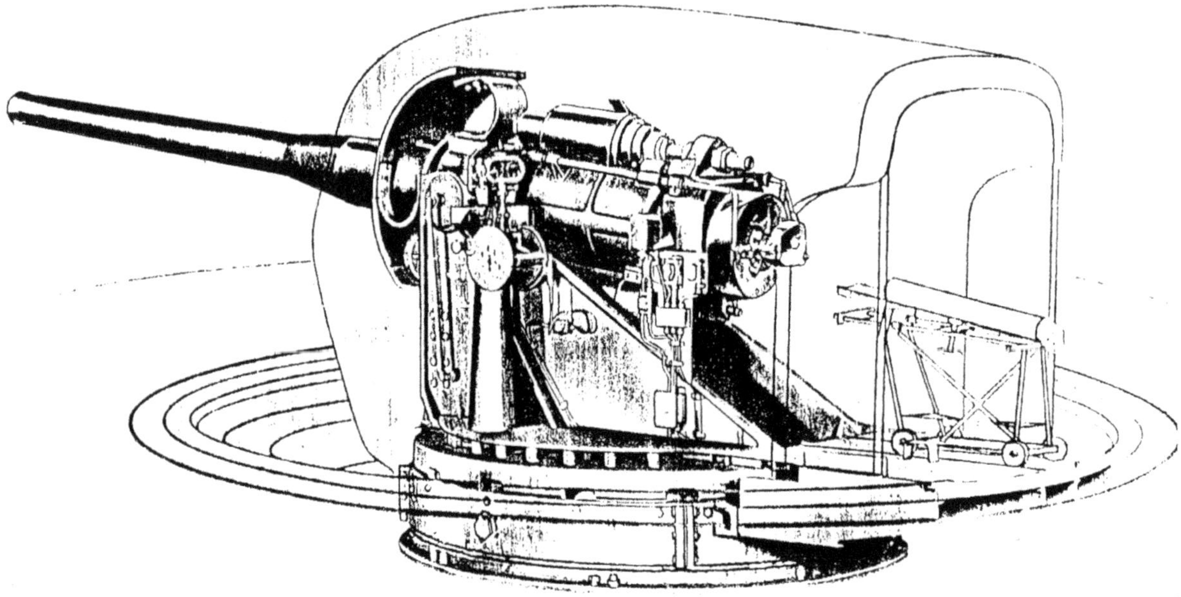

Interior view drawing of the 6-inch gun Models 1903A1/1905A1 in a shielded M4 Barbette Carriage.

The 1940 Program's primary intermediate gun was to be a new 6-inch barbette mount. Study had shown this type of fixed mount conclusively better than any mobile 155mm arrangement, even though the latter was more available and had much in common with field artillery pieces already under procurement. Fortunately, abundant quantities of 6-in, 50-caliber Models 1903 and 1905 were available. These were tubes of still relatively long-life that had been removed from their disappearing carriages in 1917 for wheeled field carriages that never made it into service before the end of the First World War. They were in storage and readily available for a new mount designed to take them. They were built-up guns with their original interrupted thread breechblock. Firing mechanism was for either percussion or electric primers. It fired separate-loading ammunition of AP and HE types. As prepared for new carriages these tubes were designated M1903A1 and M1905A1, later with gas ejection system they became M1903A2 and M1905A2.

The new barbette carriage developed in the late 1930s was a relatively simple, but thoroughly modern carriage. It had 360-degree traverse and allowed -5 to +47 degree elevation. Recoil was braked by a hydropneumatic cylinder stop the tube, and two counter recoil spring cylinders. A wrap-around "tunnel" shield of 6-inch thickness protected all but the open rear from fragments. The original carriage was known as the M1, with gas-ejection equipment it became the M2. First production orders were placed on May 16, 1941. Both of these carriages were for the M1903 or M1905 guns. Carriages for the new T-2 gun were respectively the M3 or M4. On-carriage sighting was an M31 telescopic sight. A total of 109 carriages were made, involving Watertown and three separate private manufacturers.

A standardized emplacement was eventually to be constructed and was later known as the "200-series" batteries. It featured two of the new barbettes; each on a circular apron and pit separated 210-feet from each other. Intervening was a protected, shared magazine with powder and shell storage, a power and radiator plant, and a plotting and radio room. Eighty-nine such batteries, a few modified as needed to fit local terrain, were authorized in 1940-1941 in the continental U.S. and overseas territories and bases.

Eventually the stock of older tubes was exhausted and to finish arming the authorized batteries and provide spare tubes (unfortunately the Model 1903s could not be relined) a new gun tube was developed. The new gun was never standardized (it was to be the M1); it went into production simply as the T2 6-inch seacoast gun. A pilot tube was ordered in August 1940, and 40 production tubes on May 16, 1941. Orders were increased in 1941-42 and then decreased in 1943-44 as field artillery production took precedent over new seacoast guns. Ultimately just 31 new 6-inch tubes were produced at Watervliet Arsenal and Erie Forge and Steel Co. Late in the program (1944-46) the lack of new tubes created real problems. The Army considered adapting spare navy tubes to the M1 carriage, but they had different ballistics that would have caused additional problems. In the final years there was some swapping of tubes from one battery to another to keep a dwindling number of batteries in service.

Ultimately in 1948-49 the guns and carriages were disposed of with the end of active coast artillery. Almost all of the 6-inch material was scrapped, but four guns mounted at Argentia, Newfoundland and two at Fisherman's Island, Virginia were overlooked. Four of these guns have been preserved and put on public display in 200-series batteries in fort parks. Wrecked shields and gun parts from another four still exist on site at Kodiak and Dutch Harbor, Alaska.

6-inch Guns M1903A1 or A2, M1905A1 or A2, and T2
On 6-inch Barbette Carriages M1, M2, M3, and M4

Gun Characteristics
Bore size: 6-inch
Construction: Built-up steel
Tube length overall: 310-inches
Nominal caliber: 50
Tube and breech weight: 22,000-pounds (20,550 for T2)
Usual carriage: 6-inch barbette carriage M1, M2, M3 or M4
Weight of gun and carriage: 176,000-pounds
Maximum elevation: 47-degrees
Loading: Separate
Shell weight: 105-pounds
Maximum muzzle velocity: 2,800-fps
Maximum range: 27,500-yards

Builders: Guns: Watervliet Arsenal
Carriages: Watertown Arsenal, Erie Forge and Steel, Wellman Engineering, York Ice Machinery Co.

Number built: M1903 & M1905 guns - 108 used
T-2 guns, 32 made

Ordnance drawing of the shielded 6-inch gun Model 1903/05A1 on the new M1 or M2 Barbette Carriage.

Photograph of a 6-inch M1T2 gun in M4 Barbette Carriage of Battery Construction No. 249 at Camp Hayden Striped Peak reservation in 1945.

Two surviving guns and mounts from an isolated battery in Newfoundland were relocated to a battery at Fort Columbia State Park, WA.

8-inch Gun Mk VI Mod 3A2 On 8-inch Barbette Carriage M1

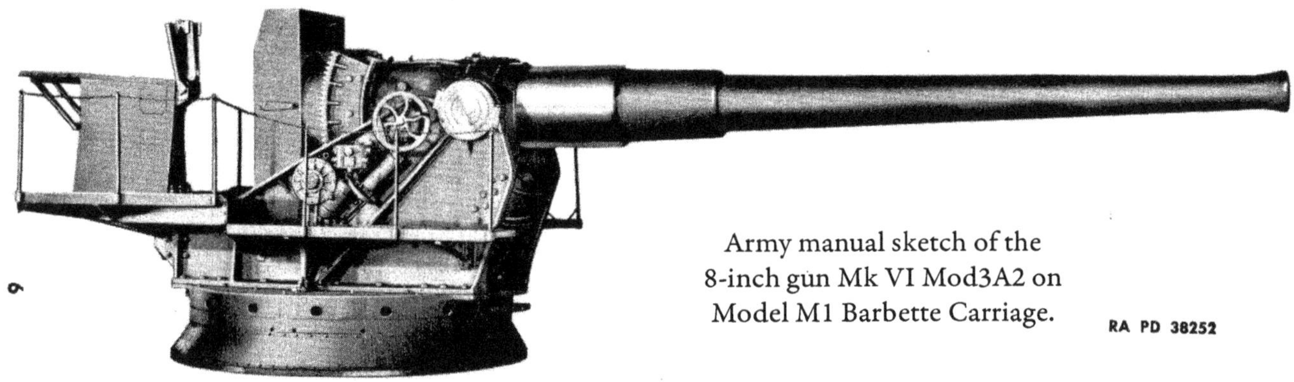

Army manual sketch of the 8-inch gun Mk VI Mod3A2 on Model M1 Barbette Carriage.

RA PD 38252

To supplement the 6-inch intermediate batteries in the 1940s program, a relatively small number of 8-inch batteries were included in the project. It was recognized that in some defenses not being equipped with new 16-inch batteries, that an intermediate armament was desirable. Places like Kodiak, Dutch Harbor, and Roosevelt Roads had to rely on 8-inch guns as their heaviest defenses. Fortunately, again like the 6-inch program, a number of useful gun tubes already existed. Ex-navy 8-inch/45-caliber tubes removed from pre-dreadnaughts when they were scrapped in the 1920s had been offered to and accepted by the Army. Initially 48 Mk Vs and Mk VIs were provided in June 1924. This total of 74 guns were obtained by the army, using them both in a new railroad mount and as a fixed coast artillery weapon.

The gun was a conventional tube, jacket and four-hooped, built-up nickel steel gun tube. In Army hands it was relined and had a modified powder chamber. The Army preferred to use a lighter projectile (240 vs. 260-lbs.), than the offered Navy rounds. The barbette carriage was of conventional design, it used hydro-pneumatic recoil provided by two cylinders and counter recoil used one pneumatic cylinder. Elevation of -5 to +45-degrees was obtained by manual gears. The gun could be operated and fired manually, an advantage in designing emplacements as it required only a minimal electric power plant. The first two new 8-inch barbette carriages M1 were ordered in February 1937, soon followed by another pair. A substantial run of twenty was added in June 1941 to be made by Baldwin Locomotive. Altogether 24 carriages were ordered and delivered by mid-1943.

The first emplacement designed for the 8-inch was an elaborate, fully-casemated battery emplaced in the defenses of Narragansett Bay. Quickly it was realized that the expense and effort was not worth it for this sized gun, and another prototype emplacement at Fort Rosecrans in San Diego was much more practical. Subsequent emplacements were similar to this one (except in Hawaii, where all emplacements in the Second World War seem to have been unique). A pair of guns, separated by 240-foot, were emplaced on open concrete pads. The intervening, protected structure provided magazines, power, and support rooms. Work was initiated to design a new splinter shield for the 8-inch barbette. Prototypes of this shield were evaluated at Aberdeen Proving Ground, but it was never put into production. The guns emplaced were protected solely with camouflage. Besides the first two batteries transferred to troops in 1942, the other half dozen emplacements armed in Alaska, Roosevelt Roads and Hawaii were completed in 1943-44. Three additional batteries on Oahu were begun but never completed.

After the war, in common with most of the rest of the final program, the batteries were disarmed and retired. The guns were authorized for salvaging in August, 1948. No intact carriages have survived, but substantial parts of the tubes are still at several Alaskan locations of the Kodiak defenses and are on public display.

8-inch Gun Mk VI Mod 3A2
On 8-inch Barbette Carriage M1
Gun Characteristics

Bore size: 8-inch
Construction: Built-up nickel steel
Tube length overall: 369-inches
Nominal caliber: 45 caliber
Tube and breech weight: 42,000-pounds
Usual carriage: 8-inch Barbette Carriage M1
Weight of gun and carriage: 103,000-pounds
Maximum elevation: 45-degrees

Loading: Separate
Shell weight: 240-pounds
Maximum muzzle velocity: 2,840-fps
Maximum range: 35,200-yards HE round)

Builders: Naval guns built by the
Naval Gun Factory, Watervliet Arsenal
Carriages by Watertown Arsenal,
Baldwin Locomotive

Number built: 74 guns acquired by the U.S. Army
24 carriages (No. 1-24)

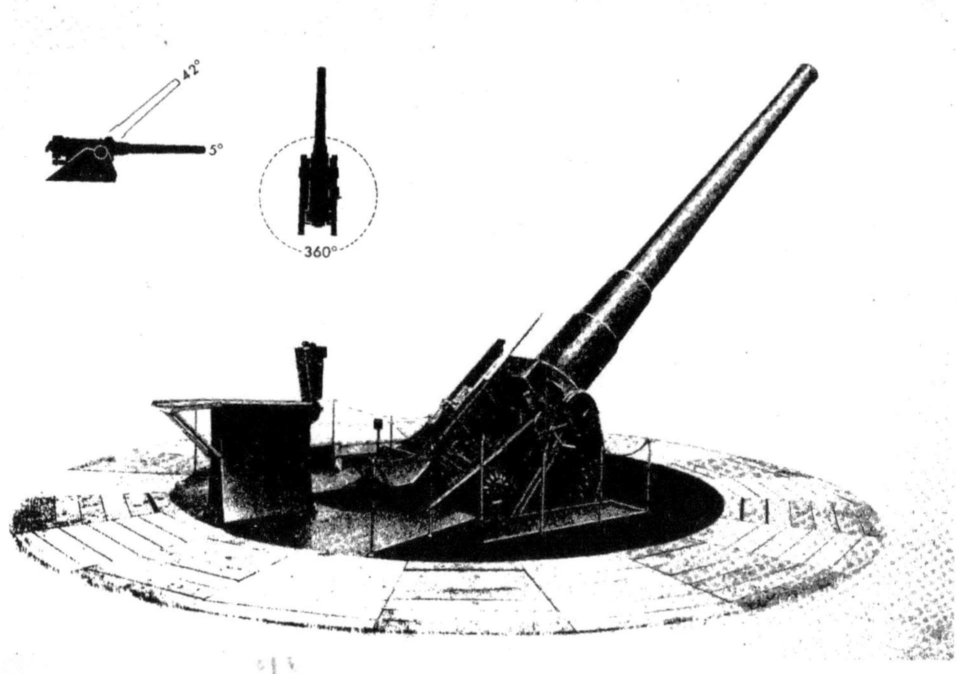

Army sketch with ¾ view of an emplaced 8-inch gun Mk VI Mod3A2 on M1 Barbette carriage.

Proofing ground photo of the intended shield for the seacoast 8-inch gun Mk VI. These shields were never put into production, much less issued.

Army Technical Manual illustration photograph of the 8-inch gun Mk VI on M1 Barbette Carriage mounted in a casemated emplacement. Only a single emplacement for two guns was ever built for this configuration—Battery Reilly at Fort Church of the Narragansett defenses.

Two views of a 8-inch M1 barbette gun in Battery Construction Number 403 at Fort J.H. Smith, Kodiak, AK

Most of the Second World War new 8-inch batteries consisted of covered magazines and open emplacements for barbette carriages. This is a March 1945 photograph of Battery Construction No. 405 near Fort Hase, Kaneohe Bay, Oahu.

No intact carriages survive for the 8-inch Model 1 barbette carriage, however major sections of the tubes and some lower carriage parts exist at Alaskan posts as a result of on-site demolition postwar. These are the remains of Battery No. 404 at Fort Abercrombie, Kodiak, Alaska.

16-inch Gun Mk II Mod 1
On 16-inch Barbette Carriages Model 1919 M1, M2, M3, M4, and M5

Army illustration of the 16-inch Mk II M1 gun on a Model 1919M4 Barbette Carriage in a casemated emplacement.

As the result of the Washington Naval Treaty of 1922, the Navy offered the Army a large number of 16-inch, 50-caliber Mk II gun tubes that had been completed or at least were partially finished forgings. These guns would not be needed for the cancelled six battleships of the South Dakota and six battle cruisers of the Lexington classes. In several lots between 1923 and 1941, 112 guns were transferred. While not quite as powerful as the Army's own Model 1919 16-inch gun, the cost savings were irresistible. Quick modification was made to the design of the current barbette carriage and a new ballistic table composed. This gun became the Army's primary weapon for the final decades of coast artillery. The tube was of nickel alloy

A February 1938 Watertown Arsenal photograph of a 16-inch Mk II gun with the protective shield of a M1919 barbette carriage with the two-inch thick shield.

steel and built-up (the Army gun was wire-wound). After relining for coast artillery use it had 96 grooves and right-hand uniform rifling of 1 in 32. It retained the typical Navy interrupted-screw breech.

The new carriage designed for this gun was simple modification of the similar Model 1919 type, originally made for the Army's own 16-inch gun. The "thinner" gun required side frames to be placed closer together. The base ring was bolted together and emplaced on a concrete foundation. Atop the ring were the racer ring and traversing rollers. The top carriage had two side frames with trunnion beds to hold the cradle. The cradle had trunnion extensions and completely encircled the gun tube. A single recoil cylinder and piston were underneath the cradle, and three recuperator cylinders were mounted atop the cradle. Electrically powered traverse could rotate the gun 360-degrees and powered elevation by rack could deliver elevation of -3 to +47-degrees. The heavy shell weight meant power loading and ramming—generally at a 4-degree elevation. They were equipped with compressed-air gas ejector systems.

The M1919M1 modified carriage was used with the first navy-type 16-inch guns (carriages No. 7-9 and 11-13) with orders starting in 1921. Subsequently the carriages M1919M2 (No. 10 and 14-16) and M3 (No. 17-18) carriages were produced, which were meant for the casemated emplacements with 2-inch thick shields. When emplaced in casemates they could achieve a maximum elevation of 47-degrees. The subsequent M4 carriages (No. 19-45) had a thicker, cast steel 4-inch thick shield and more powerful traversing and ramming electrical mechanisms. The ultimate M1919M5 (No. 46-51 and 55-61 ordered in 1942) had a similar shield with improved azimuth and elevation indicators, and a different anti-friction roller device. These last two types could achieve elevation of

16-inch Gun Mk II Mod 1
On 16-inch Barbette Carriage M1919M1, M2, M3, M4, M5

Gun Characteristics
Bore size: 16-inch
Construction: Alloy steel, built-up
Tube length overall: 821-inches
Nominal caliber: 50 caliber
Tube and breech weight: 307,185-pounds
Usual carriage: 16-inch Barbette Carriage M1919 M1 – M5
Weight of gun and carriage: 972,500-pounds (1,172-500 with shield)
Maximum elevation: 46-degrees
Loading: Separate
Shell weight: 2,240-pound (AP Mk XII)
Maximum muzzle velocity: 2,750-fps
Maximum range: 45,150-yards

Builders: Gun: Washington Navy Yard
Carriages: Watertown Arsenal, Wellman Engineering, Mesta Machine Work

Number built: 112 navy MkII guns or partial assemblies acquired by U.S. Army, 52 carriages

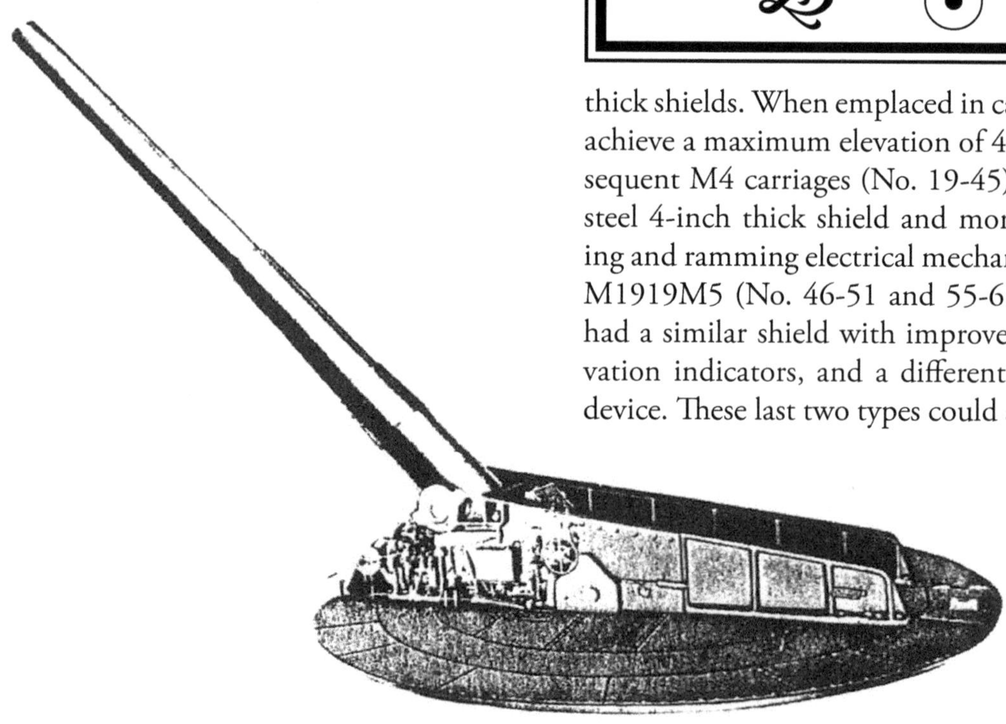

-5 to +46-degrees. Initially the carriages were all built at Watertown Arsenal, but later in the war commercial firms Wellman Engineering and Mesta Machine Works were used. Altogether 52 barbette carriages were completed to take the Mk II gun.

Six of the new navy-type guns were emplaced in the interwar period—two in Hawaii and four in Panama. They used the completely open, 360-degree traverse plan of the 1915 Board of Review system. Eventually four of these guns were given casemate cover in the 1940s.

Initially authorized in 1934, though not funded until fiscal years 1937, 1939 and 1940, the subsequent 35 batteries were to be designed and built as two-gun casemated types. After program cutbacks in 1942, only 23 of the new casemated batteries were structurally completed by 1945. While not all the armament was installed, in general it was all fabricated and prepared for installation.

Five gun tubes Mk II and Mk III still exist. Three were used in various tests and modifications at the Naval Surface Warfare Center in Dahlgren, Virginia and are still at the facility, in an area generally restricted to public visits. One Mk II gun is on display at the U.S. Navy Museum at the Washington Naval Yard. And, one gun (Mk III No. 138) on proof barbette carriage (originally built for the army M1919 gun) is on public display on the grounds near an entrance to the Aberdeen Proving Grounds in Maryland.

Aberdeen Proving Ground photograph of the 16-inch Mk II gun on a M1919M1 barbette carriage taken in 1921.

Battery Richmond P. Davis with its 16-inch Mk II gun and M1919M2 barbette carriage at Fort Funston as part of the San Francisco defenses. The battery was finished and turned over for coast artillery service in early 1939.

One of two casemated emplacements built for the 16-inch Mk II gun on M1919 barbette carriage with the 4-inch thick cast iron shield at Camp Hero, Montauk Point, New York. Photograph from late in WW2.

Five of the 16-inch Mk II guns have survived, all still in government hands. This MkIII gun on a Model 1919 barbette proof carriage, was restored and put on display at the old Aberdeen Proving Ground Museum. While the museum has closed, the very heavy gun and mount remain in place.

Emergency and Temporary Seacoast Guns
Of the Spanish American War

When the war with Spain suddenly began in 1898, something near a panic spread within the military services. It was assumed that the Spanish Navy, mostly still in Spanish waters, would sail for Cuba, perhaps stopping along the way to bombard or harass American eastern seaboard cities. The Navy redeployed ancient Civil War monitors to several coastal cities. The Army was concerned both about its lack of rapid-fire light guns and the vulnerability of several secondary ports and anchorages. It hastened to purchase available guns in Europe (giving rise to the British 4.7 and 6-inch guns that were later incorporated into the permanent system) and to appropriate a mixture of available guns in the country for temporary defenses.

Three sources were used for these weapons. A number of obsolete seacoast cannons from the Civil War were hastily emplaced in earthen batteries. Most of these were at minor ports along the South Atlantic coast. Some were also armed with standard Army 5 and 7-inch siege guns or howitzers, but soon removed when these were recalled for use by the expeditionary force bound for Cuba and Puerto Rico. To help with the rapid-fire problem, a 4.7-inch test gun at the proving ground was mounted at Sandy Hook. Responding to an Army inquiry of ordnance manufacturers, four naval-type 4-inch deck guns were found available at the American Ordnance Company in Bridgeport, Connecticut and quickly purchased and deployed them in two fairly conventional emplacements on the East Coast.

Finally in 1898 the production of 8-inch Model 1888 seacoast tubes had for the most part been concluded in runs both by the Army's Watervliet Arsenal and contractors Bethlehem Steel and West Point Foundry. However carriage production of the recent Model 1896 disappearing limited-fire carriage had lagged. The Army had a couple of dozen perfectly usable, modern 8-inch gun tubes lacking carriages. As a temporary expedient it quickly implemented a project to modify 21 older 15-inch Rodman front-pintle barbette carriages to carry the Model 1888 gun. In surprisingly short time—though not before the very few months of a genuine emergency had passed—these were modified, distributed to newly built emplacements, and mounted for service.

Of these guns, just the four 4-inch weapons were actually incorporated into the prevailing Endicott System. The single 4.7-inch Schneider gun and all the 8-inch guns on Rodman carriages were dismounted in the ten years following the Spanish American War. Most of the "modern" 8-inch Model 1888 tubes were then remounted on disappearing or barbette carriages as these became available. No examples of any of these gun mounts still exist—though there are surviving examples of the common 15-inch Rodman carriage as unmodified and used with its 15-inch Rodman gun at a number of national and state parks. However a good number of the emplacements for these expedient seacoast batteries carriage still exist in public parks.

Emplacing the 4-inch guns at Fort Warren, Boston

4-inch Driggs-Schroeder RF Gun On 4-inch RF Pedestal Mount

Catalogue illustration of the 4-inch Driggs-Schroeder pedestal-mounted gun.

Responding to a War Department inquiry, the American Ordnance Co, of Bridgeport, Connecticut found they had four navy-type 4-inch rapid-fire guns in uncommitted inventory. This firm had a long history as both a Navy and Army supplier. They offered to sell these guns for $7,500 each (along with 25 rounds common and 25 rounds of shrapnel ammunition) to the Army on April 9, 1898. On the 19th this was accepted, though at a slightly higher price to include 250 rounds per gun. Disappointed that simultaneous efforts to secure 4.7-inch guns from Europe (curiously through the same firm acting as supply agent) were not yielding enough guns, the department decided to acquire the four offered guns as substitutes for a like number of the 4.7s.

The gun and pedestal were typical naval types American Ordnance had already been delivering. In fact the Army had actually tested a slightly earlier version in 1896 and found it adequate but did not order any at the time. It was based on and described as a Driggs-Schroeder gun in reference to the major patented breech aspects. It had a built-up steel barrel of 40-caliber length, and with 30 grooves and rifling of 0 to 1 in 25. The breech was an interrupted screw type of the Driggs-Schroeder patented mechanism. The gun was carried on a cage stand with pivot, cradle and sleeve to hold the tube. Recoil was checked by a patented Driggs hydraulic recoil cylinder. The stand had a "D-shaped" hold down bolt pattern, typical of broadside shipboard mounts. It utilized fixed, cased ammunition, with a 33-pound projectile. The carriage allowed elevation from -10 to +15 degrees. The gun was equipped with a prominent shoulder rest. Initially they were delivered without any shields. These had to be ordered separately and were belatedly procured and then mounted in early 1901.

Due to the urgency of the war situation, the Army immediately authorized the emplacements for these guns. They were assigned batteries in locations that otherwise would not be armed due to the disappointing supply of Armstrong 4.7-inch guns. A two-gun battery (Battery White) was built at Fort Washington on the Potomac and another (Battery Plunkett) at Fort Warren in Boston harbor. Orders for construction were issued on April 30th. They closely followed the general dimensions of the 5-inch masking parapet type. The type of mount provided would not allow a traverse of over 180-degrees. Within a year the guns were in service.

The two emplacements, despite their armament of peculiar size and background, were retained within the approved armament of the Endicott System. As mentioned above they had shields added in 1901, but there appears to have been no other significant modifications to the armament. It is not clear how seriously they were maintained or exercised in subsequent years, several annual ordnance returns from the pre-war years mention the batteries as out of service. With non-standard guns they were not selected for alternate use or relocation during the First World War. However right after that conflict they were declared obsolete on June 15, 1919.

They soon were disarmed. At Fort Warren the guns and carriages survived for at least another twenty years as post ornaments and saluting guns. Eventually the weapons were scrapped. None of the four guns survive, but there are a few early navy 4-inch guns of similar characteristics displayed at war memorials and service clubs. Both of the two emplacements for the 4-inch Driggs-Schroeder RF gun exist at their respective fort parks.

4-inch Driggs-Schroeder RF Gun On 4-inch RF Pedestal Mount

Gun Characteristics
Bore size: 4-inch
Construction: Steel, built-up
Tube length overall: 166.5-inches
Nominal caliber: 40 caliber
Tube and breech weight: 3,613-pounds
Usual carriage: 4-inch RF Pedestal Mount
Weight of gun and carriage: 8,500-pounds
Maximum elevation: 15-degrees
Loading: Fixed
Shell weight: 33-pounds
Maximum muzzle velocity: 2,000-fps
Maximum range: 8,864-yards

Builders: American Ordnance Company

Number built: 4 guns and carriages

Few photographs survive of the 4-inch Driggs-Schroeder guns in their emplacement. This relatively poor image does show the guns of Battery Plunkett at Fort Warren, Boston about 1917.

The two 4-inch guns of Battery Plunkett were moved in the 1920s to become gate guard trophies at Fort Warren.

4.7-inch Schneider RF Gun
On Schneider Carriage for 4.7-inch Gun

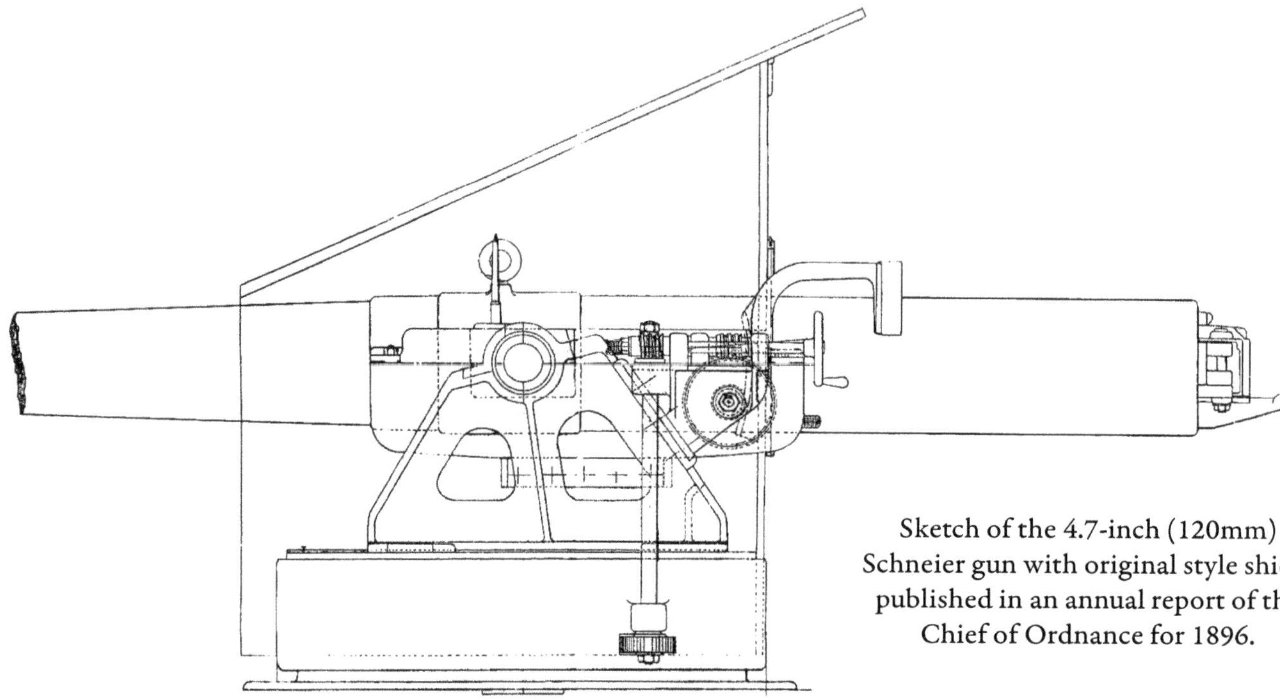

Sketch of the 4.7-inch (120mm) Schneier gun with original style shield published in an annual report of the Chief of Ordnance for 1896.

The U.S. Army's Board of Ordnance and Fortification acquired several examples of modern European rapid-fire guns in the last years of the 19th century. The entire category of light guns that could be rapidly reloaded, aimed, and fired had been previously neglected by the Army. At least it attempted to learn from what other nations were doing by buying single examples of guns from Armstrong, Vickers, and Schneider (German firms like Krupps were anxious to take a sales contract, but reluctant to sell or provide single examples that could be easily copied by foreign governments). One of the guns acquired in 1894 was a 4.7-inch Schneider RF gun and pedestal. Schneider referred to the gun as a 12-cm bore, but all American correspondence translated this to 4.7-inch.

The gun (Schneider's serial No. 2599) was bought from the French firm in July 1893 for $8000 for gun and mount and another $4000 for 100 rounds of common cast iron ammunition. The gun in fact had been on display at the Chicago World's Fair in the Schneider display, and was already in the country. The Board used the appropriation of February 18, 1893 specifically providing funds to purchase such armament for evaluation and testing purposes.

The gun had a steel, built-up tube with jacket and several hoops. It was equipped with a rather complicated interrupted-screw breech needing two separate levers (one of the things the evaluation criticized). It used a fixed round and wasn't completely semi-automatic—the extractor just loosened the spent case which had to be removed by hand rather than ejected. The mount had a cradle to hold the gun and a low chassis with revolving rollers on a racer. Two hydraulic brakes contained the recoil (but only allowed an 8.3-inch recoil), two recuperators with ten springs each returned the gun to firing position. On its circular pedestal carriage it could manage -5 to +20-degree elevation. The shield was assembled from several pieces, and had 2.7-inches of thickness in the front and sides. It could fire shells of either ordinary cast iron, steel, or shrapnel.

The gun, with a French naval officer in attendance, was subjected to firing trials at the Sandy Hook Proving Ground on November 19, 1894. The army's evaluation was that while the gun had no fundamental problems and worked as promised, it was awkward to use and lacked some of the details for safety of American designs. It was not recommended for subsequent adoption or additional acquisition.

At the start of the Spanish American war the gun and its mount were still at the proving ground, though its evaluation testing had been concluded.

The thought of dreaded Spanish destroyers romping through American harbors stimulated the emplacement of quickly acquired RF guns. One easy choice was to place the unused Schneider 4.7-inch gun in a defensive position at Sandy Hook's Fort Hancock. A relatively simple single gun platform and adjacent magazine was constructed from April 25 to 30th, 1898. Unreinforced concrete readily at hand was used for the emplacement. It was located on the northern line of light guns running east-west, not far from the single emplacement of 5-inch Battery Engle. The gun was mounted as soon as the emplacement was completed. It was never named or formally transferred to the Coast Artillery.

While no doubt manned during the early phases of the war, it is unclear how seriously the garrison subsequently took the one-of-a-kind gun in such a minimal emplacement. Exact date of removal is unknown, though the emplacement is noted as unarmed in a 1910 report. The gun was subsequently scrapped, and the emplacement was destroyed sometime later.

4.7-inch Schneider RF Gun
On Schneider Carriage for 4.7-inch Gun

Gun Characteristics
Bore size: 4.7-inch (12-cm)
Construction: Steel, built-up
Tube length overall: 236.2-inches
Nominal caliber: 50 caliber
Tube and breech weight: 6,400-pounds
Usual carriage: Schneider 12-cm Pedestal Carriage
Weight of gun and carriage: 21,390-pounds including shield
Maximum elevation: 20-degrees
Loading: Fixed
Shell weight: 48.4-pounds
Maximum muzzle velocity: 2,656-fps (with smokeless powder)
Maximum range: n.a.

Builders: Schneider et. Comp.

Number built: 1 acquired 1896 (serial No. 2599)

Photograph of the Schneider 4.7-inch gun on trials at Sandy Hook. At the start of the Spanish-American War this gun was temporarily emplaced in a simple position at adjacent Fort Hancock.

8-inch Gun Model 1888
On Reinforced SB 15-inch Carriage

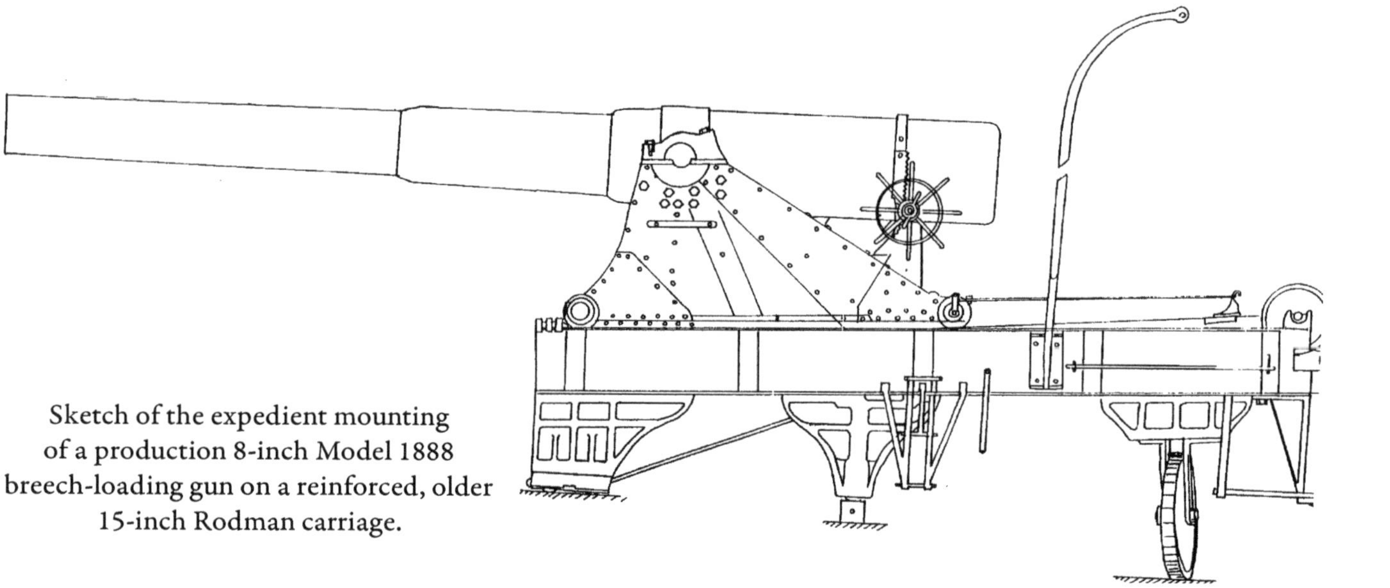

Sketch of the expedient mounting of a production 8-inch Model 1888 breech-loading gun on a reinforced, older 15-inch Rodman carriage.

The preeminent Civil War coast defense weapon was the Model 1861 15-inch Rodman cast iron cannon. A smoothbore weapon firing a round, 440-pound shot or lighter shell, it was mounted on a front or center pintle (for situations that required all-round fire) carriage. The carriage had a heavy pair of parallel side rails with transoms. Upon the top of the rails rode a top carriage of side supports with trunnion beds. Ammunition was lifted by an attached crane. After the Civil War, in the 1870s and 1880s, service carriages were strengthened to compensate for the stress caused by use of improved propelling charges. The most noticeable change was the addition of a pair of hydraulic recoil buffers (cylinders), located underneath the carriage either to the front or rear depending on the date of modification. The Ordnance Department received funds to modify a handful of carriages each year well into the early 1890s. Over 300 of the big Rodman cannon and its carriage had been manufactured and mounted through the mid-1870s, and a smaller 10-inch version was also very common. They were mounted and continued to serve into the modern era, until replaced with more modern ordnance in the early 1900s.

At the start of the Spanish American War production of Model 1888 8-inch guns (see that section) had outstripped the production of barbette and disappearing carriages. Two dozen gun tubes had been accepted and proofed but were without carriages. New carriage orders (mostly for the most modern M1896 disappearing carriage) were quickly placed, but as it would take at least a year to get these, an interim step was taken. The Department quickly adapted the standard 8-inch gun to fit some of the strengthened Model 1891 15-inch Rodman carriages—the gun being of relatively the same weight and exerting about the same stress pressure as the 15-inch. Relatively simple kits of parts were required to convert the carriage—primarily strengthening braces. The Robert Poole & Sons firm of Baltimore was given orders to produce 21 modification kits. They were to deliver one set per day starting May 1st, 1898. A couple of years later one more strengthened carriage was fabricated, to take a damaged mortar for powder testing—this has led to a report of 22 such carriages made in some accounts.

The engineers soon set about designing emplacements for the 21 guns and carriages. Batteries for one or two guns were sprinkled around many East and Gulf Coast ports (Kennebec, Portsmouth, Newport, Long Island Sound, New York, Hilton Head, Fernandina, St. John's River, Tampa, Mobile, New Orleans, Sabine River). These were either reinforcements for existing defenses, or otherwise small ports or rivers not covered in existing fortification plans. Most of the batteries were simple concrete platforms with the usual pair of rolled iron circular traverse rails and 6-inch front pintle on a raised block. An adjacent single magazine was built in positions that had no other structures. In some locations advantage was taken to place the gun in an existing Rodman emplacement from previous coast defense generations, and use was made of the magazine already adjacent.

Most emplacements and the mounting of guns took place in late 1898, frankly after the emergency of the first three months of the war had passed. Only those emplacements in fully-manned forts were actively placed into service. Starting in late 1898 the guns were gradually shipped away for use on their intended modern type carriages. By 1910 they were all disarmed and the carriages scrapped. While several of the emplacements still exist, there is little other physical or documentary evidence of these guns and emplacements. There was no ordnance pamphlet or description prepared, there are no official Army photographs of such a gun—and just a single, poor, privately taken print has to date been discovered.

8-in Gun Model 1888
On Carriage for 8-in BL Rifle Converted from Carriage for 15-in SB

Gun Characteristics
Bore size: 8-inch
Construction: Steel, built-up
Tube length overall: 278.5-inches
Nominal caliber: 32 caliber
Tube and breech weight: 32,210-pounds
Usual carriage: Converted Carriage for 15-inch SB
Weight of gun and carriage: n.a.
Maximum elevation: 12-degrees (assumed)
Loading: Separate
Shell weight: 316-pounds (short cap)
Maximum muzzle velocity: 2,200-fps
Maximum range: 11,019-yards (assumed with 12-degree elevation)

Builders: Guns: Watervliet Arsenal,
West Point Foundry, Bethlehem Steel
Carriages: Modification parts by
Robert Poole & Sons

Number built: 21 carriages modified

Only a single, poor-resolution photograph is known of the 21 emplaced 8-inch/Rodman combination. This shows one of the two guns about 1899 at Jerrys Point, New Hampshire (later to become Fort Stark).

Navy Temporary Coastal Guns

With American service policy, the U.S. Army always had primary responsibility for coastal defense. However, there were several instances where the U.S. Navy or its Marine Corps had to assume at least temporary jurisdiction over a defendable spot. In the earliest years of the 20th Century the Navy developed and deployed the concept of keeping ready a new emergency base. Known as the Advance Base Force, there were two depositories for all the equipment needed to quickly establish a new coaling/service naval station. One was kept for use in the Atlantic (Philadelphia) and the other for the Pacific either at Olongapo in the Philippines or Mare Island near San Francisco. Included in the supplies were a variety of excess naval deck guns and mountings to allow them to be quickly emplaced for defensive purposes. Some of the Philippine material was ultimately assigned to the defense of Guam from 1909-1916.

During and right after the First World War the Marine Corps was tasked to defensive assignments. They took and manned guns in Cuba, the Virgin Islands, and the Azores. Then in the late 1930s the Fleet Marine Force created garrison units known as Defense Battalions. Designed to provide anti-air, anti-boat, and anti-ship capacity to relatively small island bases, most included a fixed, seacoast component. These guns (mostly 5-inch, but on occasion 6-inch and even 7-inch) were emplaced, manned and even fought on places such as Wake, Midway, Johnston, Palmyra, Canton, Bora Bora, and Samoa Island. Stateside and in Hawaii the marines had similar weapons deployed as training batteries that had useful seacoast coverage.

During the Second World War, particularly in the panic of defensive reaction immediately following the Pearl Harbor attack, the Navy transferred to the Army a number of excess or surplus guns. While some were reserved for large-scale use (particularly useful quantities of larger guns) some of the intermediate types were quickly emplaced by the Army for defense of harbors and foreign bases that were not adequately covered by specific defense projects. A relatively large number of 4, 5, 6, and 7-inch guns found use in these places. Often, they only served for a year or two and had only the simplest of battery emplacements. Some never left the custodianship of the Marine Corps; others had been transferred to the Army during the interwar period or were transferred to them for operational manning once emplaced by the navy. Surviving records are at best spotty on some of these frankly transient assignment, and only a partial accounting of exact gun types, models, or even numbers emplaced still exists.

The Hawaiian Islands stand out as an extreme example of the use of emergency or temporary batteries. Attempts to quickly augment both the seacoast and anti-aircraft defenses of the island chain led to the use of a bewildering number of types and sizes of mostly naval guns. Some of those (5, 7 and 8-inch) are covered in their respective sections of this book. Also use was made of 3-inch and 5-inch/25 and 5-inch/38 AA guns, just about unique to Oahu or Hawaii for their use ashore. Not a few of these guns were actually salvaged off sunk or damaged ships from the Pearl Harbor attack and thus readily available. As their performance remained unchanged and is readily available from several published and online sources, they are not covered here. Finally, the use of the turrets from USS *Lexington*, *Saratoga*, and *Arizona* ashore are such unique American examples of shore fortifications that they are covered in the separate "Hawaiian Turrets" section of this book.

Emplacing a Navy 6-inch gun on pedestal mount in Samoa during the Second World War.

4-inch 50-caliber Gun Mk IX On Navy 4-inch Mount Mk XII

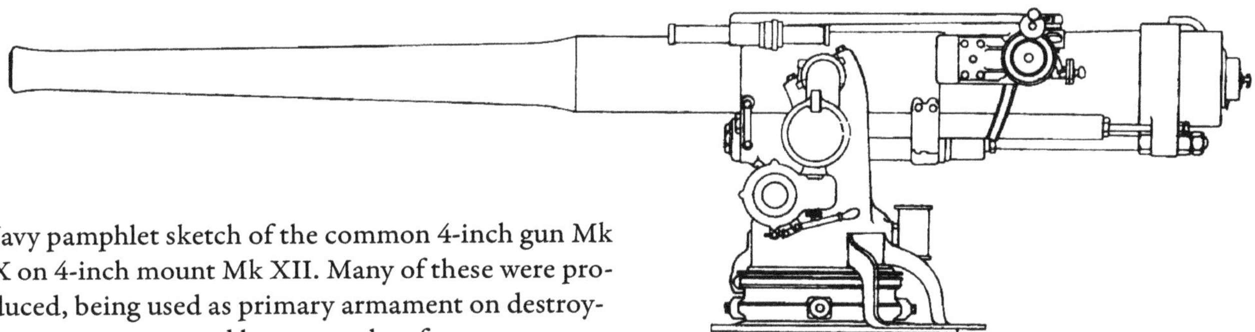

Navy pamphlet sketch of the common 4-inch gun Mk IX on 4-inch mount Mk XII. Many of these were produced, being used as primary armament on destroyers and large patrol craft.

This was a common Navy gun used in very large numbers on First World War classes of destroyers and certain other small warships and auxiliaries. Virtually all but a few experimental armament arrangements for the numerous "4-piper" classes of destroyers used the Mk IX 4-inch gun and Mk XII pedestal mounting. The gun was a nickel steel, built up tube with a full length jacket. The end of the tube had a distinct muzzle swell. An interrupted screw breech mechanism of the Smith-Asbury type swung to the right side. The tube had 30 grooves and rifling of 0 to 1/31.17 near the muzzle. The gun fired fixed rounds with a projectile weight of 33-pounds. The pedestal was a simple Navy type. The usual stand held a yoke for the slide with the gun supported on the slide by two trunnion arms. Recoil was checked with a hydraulic recoil cylinder. Elevation and traverse was with gears operated by hand. The mount could deliver -15 to +20-degrees of elevation. Maximum range was for 15,600-yards. Usually, the gun was equipped with a Mark 26 telescopic sight.

As the Navy in the 1930s embraced the 5-inch gun as its favored intermediate weapon, a considerable number of excess 4-inch guns accumulated at various naval stations. Some were contributed as aid to allied nations. The Dutch received guns for both their East and West Indies defenses. Following the Pearl Harbor attack, 25 of these guns were made available in Hawaii to the Army for use as temporary coast defense weapons. In early 1942 several batteries of two or three guns were built on Oahu: the construction work being done by the local 34th Engineer Regiment. Others went to Hilo Harbor on Hawaii and the islands of Kauai and Maui. For the most part these were the usual simple emplacements—circular concrete gun pads with embedded hold-down bolts and adjacent ammunition magazines. In one case (Battery Dodge, Fort Ruger) the guns were emplaced in an older, abandoned 4.7-inch emplacement dating from 25 years previous. At least four guns were mounted at another island base of the Southern Air Ferry Route. Canton Island received these guns also in early 1942.

As might be expected, they did not last too long. As more modern armament was completed on Oahu, and as the war moved further to the west, these batteries were gradually abandoned. The Navy requested the return of nine of the guns in October 1943. They still had a use for them on small auxiliaries and for arming transports and freighters. Most of the remaining guns were pulled and the armament scrapped by the end of 1944. Today there are no surviving, armed emplacements; even the emplacements themselves have just a few fragmentary remains. No photographs of an emplaced 4-inch Mk IX in American military service has been located. There are, however, several good examples of the Mk IX 4-inch gun on public display at navy bases, museums and a few war memorials.

4-inch Gun Mk IX
On Navy 4-inch Mount Mk XII

Gun Characteristics
Bore size: 4-inch
Construction: Nickel steel, built-up
Tube length overall: 200.0-inches
Nominal caliber: 50 caliber
Tube and breech weight: 5,900-pounds
Usual carriage: Navy 4-inch Pedestal Mount Mk XII
Weight of gun and carriage: 17,200-pounds
Maximum elevation: 20-degrees
Loading: Fixed
Shell weight: 33-pounds
Maximum muzzle velocity: 2,900-fps
Maximum range:15,600-yards

Builders: Naval Gun Factory and private contractors

Limited numbers of the 4-inch gun Mk IX were made available by the Navy for use of the Army in Hawaii and for Lend-lease to the Netherlands as shore-mounted seacoast guns. While none of the coast artillery guns still exist, there are several of the naval mount itself on display. This one was found in Bristol, Rhode Island.

5-inch 51-caliber Gun Mk VIII and Mk XV On Navy 5-inch Mount Mk VIII and XIII

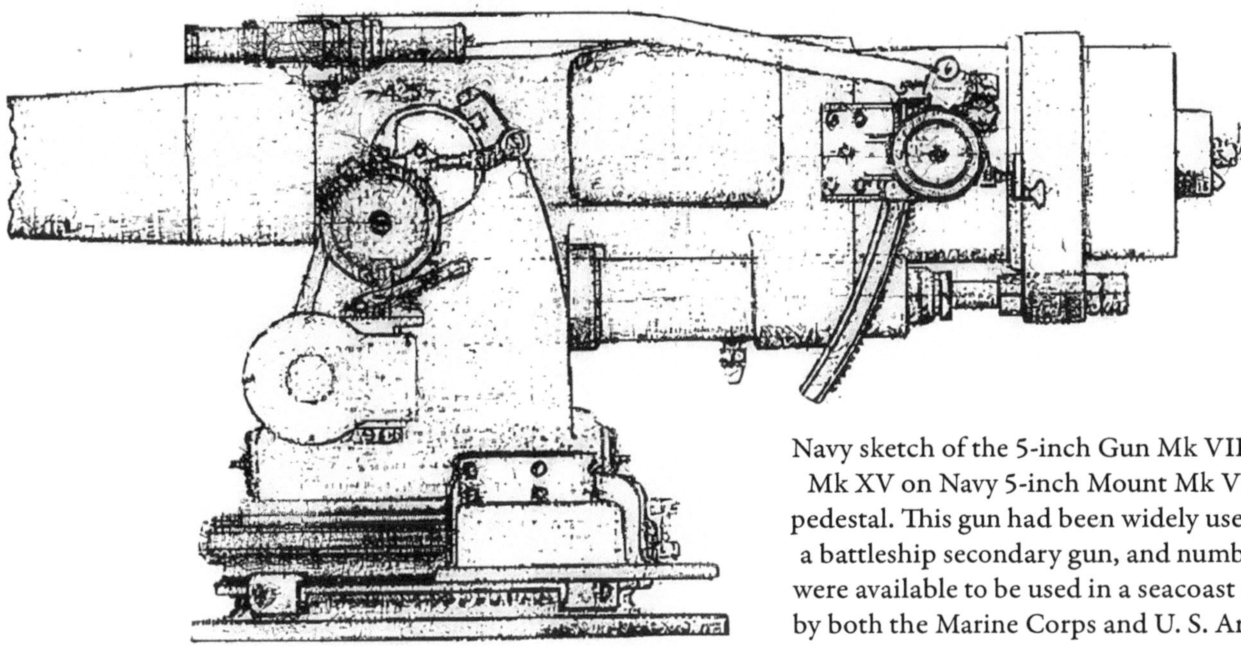

Navy sketch of the 5-inch Gun Mk VIII or Mk XV on Navy 5-inch Mount Mk VIII pedestal. This gun had been widely used as a battleship secondary gun, and numbers were available to be used in a seacoast role by both the Marine Corps and U. S. Army.

These guns were a widely used as a broadside battery on Navy battleships from the USS *Florida*-class through the *Colorado*-class. A general reduction in the number of guns carried and scrapping of several of the earlier units had resulted in an inventory of this relatively powerful light gun. Most had never fired extensively and had useful bore life remaining. The Navy selected this gun for temporary shore use initially with the final iteration of the Advance Base Force and finally, right before the Second World War, with the Marine Corps defense battalions.

The gun was a conventional built-up steel gun of 51-calibers. It had the Welin-type interrupted screw breech mechanism. The carriage was also conventional. It was a pedestal where the gun tube fit into and recoiled in a sleeve that in turn had trunnions to be held by a yoke attached to the pedestal. The carriage had hydraulic recoil cylinders and counter recoil spring mechanism attached underneath the barrel. The mount allowed elevation from -10 to +20 degrees. A distinctive prismatic sight was carried above the tube, offset on a yoke to allow the pointer sitting on the left-hand side of the gun to keep his eye continuously on the target while also controlling elevating and traversing. The yoke's shape is distinctive and useful in identifying period photographs. The mount was usually anchored to a 2-inch thick steel plate in turn laid on a 2-inch thick oak planking. While the naval mount often had a curved shield, the gun was emplaced on shore invariably without it. These were RF guns, using fixed ammunition.

The marines created a new unit called the Defense Battalion in the late 1930s. Deployed within the active Fleet Marine Force, it was a unit of roughly 1000-men designed to defend an island base. While the table of organization could and did vary, at the start of the war it generally had an anti-aircraft component (with 12 3-in/50 AA guns) and a seacoast unit with six guns (three 2-gun batteries) utilizing 5-in/51 Mk XV guns. Several battalions were immediately organized, and between 1940-42 sent to Wake, Midway, Johnston, Palmyra, and Guantanamo in Cuba. Some of these did see action; the defense of Wake in December 1941 being particularly notable. Later defense battalions manned 5-inch guns in the Solomons (including Guadalcanal). Training batteries in San Diego and North Carolina were sited to provide secondary seacoast coverage.

Several temporary batteries were also used on Oahu in Hawaii. Eleven guns and mounts were immediately offered to the Army on January 5, 1942, along with 200 rounds of ammunition each. They did require modification of their sights to allow greater deflection. Another 15 guns were taken from battleships disabled during the Pearl Harbor attack, seven going to the Army and eight more to the Marines. At least six batteries, many of them 3-gun emplacements, were sited around the island to reinforce beach defense following the December 7th attack. These were manned by Army coast defense troops, and eventually some of the island defenses set-up by the marines (like Palmyra and Johnston) were transferred to Army units.

As the war moved away from these places and American naval and air dominance increased the batteries were abandoned. The Army issued instructions for the local command in Hawaii to discontinue the 5-inch batteries as replaced (many by new 155mm batteries) starting in December 1943. In all cases the guns and mounts were removed and scrapped at the end of 1944. There are no known guns still in place, but the gun and its pedestal mount exist as war memorials or service club monuments in several cities.

5-inch Gun Mk VIII or Mk XV On Navy 5-inch Mount Mk VIII or Mk XIII

Gun Characteristics
Bore size: 5-inch
Construction: Steel, built-up
Tube length overall: 261.25-inches
Nominal caliber: 51 caliber
Tube and breech weight: 11,250-pounds
Usual carriage: Navy 5-inch pedestal mount Mk VIII or Mk XIII
Weight of gun and carriage: 23.400-pounds
Maximum elevation: 20-degrees
Loading: Separate
Shell weight: 50-pounds
Maximum muzzle velocity: 3,150-fps
Maximum range: 17,100-yards

Builders: Naval Gun Factory and several private contractors

Number built: At least 20 supplied to U.S. Army for coast defense; At least 60 used by Marine Corps for coast defense

A Marine Corps 5-inch/51-caliber Mk VIII gun and pedestal erected in a training mode.

The 5-inch gun Mk VIII/XV was a primary armament of the Marine Defense Battalions deployed to small overseas garrison in 1940-42. This one, apparently "named" in honor of the Corps hero of the Wake Island defense, is at Johnson Island in 1942.

In a few locations the 5-inch Mk VIII was emplaced by the Army, particularly to protect beaches on Oahu and the east coast. This is the battery placed near Cape Lookout, Beaufort, North Carolina during 1943.

Many Navy 5-inch tubes and pedestals were contributed postwar for civic memorials and can be seen today. This gun, formerly a broadside gun from USS Colorado, is at the Veterans Memorial Museum in Chehalis, Washington.

6-inch Gun Mk III
On Navy 6-inch Mount Mk IV

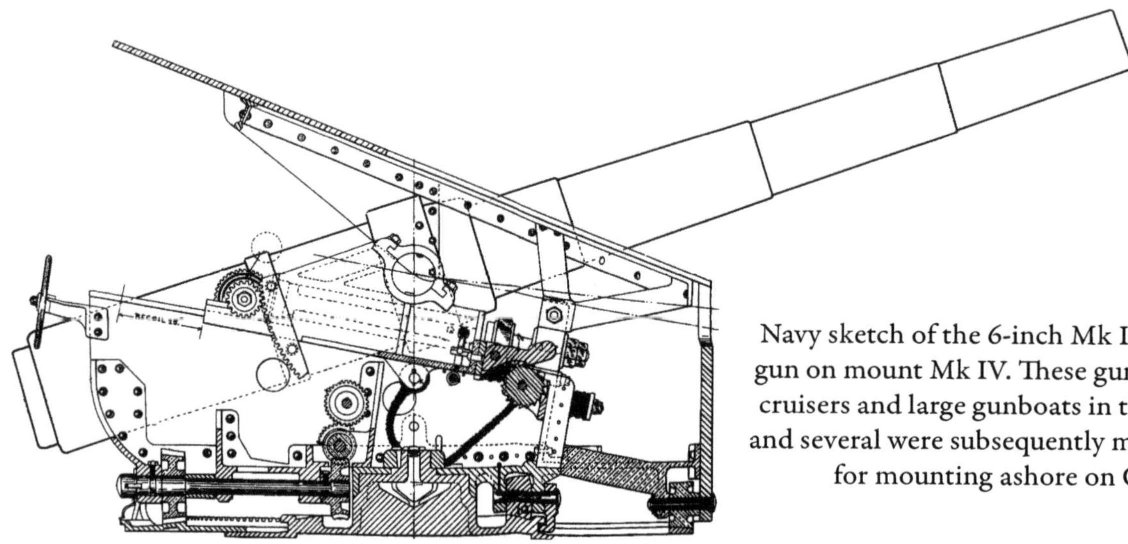

Navy sketch of the 6-inch Mk III 30-caliber gun on mount Mk IV. These guns had armed cruisers and large gunboats in the 1880-90s, and several were subsequently made available for mounting ashore on Guam.

This one of the first modern breech-loading guns developed by the U.S. Navy during the late 19th century. Only a single Mk I gun had been made (to arm USS *Dolphin*), and the subsequent very similar Mk II and Mk III guns were used for secondary armament on some of the first battleships and to arm almost all of the first generation of Navy cruisers. Over a hundred of these guns were built; most were of 30-caliber length, but a few were 35 or 40-caliber. Most were called back to the Washington Navy Yard for conversion from bag to case loading in the late 1890s. The heavy gunboats (actually these would have been classified as small cruisers in many other nations) USS *Concord*, *Bennington* and *Yorktown* with six each of these guns were slated for rearmament. When *Concord* was ordered home from her role as station ship at Guam in 1909, she left four of her 6-inch Mk III guns behind to serve as temporary defenses for that island.

The guns were conventional built-up steel tubes, with ten hoops over the tube covered with the jacket. The tube had 7-inch diameter trunnions. It had an interrupted-screw breech mechanism Mark III. The gun fired separate-loading bagged powder a shell of 105-pounds weight. The carriage was an older type designated Mk IV. It was a rather large center-pivot, gravity-return mount. It had a cast bronze top carriage, also containing the hydraulic recoil cylinders. There was no counter recoil spring, return was made by gravity moving the top carriage on inclined slides. The bulk of the carriage was a large cast steel box slide consisting of two side panels and transoms. This was held on a deck pivot supported by rollers running on a deck circle. The gun was partially enclosed with a 100-inch diameter, 2-inch thick shield with inclined top; it was attached with supports but most of its weight was supported directly on rollers on the deck tracks. Guns had mechanical traverse and elevation, which could range from -10 to +12-degrees.

The four guns, carriages and shields were mounted on specially prepared bases fabricated and shipped to Guam from the Puget Sound Navy Yard. While the guns were landed in 1909, they were not emplaced until 1911. All four were mounted around Apra Harbor. Two went to a location at the old Spanish work at Fort Santiago, the others to Orote Point. Ready magazines were erected by the guns. Each weapon was fired twice after emplacement, but apparently never again thereafter. With only a 9,000-yard range and a low rate of fire due to the use of bag powder, they were not ideally suited for the role they played. The guns lasted until dismounted in 1917 when they were replaced by Navy 6-inch Armstrong guns. Ten other 6-inch/30 guns were part of the Olongapo Advance Base Force.

Between 1908 and 1910 they were mounted around the shores of Subic Bay. Dismounted in late 1910 they stayed with the base force until retuned to the U.S. in 1914 and scrapped.

The Guam guns were also scrapped right after the First World War. However, there are several surviving examples of this gun type. The two remaining guns on USS *Concord* were removed at Puget Sound when the ship was disarmed for use as a quarantine vessel and sent to Astoria. These guns were given to the city of Seattle and still exist in a veteran's park in Chehalis, Washington. Other similar guns salvaged from the USS *Maine* are in Michigan and at the National Museum of the U.S. Navy in Washington, D.C.

6-inch Gun Mk III
On Navy 6-inch Mount Mk IV

Gun Characteristics
Bore size: 6-inch
Construction: Steel, built-up
Tube length overall: 196-inches
Nominal caliber: 30
Tube and breech weight: 10,800-pounds
Usual carriage: Navy 6-inch Mount Mk IV
Weight of gun and carriage: 21,856-pounds
Maximum elevation: 12-degrees
Loading: separate
Shell weight: 105-pounds
Maximum muzzle velocity: 2,000-fps
Maximum range: 9,000-yards

Builders: Navy Gun Factory, Washington Navy Yard

Number built: 109 built for U.S. Navy
14 made available for land emplacement use

Two nice examples of the 6-inch Mk III on pedestal mounts are on display at the Veterans Memorial Museum in Chehalis, Washington, though these were not specifically ever mounted in a seacoast role.

One of the 6-inch Mk III guns mounted to protect the navy base on Guam. Photograph dates from 1911-1912.

6-inch Gun Mk V (Armstrong) On Navy 6-inch Mount Mk VIII (Armstrong)

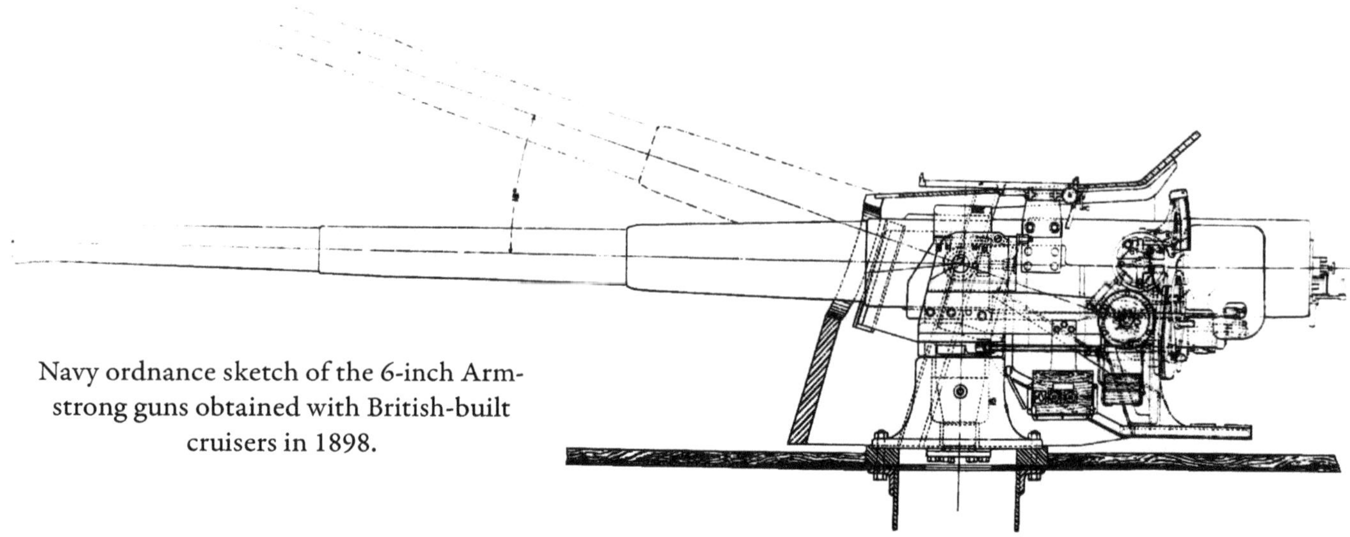

Navy ordnance sketch of the 6-inch Armstrong guns obtained with British-built cruisers in 1898.

When the U.S. purchased two Brazilian cruisers under construction at Armstrong's in England at the start of the Spanish-American War, it included the mounted armament. Each ship (to become USS *Albany* and *New Orleans*) had six 6-inch/50-caliber and four 4.7-inch Armstrong-Elswick guns. By 1904 the ships were being rearmed with American 5-inch guns, and their British RF guns went into the pool of weapons reserved for the Advance Base Force. Most of them wound up at Olongapo. Unlike many of the obsolete guns with the base force, the modern, high-performance Armstrong guns stood out as being among the best guns held there.

The guns were typical British rapid-fire guns of the period. Both Armstrong and Vickers provided nearly identical guns of this type for the Royal Navy and with a huge export market a good portion of the rest of the world's navies. It was a forged steel gun 50-calibers in length. It had the typical conical-shaped interrupted-screw breech block. The carriage was a simple forged steel pedestal. It had a bronze sleeve to hold the barrel and a single underslung hydraulic recoil cylinder and twin adjacent spring counter recoil cylinders. The sleeve had trunnions that fit into a yoke that in turn sat on a pedestal that allowed 360-degree traverse. A distinctive encompassing shield, four-inches thick, surrounded the front and sides of the gun. The guns had Armstrong markings and serial numbers, but in U.S. Navy hands were designated Mark V with new serial numbers in the American sequence. Car-

Navy 6-inch Gun Mk V (Armstrong) on Navy 6-inch Mount Mk VIII aboard USS *New Orleans*. These ex-British guns obtained when the U.S. bought Brazilian cruisers *Amazonas* and *Almirante Albreu* in 1898. When rearmed in 1905 their old armament was reserved as coastal artillery for the Marine Corps Base Forces.

riages were designated American Mark VIII also with new numbers. The gun fired a 100-pound shell using separately cased powder. A full ammunition load was acquired with the ships for these guns. Except for a shorter, 40-caliber barrel, they are very similar to the army's eight 6-inch Armstrong guns acquired by purchase in 1898 and emplaced in several coast defense batteries in the United States.

The twelve 6-inch guns were dismounted from the two cruisers in 1904 as they prepared for rearmament. Apparently, *New Orleans* left her guns in the Philippines before returning to the U.S. and *Albany* had hers removed in Puget Sound. Ten of the 6-inch were concentrated in Olongapo with the Advance Base Force. With the 1907 Japanese war scare they were mounted in temporary defenses and manned by a marine detachment pending the erection of the Army's permanent defenses on Grande Island in Subic Bay. In 1910 all were removed except a practice battery. Then in 1913 with the move of the base force to Mare Island, the ten guns were shipped to Guam to be used to replace and enhance the defenses there. Three recently dispatched Marine Corps companies emplaced four of the guns at Orote Point, two at Cabras Island, and the rest at either Tipolao Point or Mount Tenjo. One of these guns probably was responsible for the shot over the bow of the German auxiliary cruiser SMS Cormoran in April 1917 fired in an attempt to dissuade the crew from scuttling the ship.

Soon this generation of guns on Guam was replaced once again. Several 7-inch Navy guns on pedestal mounts arrived and replaced the Armstrong 6-inch guns in 1921. The removed guns were subsequently shipped back to the U.S. and soon scrapped. Only Grande Island and Guam appear to have been sites where this gun was actually emplaced.

None of the twelve original Armstrong guns still exists, though very similar examples of other Armstrong and Vickers 6-inch guns exist in museums and even original emplacements around the world.

6-inch Gun Mk V (Armstrong) On Navy 6-inch Mount Mk VIII

Gun Characteristics
Bore size: 6-inch
Construction: Steel, built-up
Tube length overall: 309.26-inches
Nominal caliber: 50 caliber
Tube and breech weight: 15,200-pounds
Usual carriage: Navy 6-inch Mount Mk VIII
Weight of gun and carriage: 36,764-pounds
Maximum elevation: 18-degrees
Loading: Separate, cased
Shell weight: 100-pounds
Maximum muzzle velocity: 2,642-fps
Maximum range: n.a.

Builders: Elswick Ordnance Com. (Armstrong)

Number built: 12 acquired by U.S. Navy in 1898

One of the Navy's 6-inch Armstrong Mk V guns deployed to Grande Island, Subic Bay in April 1909 and manned by Marines of the Advanced Base Force.

6-inch 50-caliber Guns Mk VI and Mk VIII On Navy 6-inch Mount Mk X

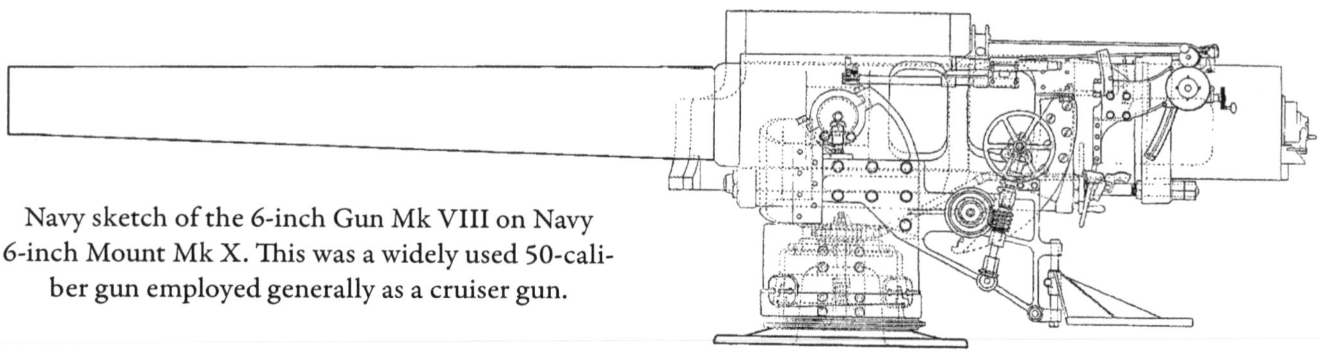

Navy sketch of the 6-inch Gun Mk VIII on Navy 6-inch Mount Mk X. This was a widely used 50-caliber gun employed generally as a cruiser gun.

These were common Navy 6-inch guns mounted on cruisers and battleships prior to the First World War. After removal from ships, large numbers remained in the Navy's inventory, and many were utilized in shore-based defense roles.

The gun was a built-up steel tube with a navy-type Welin breech. Several slightly different marks and modifications were used, but they all had the same size and performance characteristics. The carriage was the shipboard pedestal with yoke for holding the gun without trunnions in a sleeve and held down with a square-shaped bolt pattern. The breech was a typical Navy 6-sector interrupted screw type; the gun used separate, bagged powder to fire a 105-pound projectile. Some guns were mounted with a small, curved embrasure-type shield. It used on-carriage sights. The usual mount was the pedestal Mk X. This allowed elevation from -10 to +15-degrees, resulting in a maximum range of 16,000-yards. The guns could be fired by percussion but used electrical firing in most situations.

Starting in the early 1930s (twelve in 1931) the Navy transferred guns to the Army a number of times. It appears that no immediate use was made of them, but in 1942-44 the Army emplaced many of these guns in simple, temporary emplacements. These batteries usually were composed of just simple circular concrete pads with a pit and hold-down bolts for the open, unprotected mount. A magazine, usually of wood but sometimes hastily poured concrete served the gun. No crew facilities were integral with the battery, and as they used just on-carriage fire control sights, no separate spotting or plotting stations were built. Distance between guns varied with topography but appears to have varied between 130 to 160-feet.

No definitive list of 6-inch Second World War temporary batteries exists. Compiling notes from a wide variety of sources indicates that perhaps about

March 1943 Navy picture of a 6-inch gun Mk VIII on pedestal mount on Savo Island, Solomon Islands, South Pacific.

55-60 guns were emplaced in the U.S. and overseas bases. In June 1944 Army correspondence indicates that there were 24 existing batteries (of 1-4 guns each). Locations included: Beaufort Inlet (North Carolina), Los Angeles, San Francisco, Alaska (many sites, including Nome, Yakutat, Chernofski, Cold Bay, Cross Sound, Kiska and Sitka) Samoa, Espiritu Santo, Argentia, and Surinam.

During the Army's scramble to locate more serviceable gun tubes during the Second World War, the Navy again offered additional tubes of 6-inch bore. In July 1943 an additional 60 guns (and fewer mounts) were offered to the army as possible replacements for 6-inch Model 1903 guns for 1940 Program 6-inch batteries. They were accepted but never utilized, being authorized for disposal in July 1946.

In some locations the guns were removed when more modern, standard program batteries were completed. Others were either abandoned with the weaponry scrapped or they were turned over to local governments—like the four gun battery in Surinam. Several guns do seem to have survived the process; examples in Samoa, Alaska, and the four mentioned in Surinam are still in place though showing various scars of attempted scrapping. There are several other navy 6-inch/50 guns displayed as war memorials in the continental United States.

6-inch Gun Mk VIII
On Navy 6-inch Mount Mk X

Gun Characteristics
Bore size: 6-inch
Construction: Steel, built-up
Tube length overall: 300.2-inches
Nominal caliber: 50
Tube and breech weight: 8.43-8.55-tons
Usual carriage: Navy 6-inch Mount Mk X
Weight of gun and carriage: n.a.
Maximum elevation: 15-degrees
Loading: Separate
Shell weight: 105-pounds
Maximum muzzle velocity: 2,800-fps
Maximum range: 16,000-yards

Builders: Naval Gun Factory, Washington Navy Yard

Number built: Approximately 55-60 used for coast defense

Numerous 6-inch/50-caliber guns were mounted by the Army during the war, and several do still survive. Here is one (a Mk VIII M2) still at Blount's Point, American Samoa photographed in 2023.

7-inch Gun Mk II
On Navy 7-inch Mount Mk II

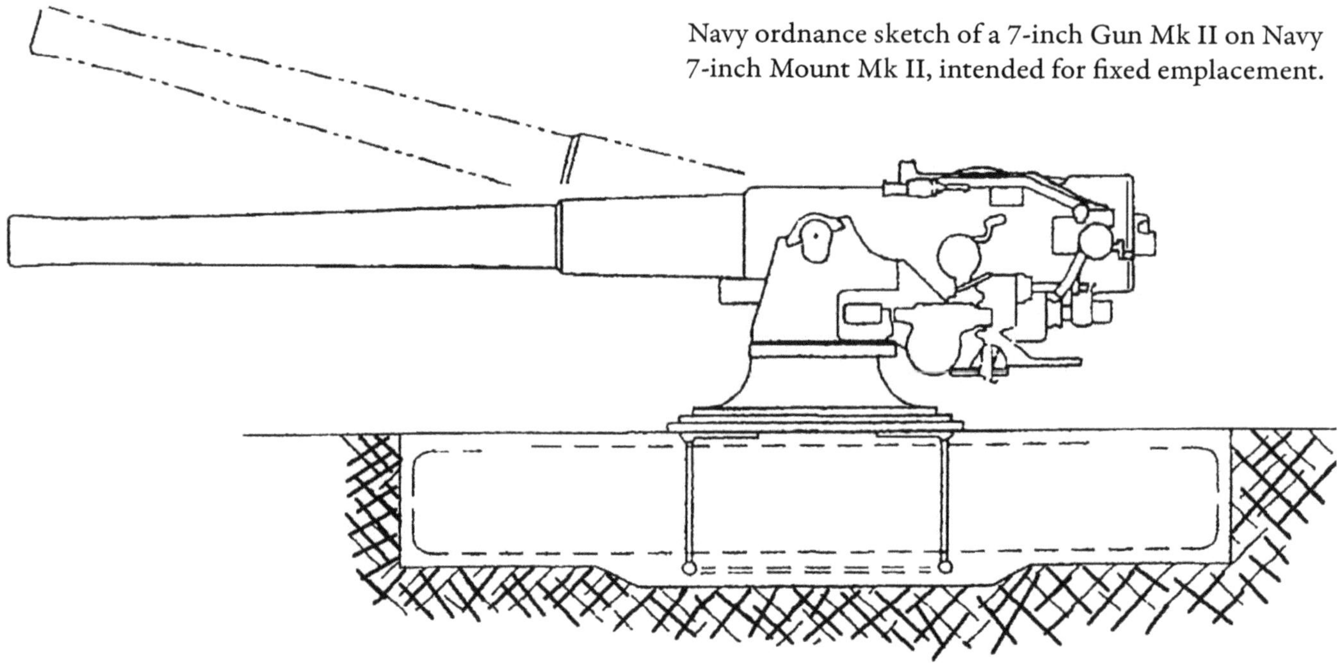

Navy ordnance sketch of a 7-inch Gun Mk II on Navy 7-inch Mount Mk II, intended for fixed emplacement.

The Navy 7-in 45-calibre Mark II gun was originally developed as an intermediate secondary gun for use in battleship casemates. Eight American battleships carried the weapon in the early 20th Century. The gun design dates from 1899-1900, production started in 1903. It was of conventional steel built-up fabrication. The guns were made by the Washington Naval Gun Factory, supplemented by contracts with the army's Watervliet Arsenal and private producer Bethlehem Steel. Never heavily used, the guns themselves were still in good condition when the First World War started. Besides a normal number of spares, additional quantities of this gun had recently become available when they were removed from the battleships to allow for better water-tight integrity in case of mine or torpedo damage. The Navy transferred a number of these guns to the Army and made others available for conversion to tracked mounts or kept them as potential shore mounted pedestal guns.

The coastal mount retained its deck pedestal Mark II. This was a conventional pedestal anchored to a square-shaped base plate. The gun was held in a cradle slide attached to the yoke. It was simply a matter of securing the bottom plate to hold-down bolts sunk either in a concrete gun block or in a heavy, buried timber platform. Shipboard casemated guns were mounted behind curved shields, but for coast defense work they were almost always open-mounted without shields. The gun fired a useful 165-pound shell with bagged powder and could reach a range of 16,500-yards with 15-degree elevation.

During the First World War 7-inch guns were used in fixed emplacements in Guantanamo, the newly acquired Virgin Islands (one at St. Thomas and three at St. Croix), and to protect the navy base in the Azores. All of these were removed soon after the war's end, except the four Azores guns that were transferred to the Portuguese government. In 1920 a total of 24 guns (18 to be emplaced, and six more for spares) were sent to Guam, replacing the Armstrong 6-inch guns already emplaced and manned by a marine detachment. Other Marine Corps 7-inch pedestal guns were with the 10th Regiment or in the Philadelphia Advance Base Force. It appears that a total of ten guns were mounted on Guam and served until removed with some ceremony in early 1933. Right before the Second World War a number of these guns and mounts were sold to the Netherlands East Indies, where some were mounted just prior to the Japanese conquest. These same naval gun tubes were also used in a caterpillar tracked mount (Mk V) and as a railway mount. Still, On October 17, 1941 the navy reported that it had 33 7-inch guns and mounts available in inventory.

During the early part of the Second World War the navy supplied eight of these guns to protect the refueling depot in Bora Bora. Another ten were mounted on Oahu and Kauai, and four were dispatched to Wake Island in the Navy's January 1942 abortive relief attempt but eventually found themselves mounted on Midway Island. Two others went to Canton Island. A training battery (Battery Zeilin) was built at Fort Rosecrans in San Diego. At least 12 7-inch guns were supplied to Brazil, and the navy still had 43 in inventory in 1944 to offer to the Army, though by then they weren't interested in acquiring further units. Somewhat surprisingly several of these pedestal mounts still exist, including all eight still in their emplacements in Bora Bora, two nice examples at Fort DeRussy in Honolulu, and even some (on pedestals mounted upside-down to allow increased elevation) Brazilian guns at military establishments around Rio de Janeiro.

7-inch Gun Mk II
On Navy 7-inch Mount Mk II

Gun Characteristics
Bore size: 7-inch
Construction: Steel, built-up
Tube length overall: 323-inches
Nominal caliber: 45 caliber
Tube and breech weight: 28,700-pounds
Usual carriage: Navy 7-inch Mount Mk II
Weight of gun and carriage: 57,245-pounds
Maximum elevation: 15-degrees
Loading: Separate
Shell weight: 165-pounds
Maximum muzzle velocity: 2,700-fps
Maximum range: 16,500-yards

Builders: Washington Navy Yard, Bethlehem Steel, Watervliet Arsenal

Number built: 112 built for U.S. Navy
78 reported available in 1920 for shore use

Bethlehem Steel brochure illustration of a 7-in L45 RF gun on pedestal mount Mk II.

Postcard from the First World War showing a 7-inch Mk II gun and pedestal emplaced to protect the new seaplane base in the Azores.

Eight 7-inch guns were emplaced and manned by Army personnel in 1942 near Bora Bora, Society Islands, South Pacific. Here is one of the guns of February 14, 1944. Incidentally, all eight of these guns survive.

Survivors of the 7-inch gun and pedestal are relatively numerous. Here is one mount originally emplaced in Battery Harbor, Sand Island Oahu and now on display at the Army's museum at Fort DeRussy, Honolulu.

Five 7-inch guns were provided during the Second World War to Mexico for coast defense. Two survivors are on display at the Mexican Naval Museum, Veracruz, Mexico

Coast Defense Fixed Emplacements

For Mobile and Railway Artillery

Following the procurement of large numbers of towed, mobile artillery and railway mounted guns by the Coast Artillery, efforts were on occasion made to develop more permanent emplacements for these types of guns. Obvious advantages of a more stable firing platform, mechanisms for more rapid traverse, and particularly the ability to tie the gun into prepared fire control observation and range plotting were appealing.

A substantial portion of the newly produced 155mm Model 1917 or 1918 GPF guns were transferred to the Coast Artillery Corps in the 1920s. Starting in the late 1920s the Corps, with assistance from Ordnance and Engineer Departments, developed fixed emplacements for these guns. After development efforts in the overseas departments, a simple center pivot block and curved track for the trails was adopted as the "Panama Mount". Numerous harbors were supplied with these inexpensive emplacements with their 155mm guns kept in storage or with their original mobile regiments. During the urgency of the early months of the war additional units were built to augment defenses at many secondary locations, including overseas island bases.

As the war progressed it was found necessary at times to use the high-speed M2 or M3 variety of 155mm GPF carriage on some of the fixed Panama Mounts. Also, in a few instances the mounts were modified to allow them to carry the new 155mm M1 and M1A1 guns, the "Long Tom". Finally, late in the war, the Army developed and produced a portable metal platform and traversing ring (sometimes called the "Kelly Mount") to serve the same purpose also for the M1 155mm guns.

An additional effort, although on a much smaller scale and limited to the island of Oahu, was made to develop a prepared, fixed emplacement for the 240mm howitzer Model 1918. Eventually over 20 such prepared gun emplacements were constructed for the 12 howitzers actually on the island. Intended to extend coverage over otherwise remote landing beaches on the island or as a component of a secure, inland, protracted-defense scheme, they served as part of the island's defense plan through most of the war.

Many 8-inch Model 1888 guns had been removed from their seacoast emplacements in 1917 and used on new railway mounts and carriages for the war in Europe. While they did not reach France in time for use, immediately after the war many of these completed mounts were organized by the Coast Artillery in special regiments. Where needed it was found that the gun and top carriage could be removed from its rail undercarriage and emplaced on a simple concrete block for a new type of fixed emplacement. Though rather ungainly in appearance and with a high profile, it was at times found to be a convenient source of relatively heavy, long-range gun during the production limitations of the 1930s. Starting in 1935 a few new battery emplacements were built, all in overseas locations.

Many of these emplacements still exist and can be viewed today. Also, there are representatives of 155mm GPF guns, and one each 240mm howitzer and 8-inch railway Model 1918 gun and top carriage. However, no guns are on public display on their original type of expedient emplacement—though at Fort MacArthur in San Pedro is a 155mm gun on a simplified, reproduction Panama Mount.

155mm Gun On Carriage Model 1917 or 1918 On Fixed "Panama Emplacement"

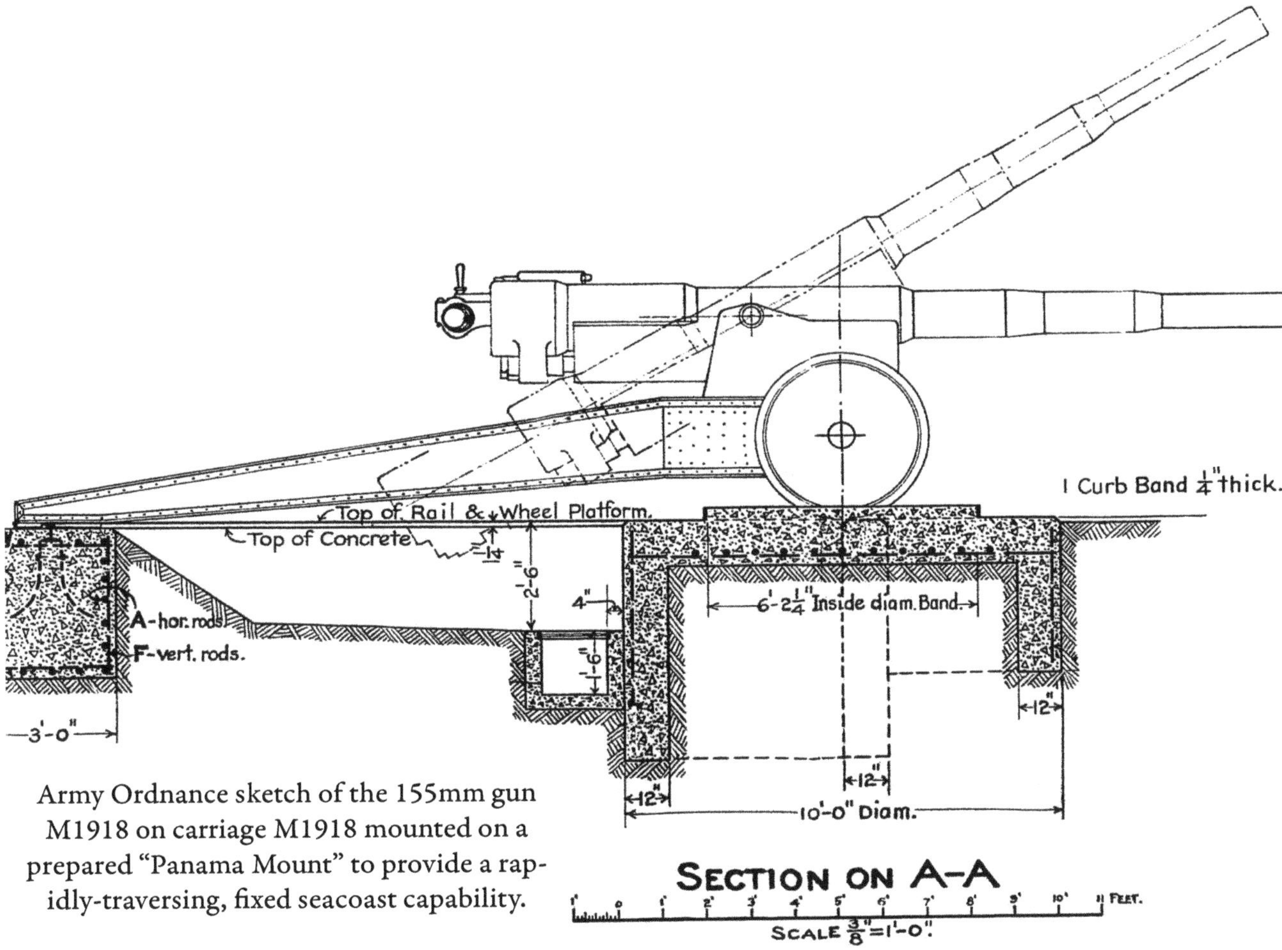

Army Ordnance sketch of the 155mm gun M1918 on carriage M1918 mounted on a prepared "Panama Mount" to provide a rapidly-traversing, fixed seacoast capability.

Following the First World War the Coast Artillery Corps was supplied with hundreds of 155mm GPF guns. The guns and carriages were designated either Model 1917 (French made) or Model 1918 (U.S. made), virtually identical in appearance, performance, and interchangeable between carriages. Most of the guns were initially assigned to truck or tractor-drawn mobile regiments. They were not deployed as active seacoast batteries but rather kept as regiments capable of being deployed to threatened locations during an emergency. During the late 1920s it was realized that pre-prepared tactical positions for the guns, tied into the fire control and command structure of existing harbor defense forts could enhance the defense, particularly for the overseas department of Hawaii, the Philippines and the Panama Canal. Development efforts to design a new type of fixed emplacement providing a stable firing platform, maximum elevation, and quick traverse was desired. In the Philippines a promising barbette-type emplacement was actually built at Fort Mills, but as it required time-consuming removal of the gun top carriage from the wheels and trails of the lower carriage it was not adopted.

The emplacement developed by the Panama Department and finalized in 1929 ultimately was favored for adoption. It featured a center circular concrete platform 10-foot in diameter, with a steel curb band. The carriage body would be placed directly atop this platform, with the wheels still free-moving and revolving around the curb. The carriage was not bolted or otherwise attached to the center of the emplacement. Outside the center platform was a concrete ring at 35-feet, 1.5-inches in diameter. In the concrete was held either a toothed or plain rail meant to support the extended trails of the carriage. Spade plates were removed from these trails and replaced with new plates engaging the traversing rail. The entire carriage, resting on its wheels on the center block and trails on the rail ring could quickly traverse left and right to follow any target. The area between the platform and rail was frequently dug out (and sometimes lined with masonry) to allow a depression for the breech recoil so the gun could fire at maximum elevation.

The mount was intended for the common 155mm gun and early, hard-wheeled carriage. It retained the same performance characteristics of the similar field gun. When emplaced as coast artillery it was usually supplied with the M8 panoramic telescope or M6A1 mount telescope.

Ultimately emplacement types were designed for complete (360-degree) or partial (270 and 180-degree) encircling rail rings. The type selected depended on the desired degree of coverage at that particular tactical site. Usually, four emplacements were built in a roughly linear configuration at about 90-foot intervals to take an entire battery of guns, though again at some sites the area or need was less and fewer emplacements were constructed. Few other structures were made permanent. Adjacent magazines were most often earth-covered wood or metal-framed buildings, and battery commander stations were either wooden or on light metal towers. In several cases fixed emplacements for 155mm guns were built adjacent to or even directly on top of abandoned emplacements from previous generations—and used structural elements from them in their design. Between 1928 and 1944 over 130 batteries with over 440 individual gun platforms were constructed in the U.S. and overseas bases. This coast defense use was the final utility for this gun; some not being disarmed until the Second World War actually ended.

155mm Guns Model 1917M1 and 1918M1 On 155mm Carriages Model 1917 and 1918 Mounted on fixed, Panama Mount

Gun Characteristics
Bore size: 155mm
Construction: Built-up alloy steel
Tube length overall: 232.87-inches
Nominal caliber: 38.2 caliber
Tube and breech weight: 8,795-pounds
Usual carriage: 155mm gun carriages Model 1917 and 1918 M1, M2
Weight of gun and carriage: 29,860-pounds
Maximum elevation: 35-degrees
Loading: Separate
Shell weight: 95-pounds (HE)
Maximum muzzle velocity: 2,418-fps
Maximum range: 20,000-yards

Builders: Guns: Puteaux, Watervliet Arsenal, Bullard Engineering Works
Carriages: Puteaux, Minneapolis Steel

Number built: Approximately 450 emplacements constructed

A high proportion of these emplacements still exist at parks, military bases, and other public and private sites. None are currently displayed with a 155mm gun, but there are numerous examples of that gun also still in existence and at museum or public display.

While many (most) concrete Panama Mount circles still exist, and dozens of 155mm Model 1918 GPF guns and carriages survive, there is currently no location that matches the two. This is a 155mm displayed as a field battery (no concrete traversing emplacement) on public display at Fort Macon State Park, Atlantic Beach, North Carolina.

A 155mm GPF gun and carriage emplaced at San Juan, Puerto Rico in 1941.

Many Panama Mounts were constructed for the Manila Bay seacoast forts. This is Battery Levagood, on the tail of Corregidor Island, Fort Mills in late 1941.

During the Second World War a temporary portable platform was developed to allow the new 155mm M1 gun (the famous "Long Tom") to be emplaced in a rapidly-traversing seacoast mode.

8-inch Model 1888 gun
On Railway Top Carriage Model 1918 In Fixed Emplacement

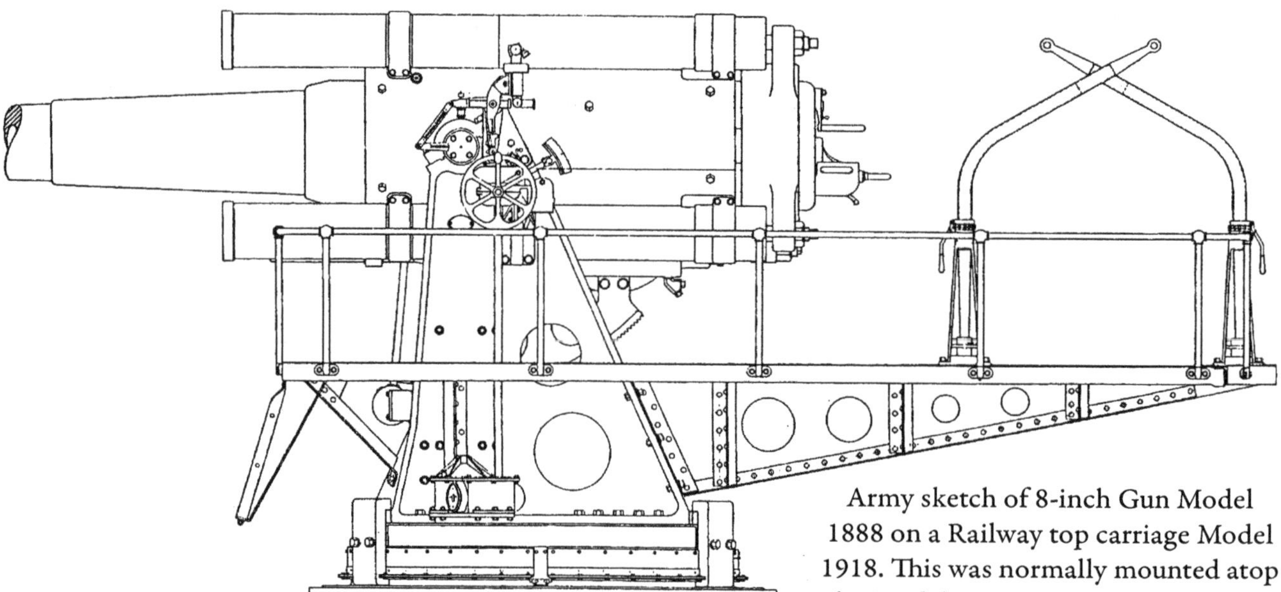

Army sketch of 8-inch Gun Model 1888 on a Railway top carriage Model 1918. This was normally mounted atop the Model 1918 Gun Railway Car, but could also be mounted directly to a fixed ground platform.

During the First World War, 50 8-inch Model 1888 guns were removed from seacoast batteries and subsequently mounted on Model 1918 Railway Mounts. The barbette top carriage Model 1918 could accommodate either of the Model 1888, 1888MI, or 1888MII guns. The carriage had the usual base ring, traversing rollers, side frames, cradle with a single hydraulic recoil cylinder mounted underneath and four distinctive counter recoil spring cylinders mounted two above and two below the barrel. It had ammunition cranes on the rear of an attached steel working platform. Essentially a complete barbette carriage, it was originally simply bolted on an appropriate railway carriage. However, it could also be bolted to a fixed emplacement ground platform. The railway equipment never made it to France, but 47 complete units were in inventory with the Coast Artillery Corps in 1919.

The gun mounted was the same model 1888 that had been previously carried on barbette and disappearing mounts. The top carriage mount had all the same ballistic performance as the railway mount. The 8-inch gun firing stress proved excessive, so developed for the railway mount and carried over for the fixed-type emplacement were new 200-pound shells and shot, capable of being fired at an increased 2,600-fps to a long 26,000-yard range. The top carriage mounted as a fixed barbette was an awkward, exposed weapon, but one with a good useful range and payload that could prove important for the interim.

The first emplacement to use the 8-inch railway top carriage was Battery Granger Adams at Black Point, Oahu. Built in 1935 it consisted of concrete-walled pits with platforms and bolts to which the carriage was bolted. Then during the war eight more of Hawaii's 8-inch railway guns were dismounted and placed onto fixed emplacements: four each at Battery Kahuku in northern Oahu and four at Battery Sylvester in southeastern Oahu. In mid-1941 seven 8-inch railway top carriage mounts (no rail cars were part of this shipment) were sent to the Philippines for use in defending straits as part of the armament for the "Inland Seas Project". The local command had desired 12-inch guns, but as no extra carriages for 12-inch guns existed, the 8-inch guns were considered adequate substitutes. Five of the guns were damaged by Japanese air action soon after the start of the war during transport—but two survived to actually be emplaced. One went to Saysain Point on western Bataan and the other to Road Junction 43 at Fort Mills, Corregidor Island. These emplacements had only concrete platforms, berms and magazines being of wood and earth. The Bataan gun saw limited action; however, both weapons were abandoned with the surrender of Filipino-American forces. Finally, several 8-inch railway guns were provided to Canada in 1940, and at least two were emplaced as fixed emplacements in Newfoundland and others on the British Columbia coast.

Curiously there are survivors. The 8-inch Model 1888 gun tube from the Corregidor emplacement still exists near the ferry arrival dock, though without its carriage. One complete top carriage and 8-inch gun also is on display at Plant Park in Tampa, Florida.

8-inch Gun Model 1888
On Railway top carriage Model 1918

Gun Characteristics
Bore size: 8-inch
Construction: Steel, built-up
Tube length overall: 278.5-inches
Nominal caliber: 32 caliber
Tube and breech weight: 32,218-pounds
Usual carriage: 8-inch Railway Top Carriage Model 1918
Weight of gun and carriage: 75,500-pounds
Maximum elevation: 42-degrees
Loading: Separate
Shell weight: 200-pounds
Maximum muzzle velocity: 2,600-fps
Maximum range: 26,000-yards

Builders: Gun: Watervliet Arsenal (56), Bethlehem Steel (25), West Point Foundry (11)
Carriages: Morgan Engineering, Harrisburg Manufacturing
American Car and Foundry Co.

Number built: 50 railway carriages constructed for 50 guns removed from fixed emplacements. At lease six gun and top carriages permanently emplaced.

8-inch gun on a M1918 Railway Carriage on display at Plant Park in Tampa, Florida

The 8-inch Model 1888 gun and Railway top carriage emplaced at Battery Granger Adams, Black Point, Oahu in the mid-1930s.

Several 8-inch railway top carriages and gun were also sent to Canada immediately before the U.S. entry into the Second World War. This unit is at Signal Hill, St. John's Newfoundland in 1941-42.

240mm Howitzer Model 1918
On 240mm Howitzer Carriage Model 1918 Emplaced on Fixed Platform

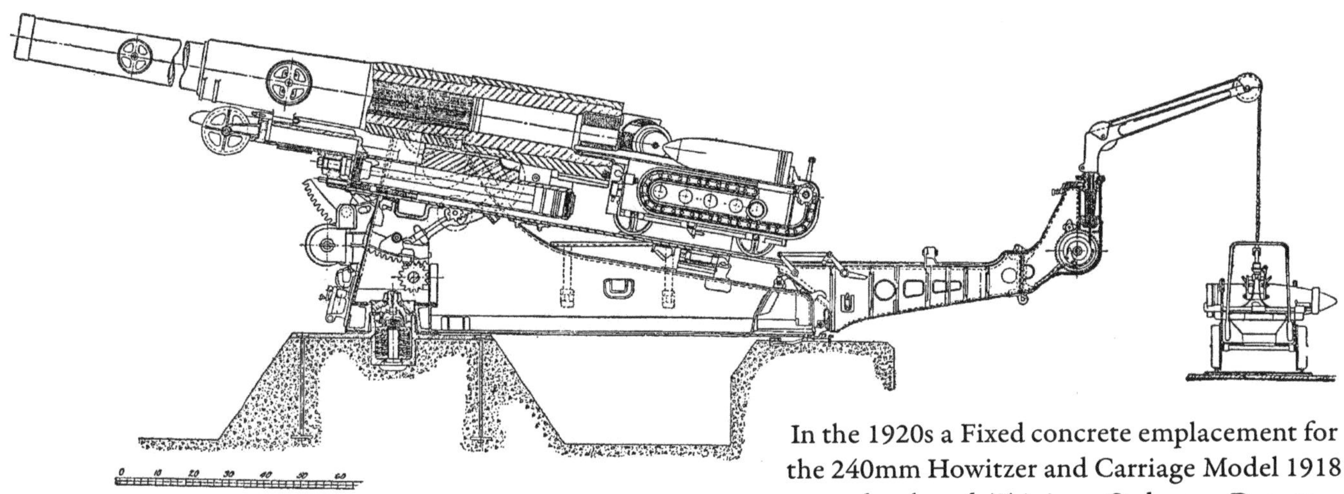

In the 1920s a Fixed concrete emplacement for the 240mm Howitzer and Carriage Model 1918 was developed. This is an Ordnance Department sketch of the howitzer on that platform.

Twelve newly built 240mm howitzers and carriages Model 1918 were diverted to the Hawaiian Department in January 1922. Originally shipped with the Philippines as a destination, the signing of the Washington Navy Treaty with its non-fortification clause caused a change in delivery. The shipment was of the howitzers only, not as an organized unit with personnel. Initially they were attached to established units and assigned defensive field positions in advance. As soon as March 1923 plans were drawn up to use these guns to augment Oahu's fixed defenses. The original equipment was not truly mobile, needing separate transportation of gun and carriage parts and a considerable delay in emplacement. Needed were prepared emplacements allowing for the rapid installation of the guns and ability to rapidly track and traverse on target. The ability to use these powerful howitzers of considerable range to project fire over all of the island's possible landing beaches was highly attractive. Starting in 1923 the department (primarily designed by Major W. A. Johnson) developed a new type of fixed emplacement similar to what would eventually become the Panama Mount for 155mm guns.

A central concrete circular pedestal was to hold the carriage's center pintle and used the pintle socket removed from the conventional platform (the balance of which was not utilized). This pedestal was surrounded by a circular trench, to allow a depression for the howitzer's full recoil upon firing. Outside of the trench was a 360-degree concrete ring holding the traversing rack, supporting the rear extension of the top carriage. The block, including center pintle and outer traversing ring had a 23-foot, 4-inch diameter. Only minor changes involving the riveting of the pintle socket plate to the top carriage was required. Even then the howitzer and its top carriage could be quickly restored to its former, mobile configuration quite easily.

In late 1923 a prototype emplacement was built at the Waimanalo Military Reservation. Several other new platforms were built in 1927, with more through 1931. This first generation of emplacements, usually composed of 2-3 individual firing platforms separated by 240-foot, were completed for a low cost of under $1000 each. These were all on remote military reservations on Oahu, not the established coast defense forts. Coverage of the more remote northern and western beaches was emphasized. Little besides the basic platforms and access roads were built, though some emplacement had simple adjacent magazines. Nine new emplacements were initially constructed; however, the guns would not be emplaced until an impending emergency required. Then in 1939-41 another twelve emplacements were built in new locations. The original number of 12 howitzers was never increased; these new positions were just for increased flexibility in ultimate location. Initially they were manned by a detachment of the 55th Artillery Regiment, a mobile unit of 155mm guns. Later they were attached to the artillery components of the 24th and 25th Infantry Divisions defending Oahu.

With full wartime deployment as of December 7, 1941, the howitzers were mounted on their assigned positions. During 1942-43 several fired during exercises, and minor improvements were made to magazines, plotting rooms, and commander stations. However, in early 1944 they were declared obsolete and dismounted. Their use did not extend postwar, and the howitzers were scrapped soon after the end of the conflict. Several of the concrete fixed emplacements still exist, though none are in easily located and at accessible locations. Only a single 240mm howitzer in field artillery mode survives, at the Fort Sill Artillery Museum.

240mm Howitzer Model 1918
On 240mm howitzer carriage Model 1918
Emplaced on fixed platform

Gun Characteristics
Bore size: 240mm
Construction: Steel, built-up
Tube length overall: 199.6-inches
Nominal caliber: 19.75 caliber
Tube and breech weight: 10,790-pounds
Usual carriage: 240mm howitzer carriage Model 1918
Weight of gun and carriage: 41,206-pounds
Maximum elevation: 60-degrees
Loading: Separate
Shell weight: 345-pounds
Maximum muzzle velocity: 1,700-fps
Maximum range: 16,390-yards

Builders: Howitzers: Watervliet Arsenal
Carriages: Watertown Arsenal, Standard Steel Car Co.

Number built: 20 fixed platform emplacements built

One of the Hawaiian Departments 240mm M1918 howitzers deployed on one of the pre-constructed, fixed concrete platforms on Oahu in the mid-1920s.

Hawaiian Turret Batteries

A unique situation was created by the urgent need to increase the seacoast defenses of Oahu, Hawaii, following the Japanese Pearl Harbor attack. The Army's need to add to the firepower of Oahu in early 1942 fortuitously occurred just as the U.S. Navy offered turrets for potential use. Undamaged in the attack and subsequent forward magazine explosion, the rear two triple 14-inch turrets from sunken USS *Arizona* were removed from the ship's wreck and were offered and accepted by the Army for emplacement. Two massive new underground emplacements (one for each turret) were constructed in 1943-1945. Repair and reassembly of the guns, turrets, and internal ammunition lifts also proved a significant challenge despite the considerable technical resources deployed. Finally, just as the war ended one emplacement was completed and test fired, and the second was structurally complete but lacked finalization of some systems. Following the start of this project, consideration was given to using the four turrets of capsized USS *Oklahoma* in a similar manner. While tentative sites were identified, the scope of the work needed, and condition of the submerged turrets eventually scrapped this plan.

Also, in January 1942 the Navy offered the command in Hawaii eight dual 8-inch mounts being removed from aircraft carriers USS *Saratoga* and *Lexington*. Once again these were accepted. The Mark IX, 55-caliber guns were designed and built in the 1920s and were effective guns with a good range capability that saw use on cruisers all through the Second World War. The gun shields were only lightly armored for splinter protection. A high priority effort was undertaken in 1942 to build four two-turret emplacements around Oahu (two covering the south coast and two the north). By the first anniversary of the Pearl Harbor attack all four emplacements were either completed or nearing that state. While technically these are more properly called mounts rather than turrets, they featured guns mounted together in a revolving, armored gun house.

Late in the Second World War the Coast Artillery Corps concluded that future heavy emplacements should consist of heavy turrets rather than casemated guns. The combination of the widest field of fire with very heavy armored protection from both ship fire and aerial bombardment was very attractive. A proposal to relocate Fort Weaver's 16-inch guns to a new turret battery near Aliamanu Crater was seriously considered, but ultimately rejected. On a more comprehensive scale was a proposal to use the partially completed 16-inch turrets from the two incomplete *Iowa*-class battleships as coastal artillery, but the day for fixed defensive artillery was ending and no project ensued.

After the war in the late 1940s the ten Hawaiian turrets were disarmed and both guns and turrets scrapped. While difficult to access, five of the six emplacements still exist.

One of the camouflaged turrets of Battery Brodie, Oahu

8-inch Gun Mk IX Mod 2
On 8-inch Navy Twin Gun Mount Mk VI

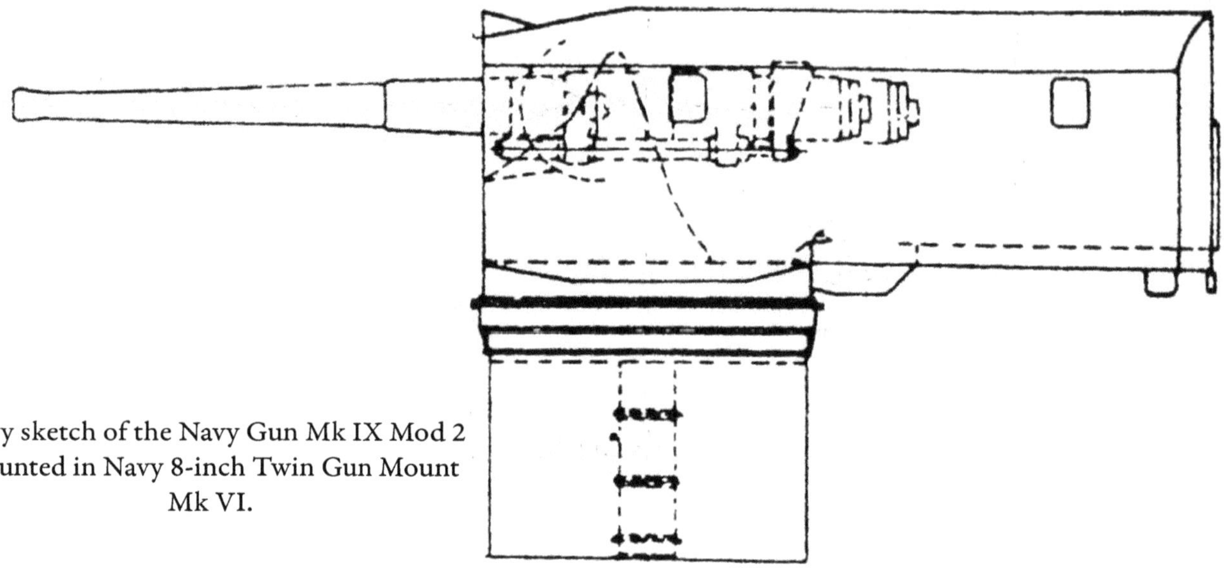

Navy sketch of the Navy Gun Mk IX Mod 2 mounted in Navy 8-inch Twin Gun Mount Mk VI.

The Navy developed new twin and triple turret mounts for a generation of ships completed in the mid-1920s. Four twin 8-inch mounts were carried by new aircraft carriers USS *Lexington* and *Saratoga.* Also, two twin and two triple mounts were assigned to the first American treaty cruisers—USS *Pensacola* and *Salt Lake City.* Plans were already made pre-war to remove the carrier armament in order to expand their anti-aircraft batteries. When damaged *Saratoga* reached Pearl Harbor in January 1942 the twin 8-inch mounts were removed; they were offered to the Army on January 17, 1942. Then a little later USS *Lexington* was similarly stripped at Pearl Harbor prior to leaving on her final mission for the battle of the Coral Sea. These eight twin mounts were offered, and eagerly accepted, by the Army for use as seacoast guns to augment the Oahu defenses.

The guns and mounts were not remarkable or unusual. The naval 8-inch Mk IX Mod 2 gun tube was a length 55-calibers, but otherwise of conventional construction and with a normal naval sliding breech. The tube was rifled with a twist of 1/35. It fired a 260-pound projectile (common, HC or AP) with separate, bagged service charge of 90-pounds. Hydropneumatic recoil mechanism was mounted underneath the barrel. The Mk VI dual mount was only lightly splinter-protected, the navy preferring to call it a mount or gun house rather than a true turret. Basically, mounted ashore still in this gun house, the guns had an elevation from -5 to +40-degrees. Initially they were accompanied by their naval Mk XVIII optical gun directors, but they were soon found unsuitable and discarded in favor of the usual Army seacoast observation and ranging system. They also had SCR-296 radar fire control equipment.

The Hawaiian engineers designed four separate emplacements, each of two dual turrets, for these guns. The sites covered sectors that were thought to have only light coverage, and with thought for survivability during a possible invasion, they were emplaced considerably inland. Each site received two turret emplacements separated by 450-foot. Two dual batteries faced northward (Battery Ricker at Browns Camp and Battery Riggs at Opaeula) and two southward (Battery Burgess at Salt Lake and Battery Kirkpatrick at Wiliwilinui Ridge). Each turret was placed over a cylindrical pit 50-feet deep containing the ammunition lift mimicking the shipboard arrangement. The magazines were concrete rooms deeply buried near the base of the cylinder. Each reservation of two batteries had a separate power station and plotting room. As the guns and turrets had been removed from active ships, they were immediately available for use. A crash excavation and construction project saw the emplacements completed within just a year of their start.

By early 1943 all four sites were in active operation by the coast defenses. They served until postwar period when these defenses were authorized for disarmament and disposal. The guns and gun houses were scrapped by 1949. No such mounts still exist, though some later naval 8-inch 55-caliber gun tubes do exist on public display. Three of the four sites at least partially still exist (Salt Lake was buried for a residential development), though they are on private property and permission for access is very difficult to obtain.

8-inch Gun Mk IX Mod 2
On Navy 8-inchTwin Gun Mount Mk VI

Gun Characteristics
Bore size: 8-inch
Construction: Steel, built-up
Tube length overall: 449-inches
Nominal caliber: 55 caliber
Tube and breech weight: 66,850-pounds
Usual carriage: Navy 8-inch twin gun mount Mk VI
Weight of gun and carriage: 374,000-pounds (twin mount with guns)
Maximum elevation: 40-degrees
Loading: Separate
Shell weight: 260-pounds
Maximum muzzle velocity: 2,800-fps
Maximum range: 31,700-yards

Builders: Naval Gun Factory

Number built: 16 guns in 8 dual turrets used for coast defense

The four 8-inch dual turrets were removed from USS *Lexington* while in Pearl Harbor in late March 1942, just prior to her April deployment and subsequent loss in the Battle of Coral Sea.

A good view of one of the ex-aircraft carrier 8-inch turrets. This one is Battery Riggs at Opaeula Military Reservation, Oahu.

14-inch Gun Mk VIII Mod 4
In 14-inch Navy Triple Gun Mount

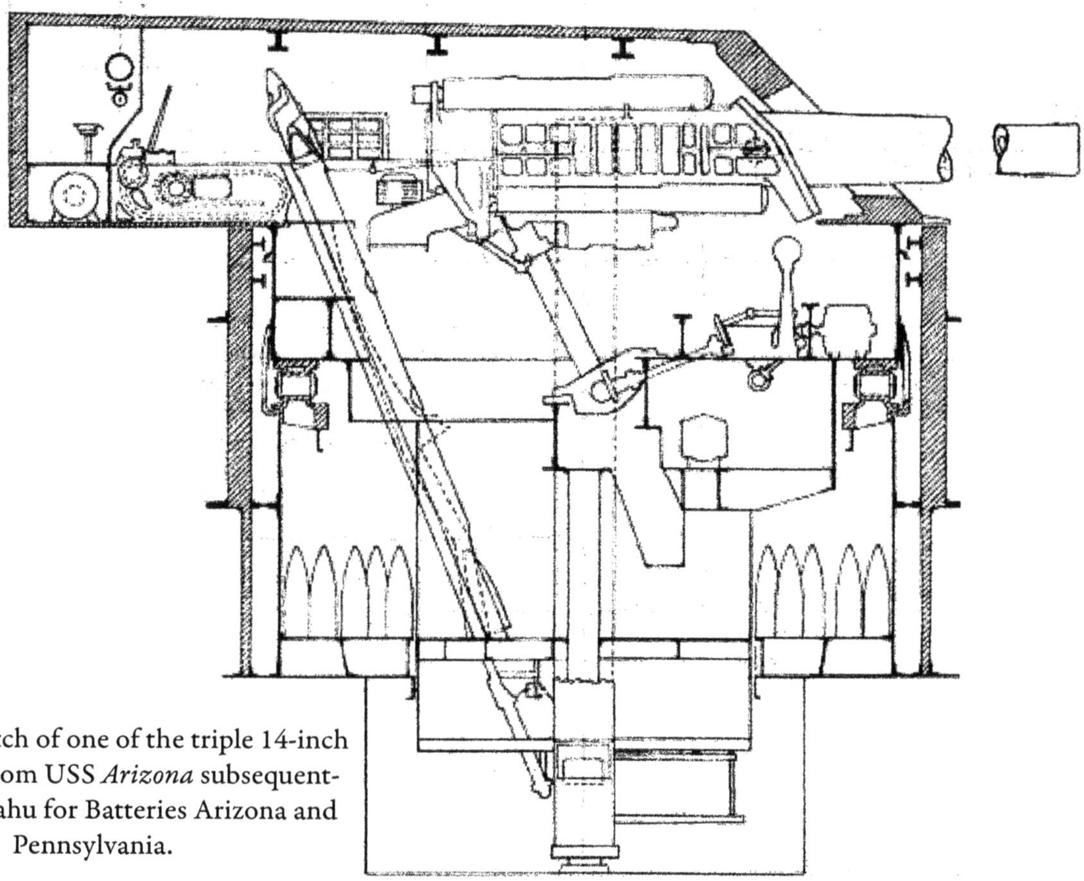

Side view sketch of one of the triple 14-inch gun turrets from USS *Arizona* subsequently used on Oahu for Batteries Arizona and Pennsylvania.

Following the Pearl Harbor attack, the Navy in early 1942 offered the Army the two rear, triple turrets off the wrecked USS *Arizona.* The forward magazine explosion had destroyed the forward two turrets, but the rear turrets were thought to be salvageable. The local Hawaiian Defense Command conditionally accepted the offer. A year was spent on a study of sites, plans, and the amount of work required to recondition the (recently submerged and hastily dismantled) turrets. On August 15, 1943 plans were approved to emplace each triple 14-inch turret in a different emplacement. Recognizing the need to enhance the seacoast defenses in the southwest and southeast of Oahu, sites were selected at Kahe and Mokapu Point.

The salvaged gun tubes were in relatively good shape. They were navy Mk VIII Mod 4 45-caliber tubes as originally produced for the battleship before the First World War. They had built-up tubes and the usual naval stepped-breech. Charges were power-rammed. The turret was a triple mount, with the guns in a single sleeve. The turret and elevation gear had been modified to allow 30-degree elevation and a nominal range of 34,300-yards with the standard 1500-pound, armored-piercing projectile. The turrets carried armor of 18-inches on the face, 10-inches on the side, 9-inches in the rear, and a roof thickness of 5-inches. Two shell and three powder hoists fed the ammunition from the magazine, the depth of the lift depending on whether the turret was an upper or lower mount.

There were significant issues in preparing the turret and mechanism. They had initially been removed from the wreck without thought for reuse, there were missing parts and dismantling cuts in the wrong places. It was found that the original plans had become obsolete with subsequent changes and at times totally new drawings had to be prepared. A joint Army-Navy team of technical experts labored hard for three years to recondition the guns, mounts, turret, and ammunition lifts for this project.

The engineers designed two massive underground emplacements. The turret sat on a steel barbette atop a concrete cylindrical shaft with underground magazines (65-feet below the surface) somewhat reproducing the arrangements shipboard. Each emplacement was to be stocked with 840 rounds. Plotting rooms, a power plant and support structures were also provided. Recognizing that the emplacements were geographically isolated, they included more gun crew features like latrines, a small galley, and first aid station. Battery crew was calculated at four officers and 157 enlisted men.

Construction of the emplacements started in the last half of 1943 and was finally complete right at the war's end. They were named for the battleship and her sister ship which provided the turrets. Battery Pennsylvania was located at Fort Hase on Kaneohe Bay and completed in August 1945. She was proof fired practically on the last day of the war. Battery Arizona at the Kahe Point Military Reservation was almost, but not quite completed when the war ended, and work was immediately suspended and she was never proof fired.

14-inch Gun Mk VIII Mod 4 On Navy 14-inch Triple Turret Mount

Gun Characteristics
Bore size: 14-inch
Construction: Nickel steel, built-up
Tube length overall: 642.45-inches
Nominal caliber: 45 caliber
Tube and breech weight: 123,200-pounds
Usual carriage: Navy 14-inch triple turret mount
Weight of gun and carriage: Approximately 1,440,000-pounds (triple mount with guns)
Maximum elevation: 30-degrees
Loading: Separate
Shell weight: 1,500-pounds
Maximum muzzle velocity: 2,800-fps
Maximum range: 34,600-yards

Builders: Navy Gun Factory, Washington Navy Yard and other contractors

Number built: Two turrets with six guns used for coast defense

The turrets and guns of these emplacements were removed in the late 1940s. The two emplacements still exist, but access to either is difficult to obtain. 14-inch guns of the same mark are on public display both at Boalsburg, Pennsylvania and Phoenix, Arizona

14-inch turrets from the USS *Arizona* being salvaged.

Battery Pennsylvania with its triple 14-inch turret just completed at Fort Hase, Mokapu Peninsula, Oahu Hawaii.

Battery Pennsylvania proof-fired the new emplacement in early August, 1945 just as the war in the Pacific was ending.

Coast Defense Dynamite Guns

The concept for these guns was the propulsion of a large, finned canister with a dynamite explosive charge by compressed air. At the time the concept was even referred to as an aerial-delivered torpedo. Several individual inventors developed small, relatively low-pressure guns in the 1880s. Army Lt. E. L. Zalinski became an early advocate for these guns, and his name is closely associated with dynamite guns. Depending on fuse type, shells would explode upon contact with an enemy warship or detonate in proximity in the water causing hull damage. The primary private developer was the Pneumatic Dynamite Gun Company of New York.

Experiencing numerous internal financial issues, that company reorganized as the Pneumatic Torpedo and Construction Company in 1894. Being successful in convincing key congressional members for support, bids for construction of the initial guns and emplacements were advertised in April of 1889. For seacoast defense two sizes were acquired, of 8 and 15-inch bore. Prototype guns of a Model 1886 were evaluated in 1887. Also, in the 1890s the firm sold representative guns to England, Italy, and Brazil. It also commercialized this type of gun for shipboard naval use. A much smaller field gun (the Sims-Dudley 2.5-inch dynamite gun) was developed in the 1890s.

Four batteries were finally begun for seacoast dynamite guns, though only two were actually completed. New York and San Francisco each were allocated a large (and very expensive) battery of dynamite guns in the early 1890s. New York was to get a battery of two 15-inch dynamite guns and one 8-inch dynamite gun. San Francisco was allotted three 15-inch dynamite guns. In 1898, mainly in order to utilize the remaining dollars still in the original 1889 appropriation, two additional batteries of dynamite guns were awarded by the War Department. Each of these was to have a single Model 1890 15-inch dynamite gun and carriage. These were at Long Island Sound's Fishers Island (Fort H. G. Wright) and Port Royal in South Carolina. The Long Island battery was armed but never technically completed, and just the gun block and adjacent steam plant for compressing air was completed in South Carolina.

Firing tests of December 1895 were undertaken by the San Francisco battery and met the specification requirements. Much of the physical gun construction was subcontracted to the West Point Foundry in Cold Spring, New York. The carriages were fabricated at the New Jersey Steel and Iron Company of Trenton, New Jersey. Shells were produced by Thomas Phillips and Son of Providence Road Island. The Pneumatic Torpedo and Construction Company also supervised various contractors for the installation of the heavy boiler and compressor equipment at the battery sites.

The technology never lived up to its promise. While in a fashion the installed guns did function, the lack of range, imprecise aiming, and high construction cost argued against any further deployment. Additionally, the engineers feared that the use of the exploding charges would jeopardize their submarine mines. Conventional guns and mortars were simply better and cheaper. Work was suspended on the two batteries under construction towards the end of the Spanish American War and they were abandoned soon after. The Board of Ordnance and Fortification declared dynamite guns obsolete in June 1901. The two constructed batteries were salvaged in 1902. After disarmament the two completed emplacements were used for other fortification purposes. The platforms for the guns and much of the substantial adjacent parados and compressed air handling rooms still exist at all four of the sites.

The Ft H.G. Wright dynamite battery

8-inch Seacoast Dynamite Gun Model 1893
On 8-inch Center Pintle Barbette Model 1893

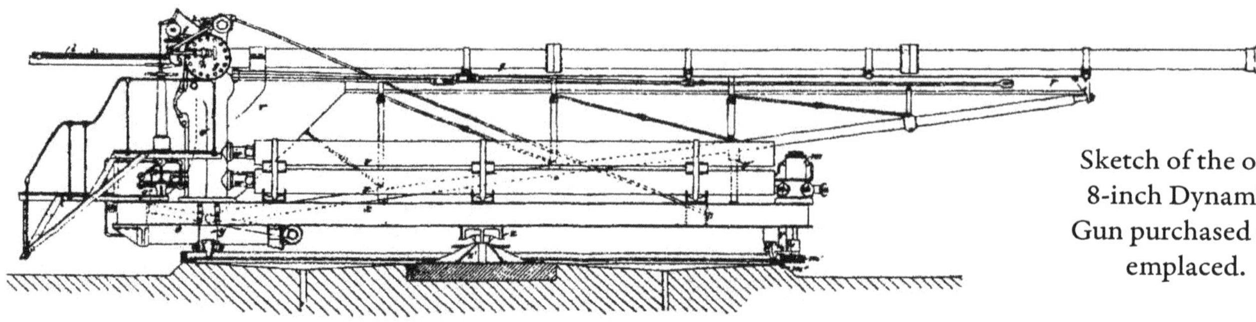

Sketch of the only 8-inch Dynamite Gun purchased and emplaced.

The 8-inch dynamite was the first attempt by the contractor to prove the gun concept with a service-sized weapon. While different in several details, it demonstrated all the same features as its later, larger companion. The initial engineering of the design was done by Babcock & Wilcox. The heavier iron parts for the gun were subcontracted to the West Point Foundry of Cold Spring, New York. This gun was completed in 1887 (as the Model 1886) and put through extensive trials mounted at Fort Lafayette, New York. In September 1887 it dramatically sank an old wooden schooner as a demonstration of its destructive power. The subsequent contract with the Pneumatic Gun Company called for two 15-inch and a single 8-inch gun. Apparently, this provision was a way to pay or reimburse the firm for its costs in developing the first two guns, including its prototype 8-inch. It would probably never have been acquired otherwise, and no subsequent thought was ever given to purchasing additional guns of this size.

The contract that was signed on March 6, 1889 with the Pneumatic Torpedo and Construction Company of New York included a single 8-inch gun and carriage. The original Model 1886 gun was modified with electric rather than complicated hydraulic power for elevating and traversing along with several other modifications to its breech plumbing. This unit (serial No. 1) now designated the Model 1893 was duly delivered and passed inspection in June 1894. The unit was immediately shipped to Sandy Hook for emplacement in the three gun battery completing there. A breakdown within the overall contract gave a price of $29,812 for the 8-inch gun and ten shells.

The gun tube was of four lengths of lightweight wrought iron tubes, with a continuous inner brass sleeve. The breech mechanism contained the inlet valve for pressurized air, stored in a firing tank. The tube was held by trunnions between cast iron uprights. At elevation of 35-degrees the 1000-pound air pressure could throw a 60-pound shell two and one-quarter miles, the 100-pound version 3000 yards. The shells were of seamless brass with either pine or metal tail and a point of cast iron. Of course, the gun was smoothbore without rifling.

The dynamite battery under construction at Sandy Hook (later named Fort Hancock) was intended for two 15-inch and one 8-inch dynamite gun. The facilities had a steam plant for compressing air and numerous pressurized air reservoirs to be supplied to all three guns. The 8-inch mount was to the flank of the two 15-inch guns, all three emplaced parallel in orientation. Construction work for the emplacement and mounting of the gun was accomplished in mid-1896. After test firing, the battery was placed into a reduced operating status. Even during the Spanish American War, it does not appear to have been manned. One of the fears by the Corps of Engineers all along was that use of these shells exploding in water would accidentally detonate the defensive minefields planted offshore. On June 5, 1901 it was declared obsolete by the Board of Ordnance and Fortification. Within a year steps were taken to dismantle and salvage the gun and carriage.

Later in the early 1920s the emplacement was used as a mining casemate. As modified it still exists, though little is recognizable as a dynamite gun emplacement. The single weapon produced was scrapped. While there are plans of the gun and carriage, photographs of it either during evaluation or as emplaced are very scarce.

8-inch Seacoast Dynamite Gun Model 1893
On Carriage Model 1893 for 8-inch Dynamite Gun

Gun Characteristics
Bore size: 8-inch
Construction: Wrought iron tube with brass lining
Tube length overall: 720-inches
Nominal caliber: 60 caliber
Tube and breech weight: n.a.
Usual carriage: 8-inch Model 1893
Dynamite Gun Carriage
Weight of gun and carriage: n.a.
Maximum elevation: 35-degrees
Loading: Shell only
Shell weight: (size to carry 100-pounds explosive)
Maximum muzzle velocity: n.a.
Maximum range: 3,000-yards

Builders: Guns: Pneumatic Torpedo and Construction Company
Carriages: Pneumatic Torpedo and Construction Company

Number built: 1 Gun (modernized pilot Model 1886) and Carriage

Navy photograph, undated and not noted for location, of an 8-inch Dynamite Gun—possibly the unit later installed by the Army at Fort Hancock. It is the only known photograph of this size dynamite gun.

15-inch Seacoast Dynamite Gun Model 1890
On 15-inch Center Pintle Barbette Model 1890

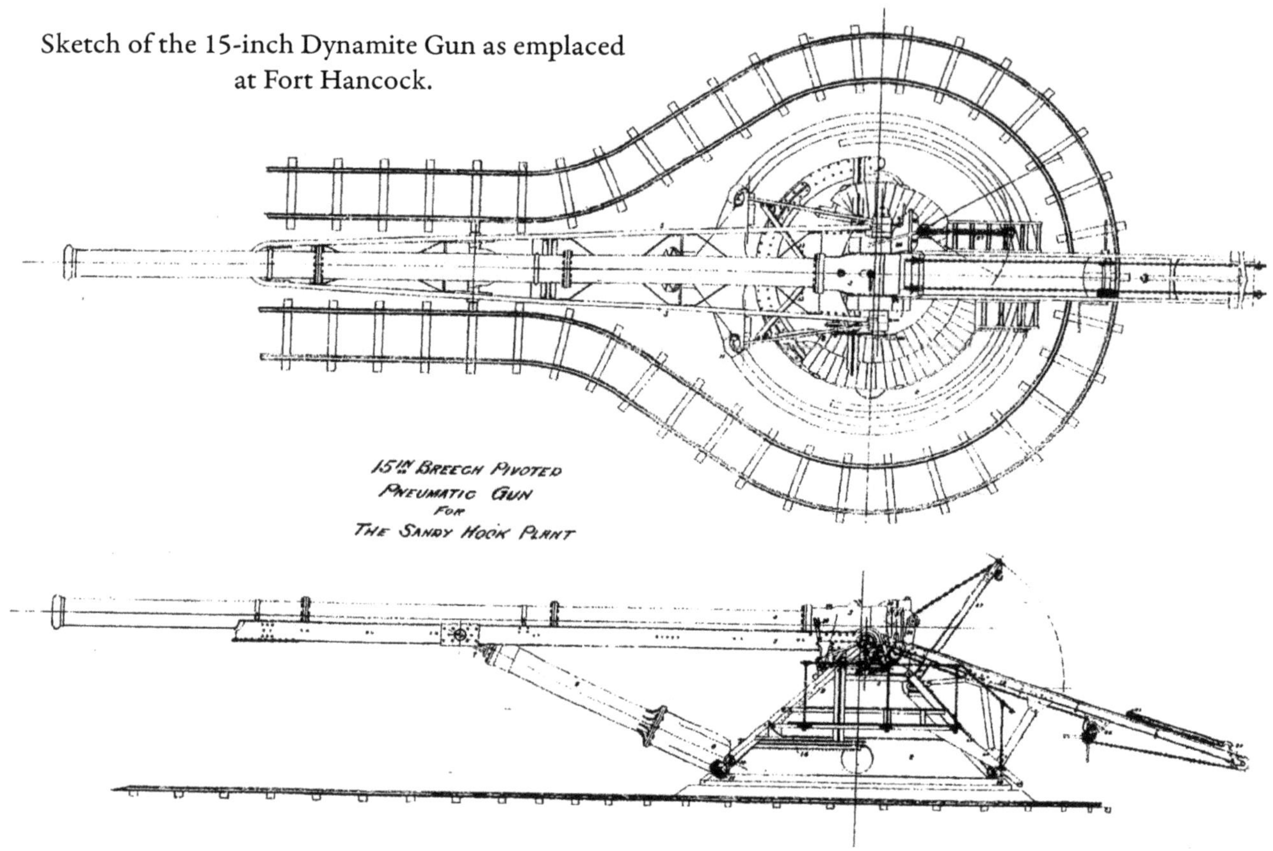

Sketch of the 15-inch Dynamite Gun as emplaced at Fort Hancock.

The Pneumatic Gun Company developed its first 15-inch dynamite gun in 1886. Initially intended for use in Italy, instead it became the prototype Model 1886 weapon evaluated for U.S. Service. This gun was tested at Fort Lafayette in a temporary emplacement. Congress began the acquisition of five 15-inch pneumatic dynamite guns in 1888. After considerable teething problems and corporate financial problems, they were completed by subcontractor West Point Foundry on behalf of the Pneumatic Gun Company in late 1895 and early 1896. Included in the first five were the original Model 1886 gun modified with new, electric training gear. All five were designated as Model 1890 types.

They were smoothbore, 40-caliber of lightly constructed wrought iron tubes with brass linings. Designed to fire at 35-degree elevation, they were mounted with trunnions on turntables allowing 360-degree traverse. Shells were 1097-pounds, 550-pounds being nitro glycerin & gun cotton explosive. It was specified to have a range of 2,000-yards with full charge, but smaller shells of 6, 8, and 10-inch diameter and longer possible range were also developed. A rate of fire of 45 rounds/hour was projected but realistically was found to be less than half that at one round per three minutes. The shells were of relatively light, cylindrical metal construction. However, they were provided a lengthy tail (the shell was over 11-feet long) to facilitate aerodynamics of flight. They had a complicated, electrical contact or immersion fuse.

The emplacement had to be equipped to generate the large volume of required compressed air. That involved a large in-battery steam plant with boilers, compressors, and heavy-pressure firing and replacement storage reservoirs (eight required per 15-inch gun). While it was possible to generate 2,000-psi pressures, usually just 1,000-psi was used for firing.

The first two large batteries were contracted for in early 1889 with the Pneumatic Gun Company. After numerous delays and a financial reorganization, the contracts were extended to the successor Pneumatic Torpedo and Construction Company in 1894. A battery for two 15-inch and a single 8-inch dynamite gun was completed at Sandy Hook (Fort Hancock) and placed in service by the end of 1896. Close behind was a congressional authorization for a battery of three 15-inch guns for San Francisco. Built on the bluff on the west side of the Presidio reservation (Fort Winfield Scott), it was constructed in 1894-96. The weapons were received and mounted in June 1897. For San Francisco during the Spanish American War additional magazines and protective parapets were added for an additional $150,000 of expense. The batteries were declared obsolete in February 1901 and scrapped by December 1904. Two batteries for slightly improved Model 1893 guns and carriages were begun later. A single 15-inch gun battery was begun at Race Point, Fort H.G. Wright, Long Island Sound in 1897. While the single gun was emplaced, completion of the berm and supporting rooms was not yet finished when work was suspended in 1898. At the same time a battery for a 15-inch gun at Hilton Head Island, defenses of Port Royal Sound made some progress, but never received any armament

No 15-inch dynamite gun still exists. In later years the Sandy Hook battery was modified and used as a mining casemate, and the San Francisco battery reused for the post power plant, fire control stations and the Harbor Entrance Command Post. As modified the emplacements still exist, as do partial remains of the South Carolina steam plant and the foundation of the Long Island site.

15-inch Seacoast Dynamite Gun Model 1890 On Carriage Model 1890 for 15-inch Dynamite Gun

Gun Characteristics
Bore size: 15-inch
Construction: Wrought iron tube with brass lining
Tube length overall: 600-inches
Nominal caliber: 40 caliber
Tube and breech weight: 40,000-pounds
Usual carriage: 15-inch Model 1890 Dynamite Gun Carriage
Weight of gun and carriage: 76,000-pounds
Maximum elevation: 35-degrees
Loading: Shell only
Shell weight: 1097-pounds with full charge (500-pounds)
Maximum muzzle velocity: n.a.
Maximum range: 4,800-yards

Builders: Guns: Pneumatic Torpedo and Construction Company
Carriages: Pneumatic Torpedo and Construction Company

Number built: Six completed guns and carriages, two others suspended during construction

– 15" PNEUMATIC DYNAMITE SEA COAST GUN –

1890 Annual Report of the Chief of Ordnance sketch of the 15-inch Dynamite Gun in its protected revetment.

The 15-inch Dynamite Gun firing.

15-INCH PNEUMATIC DYNAMITE GUN, AS INSTALLED AT SANDY HOOK, NEW YORK, AND THE ENTRANCE TO SAN FRANCISCO HARBOR.
Range, with 1,000-pound shell, 2,400 yards; with 240-pound shell, 6,000 yards.

Anti-Aircraft Artillery

First World War Anti-aircraft Guns

Only one project for an anti-aircraft gun had begun prior to America's entry into the First World War. A relatively powerful 3-inch gun on a fixed pedestal was in development. Based more or less on the previous seacoast Model 1903, the project eventually became the Model 1917 anti-aircraft gun and enjoyed a relatively successful career. However, a fixed gun was not what was needed for protection of the army in Europe. Mobile units deployable in the field were lacking, and no immediate Ordnance Department project was available to satisfy this need.

Consequently, two new projects were soon begun, one for a lighter (3-inch) and one for a heavier (4.7-inch) new mobile gun. The 3-inch gun was based on using features of the breech and the same ammunition chamber as the smaller 3-inch seacoast Model 1898 gun. The 4.7-inch project, never brought to full completion, was simply the same basic gun scaled larger. Both of these were to be trailer-mounts, being towed to their deployment sites. Good progress was made on the 3-inch Model 1918 gun, and some initial units were being completed just as the war ended. It served in the interwar years, though ultimately just as a training weapon.

To meet the immediate need of the American Expeditionary Force, other steps had to be taken. Fifty mounts were quickly improvised to carry the standard French 75mm field gun. The mount was transportable on truck but deployed on the ground for operation. American-made versions of the 75mm gun were not yet in production, so while the mounts were fabricated in the U.S., they had to be married up with guns provided by the French once in Europe. The mount was awkwardly designated the "75mm Improvised Anti-aircraft Carriage Model 1917". A number did serve in France before the war's end, but they did not survive long thereafter

The final expedient was for the mounting of a standard American Model 1916 75mm field gun on a true truck mount. In this case the existing White truck was equipped with extendable jacks to stabilize the mount for firing from the truck body. Thus, it was much more rapidly deployable than the French 75 mount described above. The design was adopted in May 1917, and 51 mounts were delivered in the fall and early winter of 1918. A total of 22 units were shipped in 1918 to France. These units persisted into the interwar period, some units assigned for training at universities, not being scrapped until 1940.

The last produced project was the original 3-inch fixed anti-aircraft gun. Designed in 1916, a total of 160 guns and fixed carriages were produced through 1919. It used an entirely new 55-caliber gun, based on the successful Model 1903 3-inch seacoast gun, married to a simple pedestal base allowing high elevation and complete traverse. The guns were produced solely to protect existing coast artillery forts and not otherwise distributed to military, government, or strategic locations. It survived in service well into the Second World War.

As previously mentioned, the development of the new mobile 3-inch AA gun continued during the war years. It was of 3-inch bore, the gun based on the 40-caliber Model 1898 seacoast gun. It was placed on a new high-angle carriage on a towed carriage from which it could be fired. Eventually designated the Model 1917 gun on Model 1918 carriage, over 600 were ordered late in the First World War, though only 120 were eventually accepted. It served in separate new coast artillery AA regiments in the interwar years; some surviving into the next war and some being contributed to allies.

None of these types were distributed as war memorial pieces, and even as museum pieces were not deemed significant enough to preserve. Of all these types only a single Model 1917/18 trailer gun survives, appropriately at the U.S. Army Air Defense Artillery Training Support Facility at Fort Sill.

75mm Gun Model 1897
On Improvised Anti-Aircraft Carriage Model 1917

Several different attempts were made to utilize the common 75mm M1897 field gun in an anti-aircraft role during the World War conflict. In this Army photograph it appears the gun is simply used on a platform allowing it to fire "up" without much capability to change elevation or traverse.

An urgent request was placed by General Pershing on behalf of the American Expeditionary Force for mobile anti-aircraft guns almost from its first arrival in France. While some 75mm AA gun mounts using the French 75mm field gun were directly acquired by purchase from the French, this early project was initiated by the Ordnance Department to come up with an expedient solution. There was a small project already underway to configure an American 3-inch Model 1905 field gun to a new carriage to provide high-angle fire against aircraft (consequently known as the 3-inch Improvised anti-aircraft carriage Model 1917). This design was reconfigured to take the Model 1897 75mm field gun then being acquired by the Army in France, and the designation changed accordingly.

The resultant carriage was essentially a disassembled three-piece base ring supporting two strong cross members mounted on the top of the circular ring. Mounted on this scaffolding and capable of traversing was a standard 75mm Model 1897 gun with its top cradle and recoil mechanism. The gun tube was exactly the same as the field gun, retaining the same length, caliber, rifling and performance. It fired a French-produced anti-aircraft shell, designated by the Americans as the AA Shell Mark III of 14.3-pounds with a 1.66-pount TNT explosive charge. Originally the mount was equipped with open sights, but later U.S. telescopic sights were attached. The gun carriage and limber were transported on 3.5-ton trailers towed by trucks. In position they could elevate from 21 to 85-degrees and had a 360-degree traverse. Once dismounted from the truck (the gun was carried upside-down), a pit had to be excavated for the carriage 109-inches in diameter and 67-inches in depth. While transportable between fixed sites, the resulting gun could hardly be called a "mobile" AA mount.

The design was completed on May 1, 1917. Almost immediately the first pilot carriage was ordered from Rock Island Arsenal on May 25, 1917. Even without testing, another 50 production units (serial No. 2-51) were ordered from Builders Iron Foundry on June 9, 1917. There were no problems in making the mounts, deliveries began that same fall. Shipment was made to France starting on March 26, 1918.

Once in France they were equipped with French-made 75mm Model 1897s and issued to troops. A total of seven anti-aircraft battalions were organized, most equipped each with four batteries of two guns each. The majority of these had the new improvised anti-aircraft gun. In service it was found that the awkward carriage framework took an excessive amount of time to erect. Some later descriptions even mentioning two weeks as necessary to properly prepare the position, mount, and prepare the gun. Hence, they were only used for static defense of important bases. They were not popular with their operators and delivered only mediocre success in their assigned role. After service they were returned to the U.S., starting in March 1919. General Pershing also directly purchased small numbers of complete French 75mm AA guns; a type using a ground-mounted, revolving pedestal with large outrigger feet.

These mounts were never considered anything but expedients, and no provision was made to incorporate them in the postwar army, in fact it is a little surprising that they were returned at all and not scrapped in France. The Secretary of War authorized that the 50 mounts be scrapped on October 1, 1919. One was kept into at least the 1930s as a museum piece at Aberdeen Proving Grounds but was probably scrapped during the Second World War. No examples survive today.

75mm Gun Model 1897
On 75mm Improvised Antiaircraft Carriage Model 1917

Gun Characteristics
Bore size: 75mm
Construction: Built-up Nickel Steel
Tube length overall: 107.13-inches
Nominal caliber: 34.5
Tube and breech weight: 1015-pounds
Usual carriage: 75mm Improvised Antiaircraft Carriage Model 1917
Weight of gun and carriage: n.a.
Maximum elevation: 85-degrees
Loading: Fixed
Shell weight: 14.3-pounds
Maximum muzzle velocity: 1,805-fps
Maximum range: 8,202-yards vertical

Builders: Guns: Various French arsenals
Carriages: Rock Island Arsenal (1)
Builders Iron Foundry (50)

Number built: 51

The American solution; an improvised design (Model 1917) utilizing a tub to hold the 75mm M1897 top carriage and allowing for quick traversing. Fifty units of this type were procured.

The French-developed specialized carriage to hold a 75mm M1897 for anti-aircraft use. It is much more sophisticated (and thus probably expensive) compared to the previous expedient.

75mm Gun Model 1916
On 75mm Anti-aircraft Truck Mount Model 1917

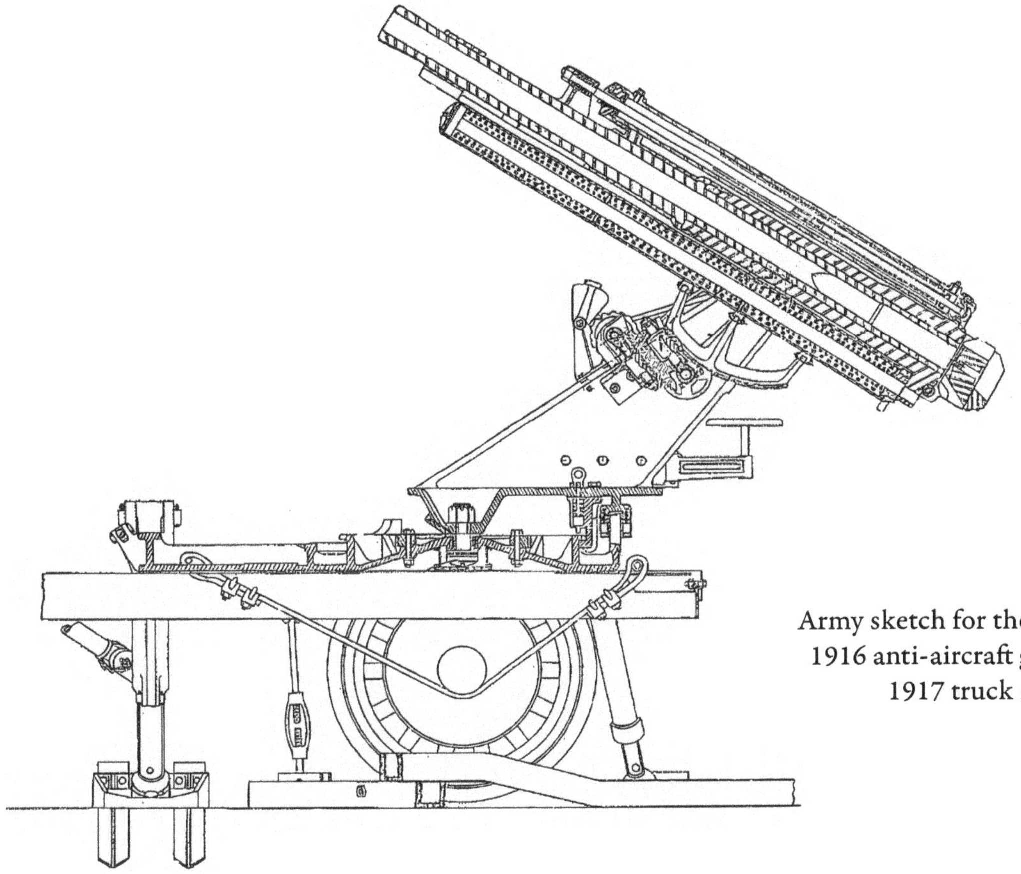

Army sketch for the 75mm Model 1916 anti-aircraft gun on Model 1917 truck mount.

Started at the same time as the 75mm Improvised Mount was this unit for a more mobile piece. It essentially used the 75mm field gun Model 1916 already entering production. The gun with its top carriage and hydrospring recoil system was emplaced on a simple, but heavy revolving base plate attached to the rear of a White motor truck. In travel the gun faced to the rear with the breech directly behind the driver. However, to allow clearance for the recoil the gun could only be fired from a position that swiveled out behind the truck. In this position it was stabilized with extendable jacks. Elevation was by worm gear and a rack operated from a hand wheel. On its truck mount the gun had a 240-degree traverse and an elevation of 31 to 80-degrees (some official sources list an 82-degree maximum elevation). The field gun mounted on its truck mount could give a vertical range elevation of about 10,000-yards and a horizontal range of 6,200-yards.

The guns came from the standard production lines at Watervliet Arsenal. It was the standard 75mm, 28.4-caliber tube with 24 grooves. The gun retained its semi-automatic breech mechanism. The recoil mechanisms were made by the New Britain Machine Co. They also were charged with assembling the gun with recoil mechanism carriage to the White 1-1/2 ton trucks. White referred to the truck type as their model 20. The trucks had a 157-inch wheel base. The gun fired fixed ammunition. Typical for these early anti-aircraft guns, it had available timed-fused high explosive and shrapnel rounds.

The design for the mount was finalized on May 1, 1917. One pilot model was assembled at Rock Island Arsenal. A production order for 51 was placed in July 1917 with New Britain Machine Co. All 51 were delivered in the fall and early winter of 1918, and a total of 22 were available in France, but not issued before the Armistice. It was recognized that this anti-aircraft gun was only an untried expedient, no follow-up orders were anticipated.

The resulting gun was not deemed successful. The limits to traverse and elevation meant that target coverage and tracking suffered. Also, it was found that the mechanisms to elevate and rotate were prone to

75mm Gun Model 1916
On 75mm Antiaircraft Truck Mount Model 1917

Gun Characteristics
Bore size: 75mm
Construction: Nickel steel, built up
Tube length overall: 90.9-inches
Nominal caliber: 28.4 caliber
Tube and breech weight: 765-pounds
Usual carriage: 75mm Antiaircraft Truck Mount Model 1917

Weight of gun and carriage: 3,300-pounds (9,500 including truck)
Maximum elevation: 80-degrees
Loading: Fixed
Shell weight: 14.3-pounds
Maximum muzzle velocity: 1750-fps
Maximum range: 10,000-yards vertical

Builders: Gun: Watervliet
Carriage and Assembly: New Britain Machine Co.

Number built: 51 + 1 pilot model

jamming. The entire rig, even using the supplied support jacks, was somewhat unstable for gun and crew. Ultimately the field gun itself did not have the muzzle velocity to truly function as a serviceable anti-aircraft gun. The guns were not issued to active units in France. After the war they were used for several years for National Guard and University ROTC training. Finally on December 27, 1934 they were declared obsolete and authorized for disposal. It appears that the last two Model 1916 AA guns on Model 1917 Truck mounts were recalled from universities as late as April 1940 and scrapped.

None of these combination Model 1916 guns on their truck-mounted carriage still exist, though a very few of the guns on their normal field mounts survive.

Proving Ground photograph of the 75mm M1916 field gun mounted on its Model 1917 truck mount. Fifty-one gun & vehicle units of this type were acquired by the Army.

3-inch Anti-aircraft Gun Model 1917
On 3-inch Anti-aircraft Mount Model 1917

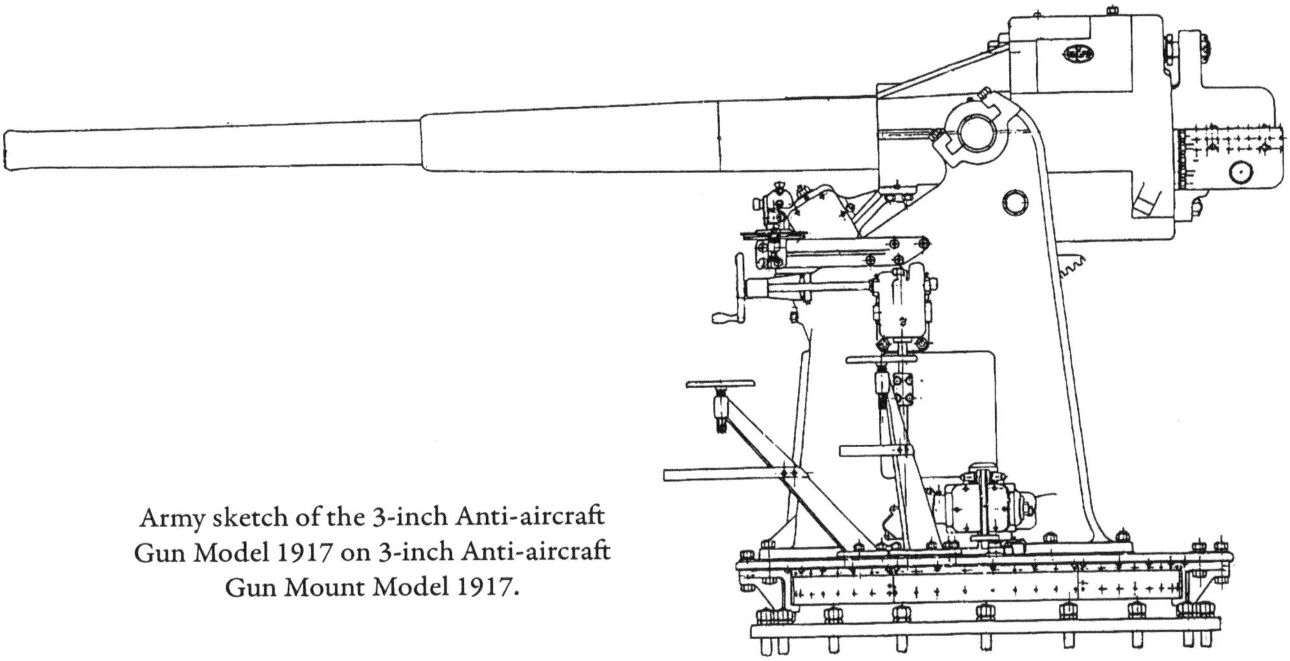

Army sketch of the 3-inch Anti-aircraft Gun Model 1917 on 3-inch Anti-aircraft Gun Mount Model 1917.

The first American fixed anti-aircraft gun design was finalized in April 1916. By the summer of that year Watertown began manufacturing a pilot mount, known as the model "E". The tube was a long 55-caliber, with a drop-block breech mechanism. Continuous-pull type of firing was used, achieved by a firing handle. The relatively simple pedestal mount allowed elevation of 0 to 90-degrees, and a full 360-degree traverse. Recoil was checked with a single hydrospring cylinder mounted above the tube, and two counter recoil spring cylinders below. The tube was balanced using a breech weight. The carriage with its pivot yoke and racer were designed to be emplaced on fixed platforms. For a foundation a concrete block 12-foot in diameter held sixteen hold down bolts emplaced in concrete in a 61.5-inch diameter circle. The gun fired a 15-pound projectile with a muzzle velocity of 2,600-fps. This gun and the subsequent M2 fixed 3-inch gun used the same ammunition, also in common with the Model 1903 seacoast gun. Fuses were set manually on the adjacent M1916 fuse setter.

Orders were placed for 160 of the new guns and pedestals; that number corresponding to an evaluation of how many anti-aircraft guns were needed to defend coast artillery forts. Watervliet Arsenal made all the guns (all forgings coming from Midvale Steel), and Watertown and Bethlehem shared the production of pedestal mounts. The first production gun was delivered in December 1917. Never envisioned as being needed for the AEF in France, the production did not receive the same priority as other projects—though 116 of the 160 were delivered through April 1919. Five guns were temporarily mounted on special front-wheel drive trucks and assigned to the chief of Coast Artillery for mobile defense of cities or depots.

During fabrication some design changes were introduced. The Model 1917M1 was generally similar but the breech ring was a separate piece from the jacket, and the locking hoop was omitted. The Model 1917MII had a different method securing the latch plate—using a lug rather than screws. For the mount, the Model 1917M1 added a second recoil cylinder above the tube, giving it a more cluttered appearance easily recognizable in period photographs. These were modifications of existing mounts; new ones not being produced. The mount Model 1917MII was also a modification and was significantly different. The trunnion seats were raised by using filler pieces, improving the elevating and traversing mechanism, and installing new roller bearings. All of the mounts could accommodate any of the model guns.

Originally guns were organized into batteries of two weapons, but later that was expanded to three. Individual concrete gun blocks were usually spaced 150-feet apart from each other. In some more permanent locations elaborate, deep concrete magazines were also constructed. They were usually manned by the same post coastal artillery unit that operated the fixed, seacoast guns rather than by separate AA regiments. Most of the more important coast defense forts in the continental U.S. and overseas departments received one or two batteries of these guns. They served through the 1920s and 1930s, well into the Second World War. In September 1943 146 of these guns were still in U.S. service (two had been lost when the Philippines fell). In January 1944 all remaining M1917 guns were reduced to Class "C" Status and not required to be manned. They were declared obsolete (probably at the end of the war in 1945) and authorized for disposal. Apparently, none made their way to memorials or museums. There are no known survivors of the Model 1917 3-inch Anti-aircraft gun.

3-inch Anti-aircraft Gun Model 1917 On 3-inch Antiaircraft Gun Mount Model 1917

Gun Characteristics
Bore size: 3-inch
Construction: Alloy steel, built-up
Tube length overall: 174.65-inches
Nominal caliber: 55 caliber
Tube and breech weight: 3,105-pounds
Usual carriage: 3-inch Antiaircraft Gun Mount Model 1917
Weight of gun and carriage: 15,280-pounds
Maximum elevation: 90-degrees (in practice limited to 85-degrees)
Loading: Fixed
Shell weight: 15-pounds
Maximum muzzle velocity: 2,600-fps (shrapnel)
Maximum range: 9,000-yards vertical

Builders: Gun: Watervliet Arsenal
Carriage: Watertown Arsenal (138) and Bethlehem Steel (22)

Number built: 160 + 1 pilot

The Model 1917 AA guns and mounts were generally assigned in batteries of two (later three) at major coast artillery forts around the country and foreign possessions. This is a late inter-war M1917 in the defenses of Chesapeake Bay, VA.

3-inch M1917 gun on a M1917 Anti-aircraft gun mount in a standard emplacement on Morrison Hill on Corregidor.

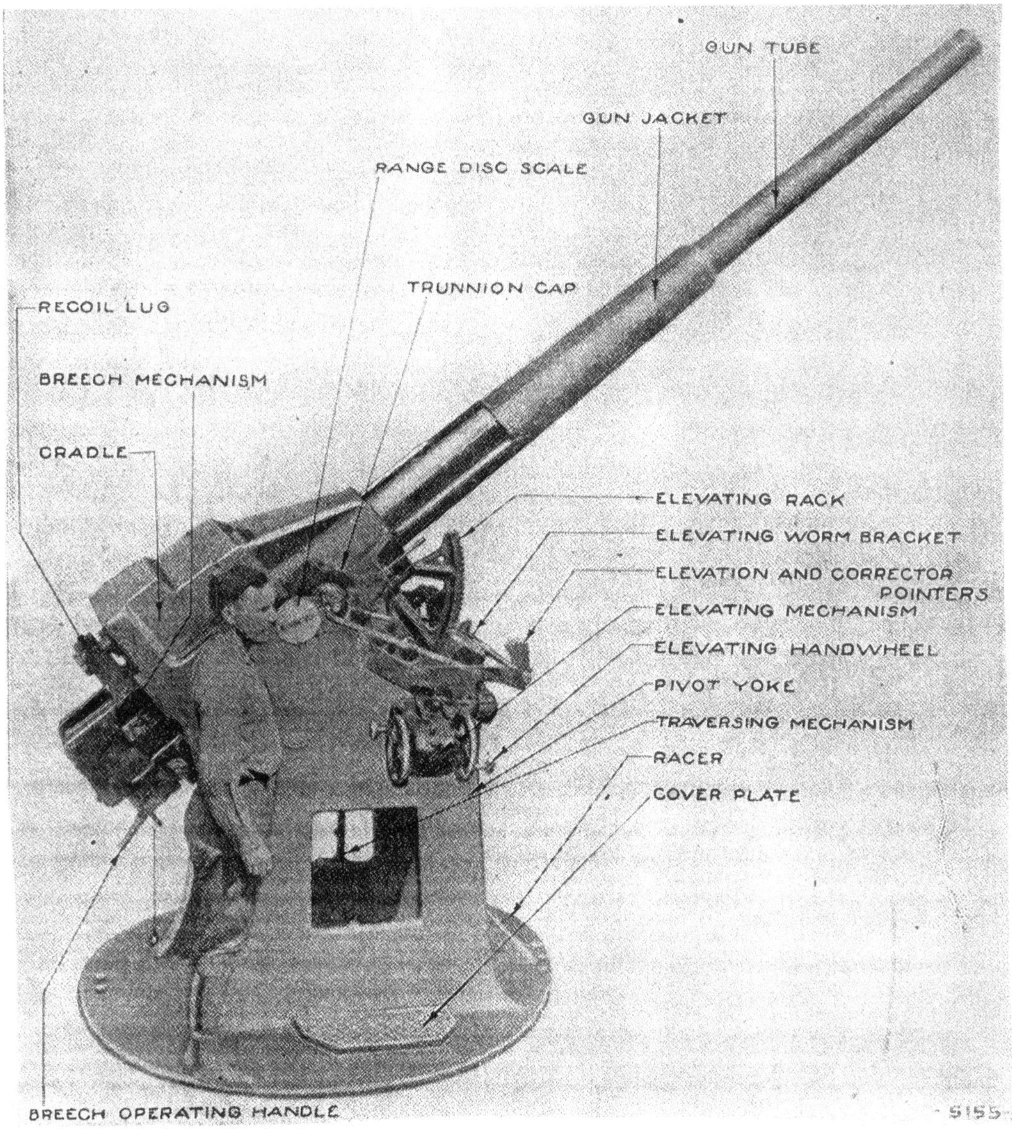

An Army Ordnance manual illustratrion of a 3-inch M1917 gun on a M1917 Anti-aircraft gun mount

3-inch Anti-aircraft Gun Model 1918 On 3-inch Trailer Mount Model 1917

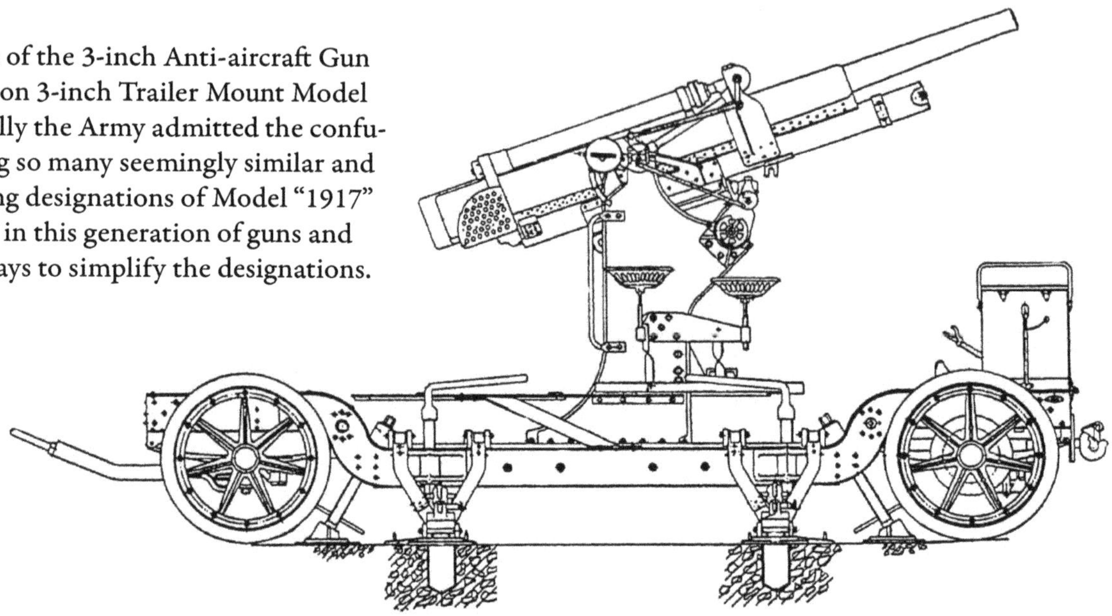

Army sketch of the 3-inch Anti-aircraft Gun Model 1918 on 3-inch Trailer Mount Model 1917. Eventually the Army admitted the confusion of having so many seemingly similar and thus confusing designations of Model "1917" and "1918" in this generation of guns and looked for ways to simplify the designations.

This was the project the Ordnance Department had the highest hopes for. The gun was designed to use already proven features. It had the same powder chamber, cartridge case and ballistics of the previous coast artillery Model 1898 gun. All of the Model 1918 guns were built new for this project. It had a 40-caliber nickel steel built-up tube, jacket and breech ring; rifling was increasing from 1 in 50 to 1 in 25. Using the fixed ammunition, it had 2,400-fps muzzle velocity with shrapnel and 2,600 with HE. The breech was a semi-automatic operating drop-block type. Recoil was braked with a variable hydrospring system for different elevations. It used sights and pointing equipment set to the target.

The carriage sat on a 4-wheeled trailer. During firing it was stabilized by extending ground jacks, though it was found to be too unstable in use. Eventually the army developed pre-designated firing positions of two parallel concrete beams set in the ground to which the trailer carriage could be secured; of course, only in certain situations could these really prove useful. An ammunition chest for 16 rounds usually was also carried by the trailer. Either a truck or tractor could pull the trailer, which had a 156-inch wheel base. The mounting allowed a full 360-degree traverse, and the gun could elevate from 10 to 85-degrees. Rate of fire was calculated to be 10-12 per minute.

After developing the pilot (delivered to the Army in August 1917), relatively large numbers of guns were ordered, in fact even prior to the pilot's completion in July 1917. Then on December 8, 1917 orders were placed for the carriage mounts. At one point 612 complete units were under contract. The guns were made by the Chalkis Manufacturing Co., the trailer by the Grant Motor Car Corp., and assembly done by the New Britain Machine Co. It appears that after orders were revised at the end of the war, only 120 Model 1918 guns and trailer mounts were actually delivered. The last delivery appears to have been in early 1920. The guns were routinely issued to new Coast Artillery Regiments (AA) being formed shortly after the end of the war. They served with units in the U.S. as well as the Philippines and Hawaii.

While not as severe as the earlier improvised mounts, there were problems with the Model 1918 carriage. As mentioned above it was never as stable as needed when firing from the trailer with jacks. The recoil system could act erratically. Loading at high angles was very difficult. Also, the trailer was fine on roads, but not in rough terrain. Still, it was the only usable mobile gun for the army in the 1920s and much of the early 1930s. In 1924 the nomenclature was changed, the mount and trailer being changed to the Model 1918 vs. their original Model 1917 designation—no doubt to avoid confusion with the fixed Model 1917 gun and mount.

These guns had a surprisingly long life in the interwar period. They fired the same ammunition as the new 3-inch M3 AA gun, which helped. In overseas possessions they proved useful as training and reserve material. In 1940 there were still 116 reported in inventory—52 overseas, 64 in the continental U.S. and there were 51 spare tubes still on hand. Though considered not nearly modern enough, such was the shortage of AA materiel that they were retained as war reserve equipment. Twenty-four of the guns and trailers were transferred to the British Middle East forces in the summer of 1941. Unlike the other First World War Program anti-aircraft guns, there is a survivor of this type. The only known existing AA gun Model 1918 on truck mount Model 1917 or 18 is with the Fort Sill U.S. Army Air Defense Artillery Training Support Facility.

3-inch Anti-aircraft Gun Model 1918
On 3-inch Trailer Mount Model 1917

Gun Characteristics
Bore size: 3-inch
Construction: Nickel Steel, built-up
Tube length overall: 129.69-inches
Nominal caliber: 40 caliber
Tube and breech weight: 1939-pounds
Usual carriage: 3-inch Trailer Mount Model 1917
Weight of gun and carriage: 13,200-pounds
Maximum elevation: 85-degrees
Loading: Fixed
Shell weight: 15-pounds
Maximum muzzle velocity: 2,400-2,600 fps
Maximum range: 8,530-yards

Builders: Gun: Chalkis Manufacturing Co.
Trailer: Grant Motor Car Corp.
Mount and Assembly: New Britain Machine Co.

Number built: 120

3-inch AA gun on a M1917 trailer mount

A 1926 photograph of the 3-inch Model 1918 AA gun and its trailer.

In contrast to the earlier examples of anti-aircraft guns, there does survive one well-preserved example at the Air Defense Artillery Training Support Facility at Fort Sill, Oklahoma.

4.7-inch Anti-aircraft Gun Model 1918
On 4.7-inch Anti-aircraft Trailer Mount Model 1918

Only four of the new Model 1918 4.7-inch Anti-aircraft gun on Model 1918 trailer carriage were ever produced. This photograph was taken in 1919 at Watertown Arsenal.

This was a First World War American project to provide a larger-sized, but still mobile anti-aircraft gun. It was started in tandem with the 3-inch Model 1918 project. Similar to most of the 3-inch gun, it was basically to be simply an enlarged version of the same to provide a higher ceiling and larger explosive charge. General Pershing requested a gun of 4-inch bore in November 1917, but a month later a 4.7-inch size was adopted for the project. The initial version was designated the 4.7-in Anti-aircraft Gun Model 1918. It had a built-up steel tube of 40-caliber length. An extended loading tray for the fixed ammunition extended out beyond the breech. Recoil was controlled by a hydraulic cylinder mounted above the tube. Development work was conducted in 1917-18. A four-wheeled trailer weighing about 10,000-pounds was adopted to carry the rather cumbersome gun and carriage, which weighed another 17,000-pounds.

Due to wartime perceived needs, production was ordered even before adequate testing could be done on the new gun. Four sets of forgings for the gun were ordered on January 31, 1918 from Standard Forgings Co. of Chicago. The guns were to be assembled and finished at Watervliet Arsenal. The trailer mounts were built by the New Britain Machine Co. As yet no pilot or prototype gun had been acquired. Quickly the number of guns ordered increased, to 260. But before any could be produced the war ended, and on December 9, 1918 orders were cut to just 80. Interestingly Watervliet was told to then suspend work until a pilot gun could be tested and approved. Apparently, that is when things fell apart. During testing of the prototype, the recoil mechanism broke. Along with other revealed defects, the gun was disappointing in terms of muzzle velocity and range. Gun construction—quite advanced on at least the initial four ordered in early 1918—was indefinitely suspended.

Efforts were undertaken to improve the gun's performance. Further development resulted in the 4.7-in Anti-aircraft gun Model 1920. With a slightly higher muzzle velocity (2,600-fps) it could achieve an extreme range of 20,030-yards. Photographs of the gun show it equipped with a large early type of muzzle brake. The carriage was rebuilt to carry two large equilibrator cylinders on its forward extension.

In the early 1920s the project changed direction. Emphasis was now on perfecting better mobility for a gun of this size, and several experimental self-propelled mounts were developed. A new mobile carriage using the Christie system of track suspension was developed and designated the Model 1921 motor carriage. The gun, complete with its muzzle brake was mounted on a flatbed between the tracks, and an open driver's position was established on the rear end of the bed. The unit was credited with a length of 272-inches and a total weight of 42,200-pounds. It was equipped with a 120 horsepower engine, though capable of moving at only slow speeds. Evaluation of a pilot unit continued through 1925.

Despite these efforts none of the original 4.7-inch Model 1918 or its improved Model 1920 were standardized, put in active production, or issued to units. The project was terminated on January 15, 1925, though some of its findings were no doubt useful when a new 120mm anti-aircraft project started in the late 1930s. None of these pilot or early pre-production units still exist.

4.7-inch Anti-aircraft Gun Model 1918 On 4.7-inch Anti-aircraft Trailer Mount Model 1918

Gun Characteristics
Bore size: 4.7-inch
Construction: Alloy steel, built-up
Tube length overall: 201.5-inches
Nominal caliber: 40
Tube and breech weight: 6,945-pounds
Usual carriage: 4.7-inch Antiaircraft Trailer Mount Model 1918
Weight of gun and carriage: 17,000-pounds
Maximum elevation: 80-degrees
Loading: Fixed
Shell weight: 45-pounds
Maximum muzzle velocity: 2,400-fps
Maximum range: 11,500-yards

Builders: Gun, Watervliet Arsenal
Carriage, New Britain Machine Co,

Number built: Probably 4

The Model 1918 4.7-inch Anti-aircraft gun at full elevation in the shop of producer New Britain Machine Co. in 1918.

Heavy Anti-aircraft Guns 1925-1955

With the return of peace and absorption of the lessons of the World War, the Army turned to further development of anti-aircraft weapons. The wartime expedients were soon discarded, and while both the 3-inch mobile (Model 1918) and fixed (Model 1917) served for many more years, clearly more capable designs were needed. Initially the Ordnance Department turned to developing improved guns for both categories of 3-inch bore. A significant effort was made through the 1920s and 1930s to develop new 3-inch guns. While several designs were perfected and trial prototypes were made, only a single project achieved a significant service status. The mobile M2/M3 3-inch anti-aircraft gun began replacing the M1918 mobile 3-inch in the late 1930s. For a period of time right before and after the start of the Second World War, it was the best weapon the Army had for anti-aircraft defense. The department also pursued several projects for new anti-aircraft guns of 75mm bore, demonstrating its equivocation about the recent adoption of standardized metric measurements.

One project for a gun heavier than 3-inch did survive between the wars, and a few were actually produced. A new 105mm gun, essentially just a scaled-up version of the 3-inch, was developed in the early 1930s. Just one prototype and 14 production guns were acquired. Twelve were deployed in new emplacements specifically to defend the Panama Canal. While the design was eclipsed by the later heavy guns, they served through the war in the canal zone.

In the mid-1930s improvements in aircraft speed and altitude sparked a compensatory need for heavier AA guns. Three-inch guns simply didn't have the size for increasing weight and shortening time of flight. Most other nations were turning to guns of larger size—this period saw the development of the British 3.7-inch and the German 88mm anti-aircraft guns. In late 1937 the service formally asked for a new caliber of anti-aircraft gun with a heavier projectile and shorter time of flight. An on-going ordnance development project for a new 90mm gun was accelerated and the gun appeared in short time. By 1941 it was in production. The M1 gun and its successor, the automated M2 were both successful. Large numbers were made during the war and supplied to many allied countries during and after the war. The same gun was also adopted by the Coast Artillery (on the shielded mount M3) for its standard harbor anti-motor torpedo boat gun.

A project for a new 120mm heavy gun was also begun in the late 1930s. More intended for fixed defenses, the concept was in many ways a perpetuation of the belief that only the largest guns could effectively deal with high-flying strategic bombers. While never given the priority for production of the 90mm project, the development proceeded during the war without major problems. The gun went into production during the war but was mostly kept in the United States and offshore possessions. Its modern operation and fire control allowed it to function well into the postwar period, not being deleted until 1960. In this mission it was joined by a final project for a fully automatic 75mm anti-aircraft gun. Development started during the war, but the M51 "Skysweeper" gun did not enter production and deployment until the early 1950s. It too served in limited numbers during the early cold war but was eventually replaced in function by surface-to-air missiles in the early 1960s, though it served in certain allied armies for yet another decade.

The survival of heavy anti-aircraft guns has been spotty. Very few of the interwar 3-inch types (and none of the 105mm) still exist. However, the wartime 90mm types have proved to be a popular memorial and museum piece and more than a few persist. The later 120 and postwar Skysweeper types were never produced in large numbers and just a scattered few of each survive.

105mm Anti-aircraft Gun M3
On 105mm Anti-aircraft Mount M1

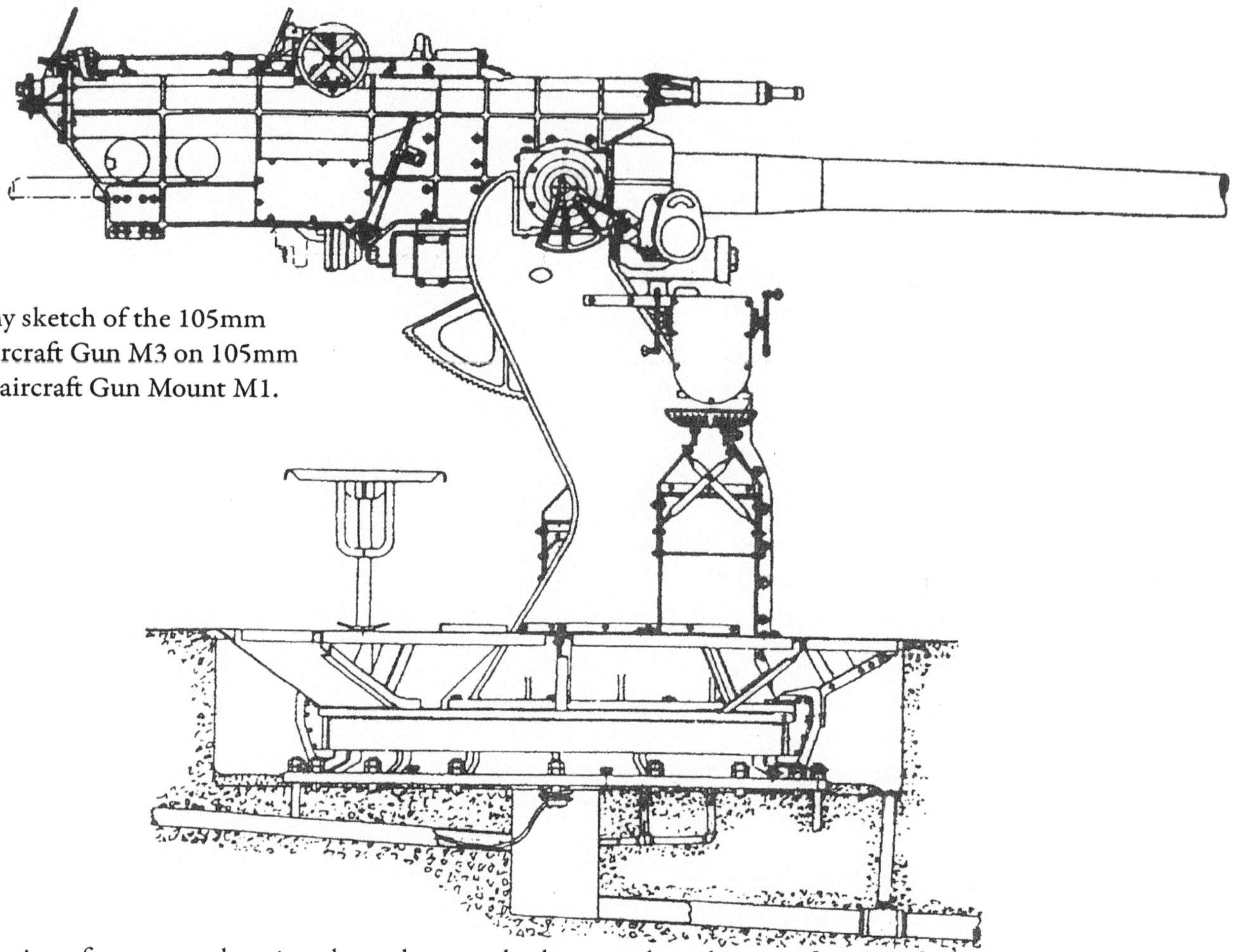

Army sketch of the 105mm Anti-aircraft Gun M3 on 105mm Anti-aircraft Gun Mount M1.

A project for a gun heavier than the standard 3-inch did not end with the abandonment of the post-war 4.7-inch gun. Eventually the Ordnance Department developed and produced a small number of 105mm guns in the late 1930s.

Development efforts began in 1924 for a new, larger anti-aircraft gun for fixed defenses and were completed in 1928 with standardized as the 105mm anti-aircraft gun M1. The gun was a longer, 60-caliber size. The barrel had 36 grooves, and rifling was a uniform twist of 1 in 30. This allowed for a larger 33-pound projectile with a useful vertical range of 12,000-yards and a relatively high muzzle velocity of 3,000-feet/second. The gun was equipped with a vertical, sliding breech block. Despite the promise of the design and continued need for something larger than the 3-inch gun, funding for production was not immediately forthcoming. When finally produced, changes in the barrel design (monobloc construction and a fixed, rather than loose liner) and a reduced muzzle velocity of 2,800-feet/second led to a new designation of the Model M3 gun. The M3 and M1 carriage were standardized in 1933. Two carriages had been produced at Watertown in 1929-30, and twelve more at Rock Island Arsenal in 1933-34.

The simple pedestal mount M1 utilized a cradle to hold the gun at a high trunnion level to allow for an elevation from -5 to + 80-degrees. Elevation and traverse were set by hand controls. The ammunition was power-rammed activated with compressed air for a relatively high rate of fire of 15-20 rounds per minute. The air compressor was recharged by the gun's recoil. A hydrospring recoil cylinder and two spring counter recoil cylinders were placed underneath the barrel. It was supplied with an M38 and an M41 high-explosive projectile. A fuse-setter was attached to the carriage base. The usual fuse was time-set.

The substantial size of the mount and unique outline of the supporting pedestal make it readily distinguishable from the 3-inch types produced at the same time. By the time the gun was finally put into production, newer designs were being developed, and the 105 was never selected for quantity production. There were some further development efforts made during the war, and some elements of the design influenced the tank gun of the same caliber.

A dozen simple emplacements were built for these guns in the Panama Canal Zone. Four batteries each were equipped with three guns. They were placed in a line; each gun emplacement was just an entirely open concrete pad with hold-down bolts. Two concrete, earth covered magazines accompanied each battery of three guns, along with a separate director station. Two batteries each of three guns were positioned on each of the Pacific and Atlantic entries of the canal. Apparently, the gun blocks were constructed in 1934, but the armament was not completed and ready for mounting until 1938.

The guns were formally declared obsolete in February 1945. Panama Canal defenses were still considered of high priority, but they were now receiving the new 120mm anti-aircraft gun. They were scrapped either late in the Second World War or shortly thereafter. No examples of the 105mm M3 gun or M1 carriage managed to survive. Several of the simple gun block emplacements in Panama still exist.

105mm Anti-aircraft Gun M3
On 105mm Anti-aircraft Gun Mount M1

Gun Characteristics
Bore size: 105mm
Construction: Alloy steel, monobloc construction
Tube length overall: 259.3-inches
Nominal caliber: 60 caliber
Tube and breech weight: 6,575-pounds
Usual carriage: 105mm Antiaircraft Gun Mount M1
Weight of gun and carriage: 19,845-pounds
Maximum elevation: 80-degrees
Loading: Fixed
Shell weight: 33-pounds
Maximum muzzle velocity: 2,800-fps
Maximum range: 42,000-feet (37,000-feet effective with time fuse)

Builders: Gun: Watervliet Arsenal
Carriage: Watertown Arsenal, Rock Island Arsenal

Number built: 14

Aberdeen Proving Ground photo of the 105mm Anti-aircraft Gun M 1927E (later standardized as the M3) in November 1926.

3-inch Anti-aircraft Gun M3
On 3-inch Anti-aircraft Mobile Mount M2

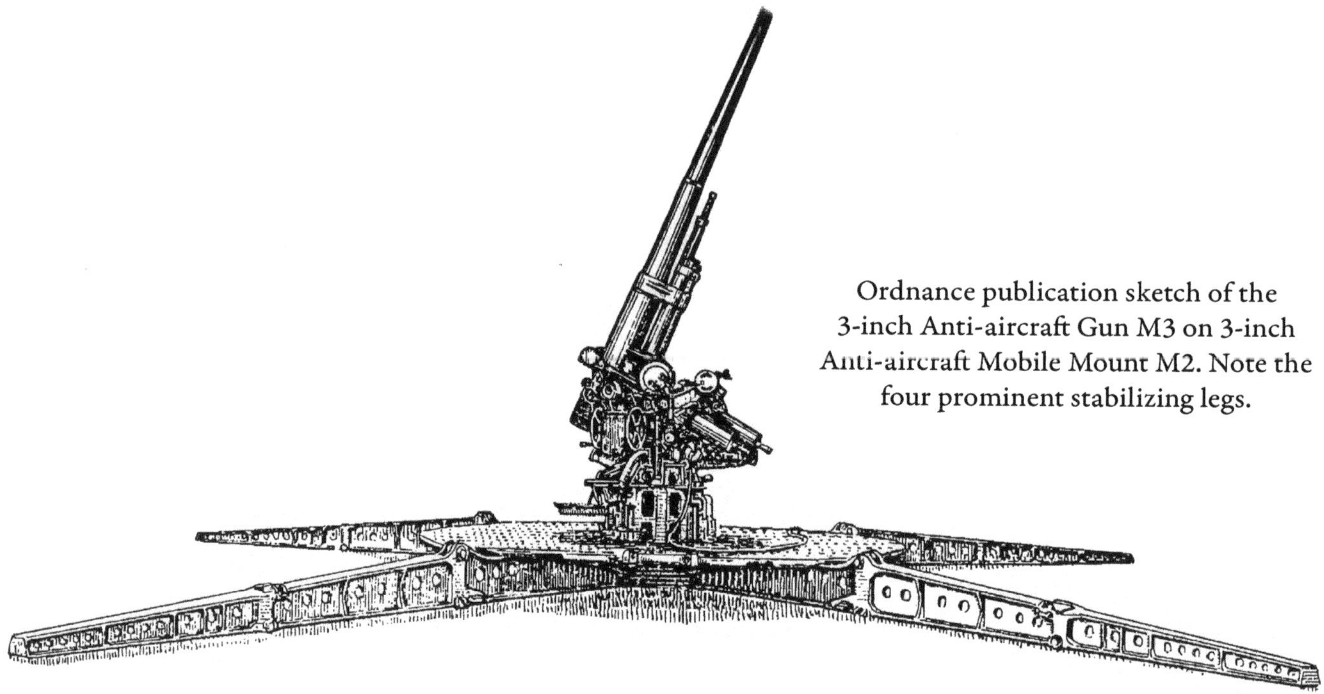

Ordnance publication sketch of the 3-inch Anti-aircraft Gun M3 on 3-inch Anti-aircraft Mobile Mount M2. Note the four prominent stabilizing legs.

The Army began to immediately develop 3-inch anti-aircraft guns of improved performance after the First World War. The first new major mobile gun standardized was the 3-inch M1923E, which emerged as the M1 in 1927. The tube was cold worked, a process known as autofrettage with a removable liner. The M3 was an incremental iteration featuring a slightly larger tube to better accommodate the replaceable liner. This tube was 50-calibers long with a screwed-on breech ring holding a drop-block breech. The barrel had 28 grooves and uniform rifling of 1 in 40. The breech mechanism was semi-automatic. It developed 2,800-fps muzzle velocity with HE rounds and 2,600-fps with shrapnel. It fired using a continuous-pull mechanism and short lanyard. Maximum rate of fire was calculated at 25-30 rounds/minute. For its time and bore size, it was considered a successful weapon.

The gun sat in a cradle with the hydropneumatic recoil cylinder mounted underneath. Trunnions of the cradle were supported by the yoke mounted close to the breech, allowing for an elevation of -1 to +80-degrees. For balance two equilibrators were provided. This mount was the M2, like its tube a progressive development of the earlier M1 type. This mount was carried on a large trailer with pneumatic tires. On the trailer were fitted large outrigger legs that held a sectioned, perforated platform. This platform, in segments, could be lowered and function as the gun's loading platform. The four outriggers had circular cut-outs, a useful characteristic for photo identification. Carriages were later either produced or modified to the M2A2 standard with electric brakes included.

Developmental pilot guns and mounts, including enough for service trials, were ordered beginning in 1932. Major orders for production M3 guns and M2 mounts occurred from 1938 to 1941 (a total of 134 were delivered in that year). On December 11, 1939 321 of these guns were reported on hand, and more were added through early 1942. A final order for another 100 3-inch anti-aircraft guns was placed in July 1940, after which production ended. It appears that the entire production of such guns was 648. The tubes were made at the Army's Watervliet Arsenal and carriages at Watertown Arsenal. They were issued almost as fast as they could be produced to the Army's regular and National Guard Mobile Coast Artillery Anti-aircraft regiments. A number were transferred to the Marine Corps for use with their defense battalions. This model of gun played a major role in the early Pacific island defenses and with units in Panama,

Hawaii, and the Philippines. Consequently, many saw combat early in the war. Even with the increasing availability of the 90mm gun, the 3-inch continued to serve through the war, those in Hawaii remained part of the active garrison into 1944.

A number of 3-inch anti-aircraft guns were made available for lend lease. Twenty were supplied to the British, but the majority (48) went to Latin American countries, particularly Brazil, Chile and Peru. Very few examples have managed to survive. At least one museum example seems to still exist and be on display each in Peru and Brazil. The only American guns that are known to have survived are two battered examples on Wake Island. Apparently, these were part of the Marine Corps defense battalion garrison, surrendered to the Japanese, and after recapture left in place as sort of wartime historical ornaments.

3-inch Anti-aircraft Gun M3
On 3-inch Anti-aircraft Mobile Mount M2

Gun Characteristics
Bore size: 3-inch
Construction: Steel, autofrettage
Tube length overall: 158.2-inches
Nominal caliber: 50 caliber
Tube and breech weight: 2,302-pounds
Usual carriage: 3-inch Antiaircraft Mount M2
Weight of gun and carriage: 16,800-pounds
Maximum elevation: 80-degrees
Loading: Fixed
Shell weight: 12.7-pounds
Maximum muzzle velocity: 2,800-fps (HE),
Maximum range: 14,200-yards (vertical) and 9,700-yards horizontal

Builders: Gun: Watervliet Arsenal
Carriage: Watertown Arsenal

Number built: Approximately 650

The M3 on M2 3-inch AA gun was the primary anti-aircraft weapon of the pre-war Army. This is a gun in 1941 at the Louisiana Maneuvers

A long line of 3-inch M3 Anti-aircraft guns belonging to the 64th Coast Artillery Regiment in Hawaii during the interwar years.

A 1940 photograph of a 3-inch Model M3 and M2A2 mount being transported by truck, note the folded-up ground platform that would be extended once the gun was in its field position.

Somewhat surprising seeming how it was once so common, few 3-inch M3 AA guns survive. This example is actually in Peru as a result of wartime Lend-lease largesse.

There are a couple of 3-inch M3 AA gun survivors still on Wake Island, originally brought there with Marine Corps defense battalions, and then not destroyed or scrapped by Japanese occupational forces. This is one photographed in 2009.

90mm Anti-aircraft Gun M1 and M1A1 On 90mm Anti-aircraft Carriage M1 and M1A1

Army Technical Manual illustration for the 90mm M1 Antiaircraft Gun on 90mm M1A1 mount.

The Chief of Coast Artillery, the branch responsible for anti-aircraft service, advocated the adoption of a gun of 90mm bore on November 27, 1937. They needed a gun of greater projectile size and shorter time of flight than the previous 3-inch standard. Desired was a pilot battery of four guns giving a 2,800-fps muzzle velocity of not more than ten tons per gun, firing a fixed round still capable of being handled by a single man. The project, which had already internally begun, rapidly accelerated. A test pilot was made and evaluated during 1938 as the T-2. After improvements the T-2 model was standardized and adopted in March 1940 as the 90mm anti-aircraft gun M1. It moved into production status almost immediately.

The tube itself was a 50-caliber, monobloc, autofrettage alloy steel tube with a vertical-sliding drop-block breech. It had a percussion firing mechanism. It was rifled at 1/32. The initial M1 gun used manual loading of a fixed round. The mount was a conventional cradle held by a frame to a rotating platform on revolving rollers. Recoil was absorbed by a hydropneumatic system under the barrel using a recoil cylinder, gas cylinder and floating piston cylinder. Counter recoil was achieved by a spring mounted above the tube. The gun had a single equilibrator to help balance it at higher elevation. That elevation could be between 0 and +80-degrees. The mount was provided with four outriggers and a single axle, 2-wheeled bogie for towing. The firing platform could be folded up and one outrigger leg used as a towing bar for transport. Still, it seems the time to properly emplace the mount was considered excessive.

Orders were placed in late 1940 and production deliveries began in June 1941. A test battery of four guns went to Fort Monroe in July, and then to Britain in October to evaluate under combat conditions. A total of 171 guns and mounts had been delivered to December 1941; though production rapidly accelerated in early 1942 with orders in January and February for 5000 more units. Five different barrel producers and six different carriage manufacturers were utilized. The war tempo soon changed from defense to offense, that and a shortage of steel among high priority projects led to cutbacks in orders as soon as November 1942.

The 90mm M1A1 gun was the same M1 gun with a mechanical rammer, designated the M8 or M8A1. It used the gun's recoil to power ramming the subsequent round. It did not prove popular with the crews and often was disconnected or removed in service. The mount M1A1 was developed to specifically work with the M2 remote control system—where the director automatically fed firing data and elevation settings to the gun. Changes in the steel used led to the M1A2 and M1A3 variants. In April 1943 the latest 90mm M1A1 was reclassified as limited standard and replaced in production with the 90mm M2 type gun. All production of the M1 and M1A1 terminated in October 1943, though many had been made and were successfully serving.

The gun was quickly issued to the dedicated Coast Artillery Regiments (AA), replacing the 3-inch mobile guns. They saw service protecting airfields, forts, industrial, and tactically important sites around the world during the war. Units were provided to the Marine Corps to replace 3-inch guns in their Defense Battalions starting in late 1942.

During the war Lend Lease transfers were made to China (28), France (210) and the USSR (251). After the war guns were supplied to Turkey, Yugoslavia, Argentina, Brazil, Greece, Pakistan and Spain. Postwar in U.S. service the gun was redesignated the M117. They lasted in American service with dedicated AAA Battalions until the early 1960s when generally replaced with HAWK missile batteries. In foreign service some guns lasted until the 1990s. They have proved popular museum and memorial guns, and numerous examples still exist.

90mm Anti-aircraft Gun M1 and M1A1 On 90mm Anti-aircraft Gun Mount M1 and M1A1

Gun Characteristics
Bore size: 90mm
Construction: Alloy steel monobloc
Tube length overall: 186.15-inches
Nominal caliber: 50 caliber
Tube and breech weight: 2,445-pounds
Usual carriage: 90mm Antiaircraft Gun Mount M1A1
Weight of gun and carriage: 19,000-pounds (M1A1)
Maximum elevation: 80-degrees
Loading: Fixed
Shell weight: 23.4-pounds
Maximum muzzle velocity: 2,700-fps
Maximum range: 11,273-yards (time fuse)

Builders: Guns: Watervliet Arsenal, Chevrolet, Wheland
Carriages: Watertown Arsenal, York Safe and Lock, Allis-Chalmers, Worthington Pump, Buckeye Traction, Aetna-Standard Engineering, Fisher Body division of General Motors

Number built: 7,010 M1 and M1A1 guns & mounts

90mm M1 of the 504th AA Bn on Okinawa June 6, 1945.

Many 90mm Anti-aircraft guns survive in museums and as public memorials. This one is on the Canadian side of Sault Ste. Marie, photographed in 2014.

90mm Anti-aircraft Gun M2
On 90mm Anti-aircraft Carriage M2

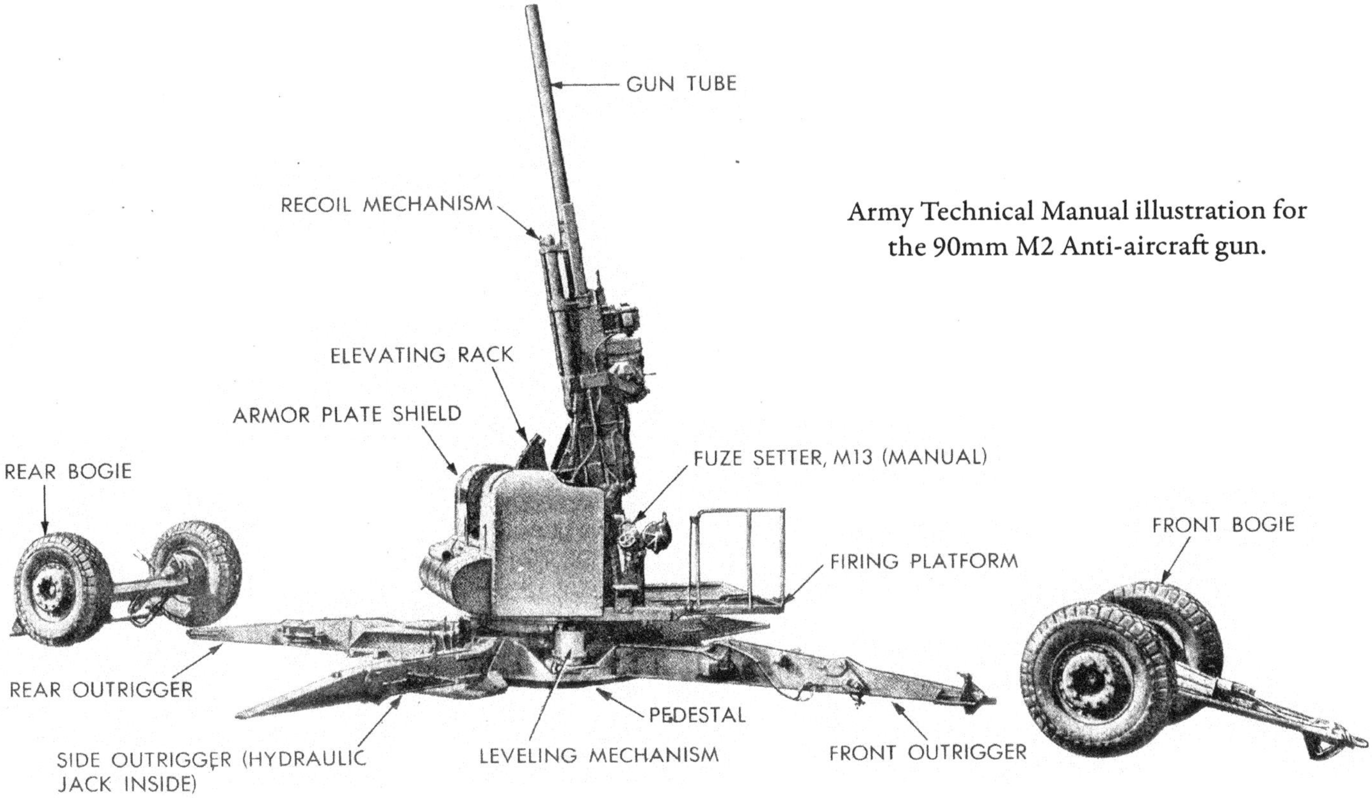

Army Technical Manual illustration for the 90mm M2 Anti-aircraft gun.

In June 1941 a project was started to increase the 90mm M1's ability to function in the anti-tank and anti-ship role. Also, the inability to fire from transport mode and continuing issues with the automatic rammer of the M1A1 needed to be addressed. Within a short time, an improved gun T4 on carriage T2 were standardized as the 90mm anti-aircraft gun M2 on carriage M2 on April 29, 1943. This new weapon succeeded the M1A1s (which was on the same date declared as limited standard) in the production lines. A total of 2,750 guns and mounts M2 were initially scheduled for production, but with the cutbacks in heavy AA needs in late 1943 the needs were reduced. Ultimately just 821 of these guns were produced through late 1944.

It used the same type tube as the M1, except the breech was attached with an interrupted-screw breech ring. The main difference in the gun itself was the adoption of a new electric-assisted automatic rammer and fuse setter (designated the M20) to replace the imperfect M8 rammer of the M1A1. The carriage saw the most dramatic improvements. The gun unit was placed on a new platform with dual-wheeled bogies on either end. Time from site arrival to deployment ready to fire was much improved, one of the desired benefits of the new design when considering its potential antitank role. Usually, the gun was to be lowered to the ground and supported by four outriggers for support. In an emergency the weapon could be fired from the wheeled transport configuration. It was equipped with two spring equilibrators between the side frames of the top carriage. Also, the new carriage, while maintaining an 80-degree elevation, could allow depression of -10 degrees to facilitate use in ground (antitank) and seacoast firing. Finally, a curved defensive shield was added in front of the mount to provide crew protection.

The gun was supplied with a full range of AA HE rounds with several fuse types and with AP and APC. The assembled carriage was usually towed by the M4 artillery tractor. The large automatic loader above the breech and distinctive shield, along with the two bogie sets of wheels make photographs of the M2 mount quite distinguishable from the earlier M1 and M1A1 types.

Perhaps the most impressive performance enhancement was with the automation of fire control. The gun, with its autoloading rammer and fuse setter was coupled directly to the new M1A1 or M3A1 height finder and director M7. Thus, fitted it could manage 20-24 rounds per minute. At the time this was considered an advanced and highly successful system. Soon it benefited also with the advent of the VT fuse, radar, and the more advanced M9 fire director. Postwar the unit was often used with the M33 fire control system with trailer and dedicated acquisition and tracking radar.

During the war the gun was introduced into combat service from 1944 and served in European and Pacific theatres with heavy Anti-aircraft Battalions. The 90mm M2 anti-aircraft gun also continued to serve postwar. The unit was reclassified as the M118 when designations changed in the early 1960s. It was in American service well through the 1960s, until generally replaced by HAWK missile batteries. Postwar the gun also served with a number of allied nations. Guns were provided to Brazil, Greece, Turkey and Japan, lasting well into the 1980s in some locations. While not as numerous as the preceding M1 or M1A1, several good examples survive at museums and as municipal memorial guns on public display.

90mm Anti-aircraft Gun M2
On 90mm Anti-aircraft Carriage Model M2

Gun Characteristics
Bore size: 90mm
Construction: Alloy steel monobloc
Tube length overall: 186.15-inches
Nominal caliber: 50 caliber
Tube and breech weight: 2,445-pounds
Usual carriage: 90mm Antiaircraft Gun Carriage M2
Weight of gun and carriage: 32,300-pounds
Maximum elevation: 80-degrees
Loading: Fixed
Shell weight: 23.4-pounds
Maximum muzzle velocity: 2,700-fps
Maximum range: 11,273-yards (time fuse)

Builders: Guns: See list for 90mm M1
Carriages: Fisher Body, Link-Belt Ordnance

Number built: 575 M2 guns and carriages through October 1944

Army Technical Manual illustration for the 90mm M2 Antiaircraft gun.

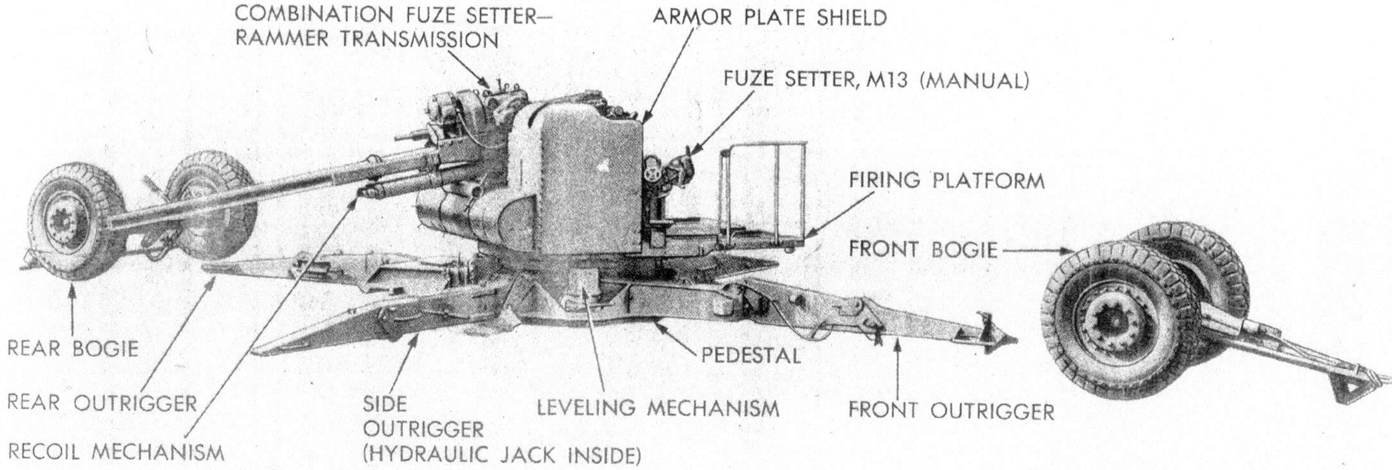

First production 90mm Anti-aircraft gun in November 1942 preparing to fire at Aberdeen Proofing Ground.

There are a couple of 90mm AA M2 in museums. This example is of the one at the Aberdeen Ordnance Museum before its closure in 2005.

120mm Anti-aircraft Gun M1 On 120mm Anti-aircraft Mount M1

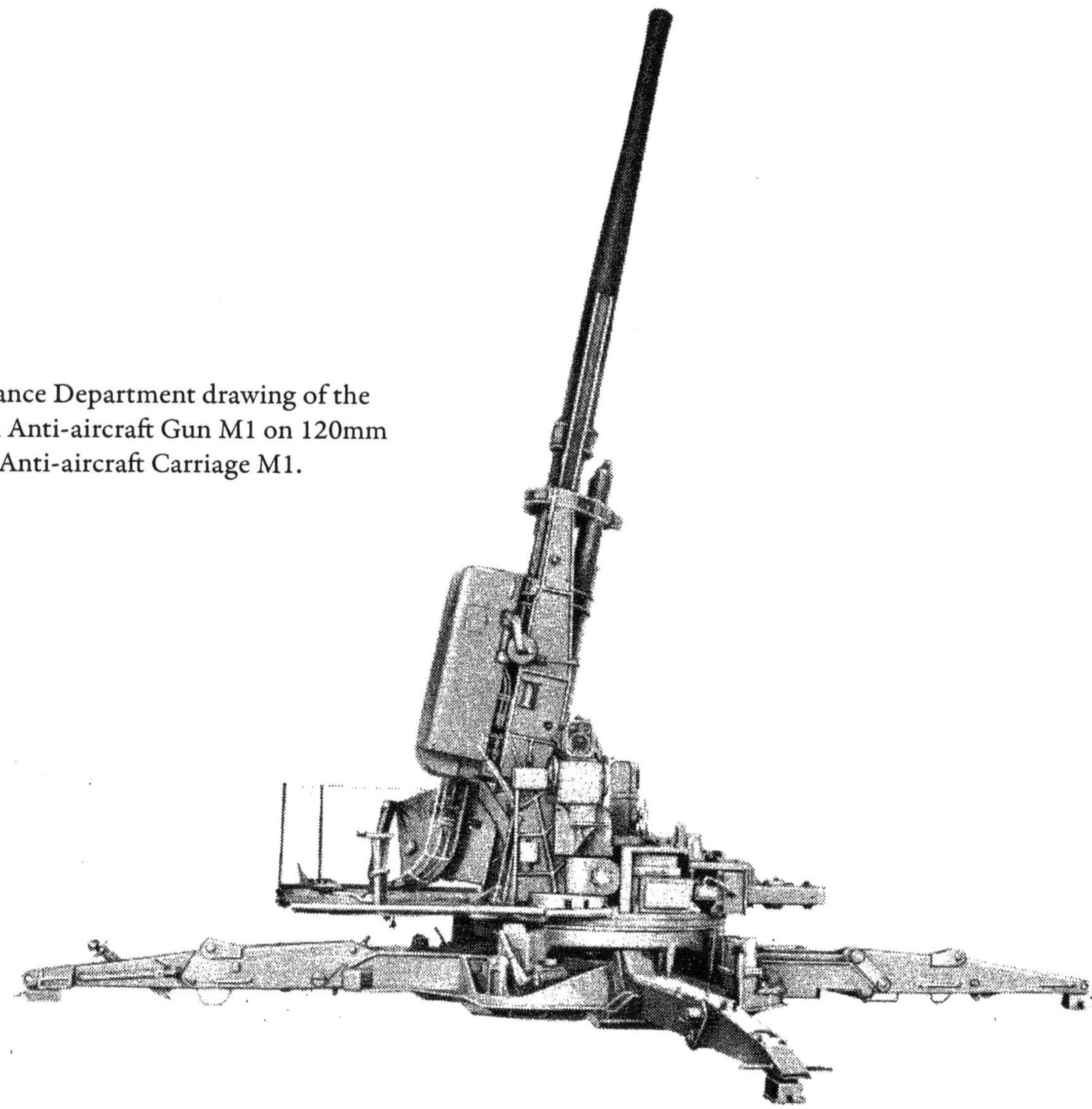

Ordnance Department drawing of the 120mm Anti-aircraft Gun M1 on 120mm Anti-aircraft Carriage M1.

This project perpetuated the First World War concept of having a very heavy anti-aircraft gun to complement a more numerous "standard" gun. Development started in 1938 and was perhaps spurred by America's own development of strategic bombers like the B-29. Initially design work referred to a 4.7-inch bore, ultimately the gun was classified with a metric 120mm bore size. A contract with Link-Belt Co., to design and manufacture prototype 4.7-inch AA gun mount (tubes would come from Watervliet) was granted on December 30, 1940. The long, 60-caliber tube was of monobloc, alloy-steel construction. The drop-block breech was accompanied by unique fuse setter- power rammer that was an integral part of the design. It fired a 50-pound projective to a maximum altitude of 19,150-yards—though effective range depended more on the abilities of fire control than the ballistics of the gun.

The usual transport method was by trailer, carried on two dual-wheeled bogies. Generally, it was towed by an M6 high-speed tractor prime mover. Set-up time was established as 25 minutes. Fifteen-ton hydraulic jacks in the outriggers lifted the gun from the bogies and lowered it to the ground. Also, the tube had to be racked into the gun from its traveling location. On account of its size it was not found easy to transport, the gun often served during and after the war in fixed emplacements. Full electric traverse and elevation of -5 to +80-degrees were possible through automatic operation from the remote director M10.

A conventional hydro-pneumatic recoil cylinder was mounted under the tube, and equilibrators were used to balance the gun for elevation. It fired separate projectiles from cased ammunition, but the automatic loader made for a high rate of fire.

The gun was standardized in October 1942 as the 120mm M1. Initial orders were placed for 200 units but quickly increased to 1450 in 1943. This optimistic requirement was subsequently pruned back twice during 1943, ultimately to an order of just 500 guns and carriages by February 1, 1944. The government funded an $18 million dollar plant at the Link Belt Co. to assemble the gun and breech mechanism. A few initial carriages were manufactured at Watertown Arsenal, but most of the mounts came from the Fisher Body Division of General Motors at their Grand Rapids Plant. This plant under contract of January 1942 began deliveries in April 1943.

The gun began to reach coast artillery anti-aircraft units in 1944. By then the need for such a weapon in either Europe or Asia was considered unnecessary. In wartime the guns did get assigned to Hawaii and the Panama Canal Zone, and a small quantity (just four) went to Northern Ireland. Apparently one battalion was deployed to the Philippines in early 1945, but it did not see combat.

During the war, only a single transfer of just four guns was made under lend-lease to the Soviet Union. Postwar the gun continued in service for a number of years. A small number were also supplied to Can-

120mm Anti-aircraft Gun M1
On 120mm Anti-aircraft Carriage M1

Gun Characteristics
Bore size: 120mm
Construction: Alloy steel, monobloc construction
Tube length overall: 311-inches
Nominal caliber: 60 caliber
Tube and breech weight: 10,675-pounds
Usual carriage: 120mm Antiaircraft Mount M1
Weight of gun and carriage: 61,500-pounds
Maximum elevation: 80-degrees
Loading: Semifixed projectile with cartridge case
Shell weight: 50-pounds
Maximum muzzle velocity: 3,100-fps
Maximum range: 19,150-yards vertical

Builders: Guns: Watervliet Arsenal,
Chain-Belt Co.
Carriages: Watertown Arsenal (pilots),
Grand Rapids Stamping Division of
Fishers Body (General Motors)

Number built: pilot units + 550 production units to April 1944

Wartime production line of the 120mm M1 Anti-aircraft gun and carriage at Fisher Body Co.

ada, where they joined the fixed anti-aircraft defenses of the Sault Ste. Marie Canal. Some development continued postwar, Watervliet Arsenal engineered an improved M1E1 gun in 1950, and a short production run ensued. The 120mm M1E1 gun served well into the postwar era in the U.S. Army. Sixty-six army or National Guard battalions were equipped with the gun. Units were deployed extensively to Japan, Western Europe, and to Korea during the Korean War. Starting in 1954 the gun was phased out in favor of the first generation of Nike missiles. It was finally deleted in 1960. Half a dozen guns on mounts are still in existence, mostly at Army or National Guard museums.

Ordnance Department drawing of the 120mm Anti-aircraft Gun M1 on 120mm Anti-aircraft Carriage M1 in transit mode.

Cold War (1957) photograph of a 120mm M1 Anti-aircraft gun and mount at the Yakima Firing Center, Washington.

A 120mm M1 Anti-aircraft gun and mount at the outdoor ordnance display at the USS *Alabama* Memorial Park.

75mm Anti-aircraft Gun System M51 (Skysweeper)

Army Technical Manual drawing from 1951 of the early 75mm M51 Skysweeper anti-aircraft gun.

The final type of heavy anti-aircraft gun developed by the U.S. Army was this advanced 75mm project. While the gun size and ballistics were not significant improvements, the gun was designed to take advantage of a fully automatic, self-sustained mounting. The intermediate-ranged gun was to depend on volume firing of VT, proximity-fused ammunition not requiring time-consuming fuse setting. The project started in in the 1944 to produce a new 75mm gun with a high, 3,000-fps muzzle velocity and automatic loader—in some ways a smaller version of the 120mm project. The first gun prototype was made in 1945 but underwent further modifications to improve performance when the initial muzzle velocity only turned out to be 2,300-fps. The more innovative carriage and autoloader took even more time to perfect. By 1948 a refocused project was fully underway.

The tube itself was the M35 (T83E1) alloy steel tube with screwed-on breech ring. It had a fully automatic breech mechanism. The gun had 28 grooves and uniform, right-handed rifling of 1/25. The breech block was a vertically sliding type; rounds were fired by percussion. The gun sat on a rotating top carriage M84 (standardized from the T69) was mounted on a lower chassis with retractable outriggers. For transportation two removable 2-wheeled suspension assemblies could be attached. The gun tube was held in the cradle by rails guiding the recoil path. Under the tube was a conventional hydropneumatic T47E2 recoil mechanism with two cylinders and recuperator system. The carriage could achieve elevation from -6 to +85-degrees, and of course traverse of 360-degrees. The front and rear bogies could be removed and hydraulic jacks used to lower the carriage to the ground. The usual prime mover was the M8 tractor.

The fixed ammunition was fed from two revolving magazines holding ten rounds each (in turn fed by hand). It fired high-explosive rounds of 12.3-pounds. The gun was credited with a rate of fire of 45 rounds/minute. Onboard the carriage was the M38 fire control system. For the first time in an anti-aircraft weapon, it had a self-contained radar (the M4 tracking radar) and the M10 analog ballistic director/computer, and M15 director. The radar, developed by Sperry Gyroscope Company, was credited with a range of 30-kilometers and could track targets moving up to 1100-kph. Also, the gun could use manual settings obtained from the M37 auxiliary sighting system.

The M51 anti-aircraft system entered limited production in the early-1950s. It was publicly announced in 1953. Production numbers are unknown, but likely were in the 500-700 range. The tubes were made at Watervliet Arsenal, the carriage and final assembly done at Aetna-Standard. Gun information was inscribed below the breech opening on the breech face, carriage data was on a plate on the right front transom of the bottom carriage. In U.S. service it was deployed to domestic and overseas bases. Army units were part of the newly organized ARACCOM (Army Anti-Aircraft Command). During the Korean War the U.S. Air Force moved to acquire M51s directly for use in protecting Korean airfields, presumably to be organized into separate Air Force battalions. However, delivery came too late for that conflict, and the guns were disposed of. For the Army In the late 1960s it was basically replaced by intermediate missiles like the HAWK system.

Some of these weapons lasted at overseas bases and with National Guard units until the early 1970s. After being phased out of U.S. service, units were made available to Greece, Turkey, and Japan. Several did find their way to memorial gun status and others to Army museums. At least 15 are on public display in the U.S., and it is likely that others are with the foreign nations supplied with the M51 (at least two are on public display in Japan).

75mm Anti-aircraft Gun System M51 75mm Gun T83E1 (M35) On 75mm Gun Mount T69

Gun Characteristics
Bore size: 75mm
Construction: Alloy steel
Tube length overall: 183.19-inches
Nominal caliber: 60 caliber
Tube and breech weight: 1,590-pounds
Usual carriage: 75mm Gun Mount T69
Weight of gun and carriage: 19,200-pounds
Maximum elevation: 85-degrees
Loading: Fixed
Shell weight: 12.30-pounds
Maximum muzzle velocity: 2,825-fps
Maximum range: 10,000-yards (vertical), 14,415-yards (horizontal)

Builders: Guns: Watervliet Arsenal
Carriages: Aetna-Standard

Number built: Unknown, probably 500-700 total

A 75 Skysweeper M51 at Ft. Bliss March 25, 1955.

Several of the postwar M51 Skysweeper guns and boxlike mount exist as museum and memorial guns. This is the unit currently at the Rock Island Arsenal Museum in Illinois.

Light Anti-aircraft Guns 1938-1945

Low-level aircraft attacks during the First World War could easily be met with machine guns and even rifle fire. However sturdier and faster aircraft emerging in the 1920s and early 1930s led to concerns among armies around the globe. The required quick traverse and rapid rate of fire to match a swift dive bomber or strafing aircraft in close proximity could not be provided with the relatively slow-firing 3-inch standard anti-aircraft cannon. A new class of weapons was required to meet this need. Automatic cannon of 20 to 40mm in bore size, firing an explosive shell at a high rate of fire seemed to fit the requirement. Foreign designs soon emerged to fill this niche. As it turned out the American 20mm and 40mm were essentially licensed versions of European guns, even the domestic 37mm was ultimately married to the British M5 (Kerrison) fire control director.

U.S. Army development efforts with light automatic cannons were seriously begun by John Browning at Colt's Patented Firearms immediately following the First World War. Starting with existing 37mm infantry guns, Browning (and the firm following his death in 1925) progressively increased the muzzle velocity and rate of fire of a series of guns. By 1928 the basic 37mm automatic cannon was standardized as the M1. The Army's Ordnance Department progressively developed a series of mountings for this gun. Finalized substantially later than the gun itself, the M3 mounting was standardized in 1938.

It was found that a highly transportable 4-wheeled trailer with stabilizing outriggers was best used to move this type of weapon. Of course, theoretically they could be mounted on fixed mounts (and were extensively in their shipboard versions), but generally for field use these guns had to travel with the field army and be quickly deployable. Similar carriages, totaling about 5000-6000 pounds with the gun and mount, were adopted for both the 37mm and 40mm guns. A small number of dual 20mm mobile mounts were provided certain Marine Corps units, using a modified 37mm trailer. Prime movers were almost always the standard military 2.5-ton truck.

The 37mm gun was put into production in 1938, ending in July of 1943. The 40mm mobile gun came essentially three years later, production being delivered from March 1942 through March 1944. Both types were made in very large numbers, primarily using contracted commercial manufacturers. Colt's Firearms was the prime producer of the 37mm, but numerous commercial firms not usually associated with armaments production made the 40mm. They armed the automatic weapon batteries of standard American mobile and semi-mobile anti-aircraft battalions. In separate AA Brigades and Regiments, they were teamed with heavier 90mm or 3-inch AA. As the war advanced and allied air superiority dominated, the issued guns saw frequent employment as ground support weapons. The 37mm was supplied to allies in relatively small amounts under Lend Lease, on the other hand the 40mm was one of the major contributions made to the British Empire and Russian armies. While the 40mm continued with the postwar army, it was eventually replaced with a range of anti-aircraft missiles.

Neither gun was favorably viewed for donation as a monument gun, though numerous examples survive in Army and private military museums. The dual mobile 20mm mount, never built in large numbers, is not known to have a surviving trailer version mount.

A Light 40mm AA of 453rd AAA Bn at Camp Perry

37mm Automatic Gun M1A2
On 37mm Automatic Gun Mount M3A1

Ordnance Department illustration for the 37mm Automatic Gun M1A2 on 37mm Automatic Gun Mount M3

This gun was the result of an almost twenty year development program. After the First World War, with the rapid development of faster, sturdier aircraft, the perceived need for a rapid-fire light cannon to fill the "gap" between machine guns and 75mm/3-inch heavy AA guns emerged. John Browning of Colt's Patented Firearms began development of an anti-aircraft gun of this bore in the early 1920s. The Army adopted it as the M1 weapon in 1927, though delays plagued the development of a suitable carriage. The M3 carriage was not standardized until 1938.

The gun was fully automatic and air-cooled. The barrel was a one piece forging or casting. It had a vertically sliding wedge breech. Firing was by a foot pedal firing mechanism. Recoil was controlled by hydrospring, and spring-type equilibrators balanced a trunnion located near the breechblock. Generally credited with a muzzle velocity of 2,600-fps, it was supplied with a variety of ammunition including HE, AP, APC, usually in ten-round clips. A change in rifling to 12 grooves of a uniform 1 in 30 twist resulted in the M1A2 final gun version.

The lower carriage was mounted on a four-wheeled trailer. The wheels were mounted on spindles offset from the axles, allowing the carriage to be lowered for firing. An outrigger on each side, perpendicular to the carriage could be extended for additional stability. The gun could be deployed in just two minutes, though setting-up the fire control equipment took ten minutes more. Generally, the gun was pointed by a remote control system. On its carriage platform it could rotate 360-degrees and fire electrically from 0 to 85-degrees elevation. Prime mover was usually the common 2 ½- ton truck.

The first substantial orders for M1 37mm guns were placed in November 1938 (carriages at Watertown, barrels at Watervliet Arsenal). Initial issue was to Army Anti-aircraft regiments, with a requirement of 24 37mm guns in each. The early M2 sighting system was replaced with a licensed version of the British Kerrison predictor (known as the M5 in U.S. usage). When the Bofors gun was standardized in April 1941, the 37mm was classified substitute standard, though large orders (4200) were placed as late as April 1942. A year later the program was suspended. Colt Firearms was the major gun producer. The gun and carriage were declared obsolete in January 1945. Figures vary by source, but about 4,048 complete units are reported as being manufactured and assembled with M3 wheeled carriages from July 1940 through end of production in July 1943. Perhaps another 2,200 were used with motor carriages (primarily the M15A1 halftrack type).

Smaller numbers were used in a combination mount with two 0.50-caliber AA machine guns, designated the M42 and M54. Most of these were mounted on halftrack GMCs, but a few were also trailer-mounted. The gun was soon considered inferior to the 40mm type, however as licensing rights for the 40mm were delayed, it continued in production in very large numbers early in the war. By 1943 production of the 40mm type (of roughly the same size and weight, and also primarily mounted on similar wheeled carriages) supplanted the 37mm. The gun did remain in service throughout the war and saw action in virtually every Army theatre.

A total of 620 37mm AA guns were transferred to other countries through lend lease, major recipients were the British Empire (432) France (100) and Brazil (48). After the war the 37mm was quickly phased out and replaced with the 40mm type. Compared to the large numbers produced, surprisingly few still exist, though there are a number of good representatives at several Army museums.

37mm Automatic Gun M1A2
On 37mm Automatic Gun Mount M3

Gun Characteristics
Bore size: 37mm
Construction: Steel
Tube length overall: 104-inches
Nominal caliber: 53.5 caliber
Tube and breech weight: 365-pounds
Usual carriage: 37mm Automatic Gun Mount M3
Weight of gun and carriage: 6,124-pounds
Maximum elevation: 85-degrees
Loading: Fixed
Shell weight: 1.34-pounds
Maximum muzzle velocity: 2,600-fps
Maximum range: vertical effective range 4,000-yards
Horizontal effective range 4,070-yards

Builders: Guns: Watervliet Arsenal,
Colt Firearms, Oldsmobile
Carriages: Watertown Arsenal, Koppers Co.,
Aetna-Standard Engineering Co.

Number built: Approximately 5541 on M3 mobile carriages

The combination 37mm + 0.50-cal Anti-aircraft mount of the 61st Coast Artillery Regiment on wheeled trailer in the defenses of Keflavik, Iceland on August 23, 1942.

A 37mm M1A2 Antiaircraft gun belonging to the 62nd Coast Artillery Regiment on maneuvers on November 25, 1941.

The 37mm M1 Anti-aircraft gun was also combined with a pair of 0.50-caliber machine guns to create the M42 mount—which was usually mounted on a halftrack, but could also be truck and ground mounted.

37mm Anti-aircraft guns in prepared positions at Fort McAndrew, Newfoundland in May 1943.

Not as many 37mm AA mounts survive today compared to their replacement 40mm mounts. A M42 mount combination 37mm AA with 0.50-cal machine-guns at the Indiana Military Museum in Vincennes, Indiana.

The U.S. Army never adopted the automatic 20mm gun for anti-aircraft service; however, the Marine Corps did. The common Navy 20mm gun Mk II was adapted to slightly modified 37mm M3 mobile trailer carriages in either single or dual mountings and issued to Marine divisions mid-war. The mount could also be detached and secured to a ground platform. This photo was taken in the South Pacific on November 4, 1943.

40mm Automatic Gun M1 On Carriages M2 and M2A1

Army Ordnance sketch of the 40mm Automatic Gun M1 for use on Carriages M2 and M2A1.

The Bofors-designed 40mm automatic anti-aircraft gun was the result of a long, careful development of that Swedish arms firm. The tube was forged alloy steel. It had a rectangular breech housing and sliding vertical breech block. Most often it used a threaded flash hider. An automatic loading mechanism fed fixed cartridges into a loading tray from which they are pushed into the chamber by a mechanical rammer. The gun was fed with 4-round clips. Ammunition was an HE or AP shell of 2-pound weight, capable rate of fire was 120 rounds per minute.

The upper carriage supported on a 360-degree base ring on bearings sat on a 4-wheeled trailer with two hinged outriggers. The carriage axles could rotate to raise the wheels when set in the firing position. Elevation to 90-degrees was possible. The original M1 British-type carriage (of which at least 1500 were made for lend-lease) was changed to the American M2 carriage with new drawbar, lunette, electric brakes and controls. Later the M2A1 carriage had gearing for faster traversing and elevation. The rig was usually accompanied with the M5A1 off-carriage director. A more compact M5 carriage was designed for airborne use, but only 200 were fabricated. The gun was a little larger and heavier than the 37mm. However, it had a higher muzzle velocity, longer range, and a heavier explosive shell. Also, its quick barrel replacement was another crucial benefit.

Even though satisfied with its internal development of the 37mm automatic gun, the Army kept careful watch over foreign developments. Trials in the late 1930s clearly showed the Swedish 40mm to be superior to virtually every other commercial offering. The widespread employment of dive bombing early in the war added emphasis to the need for a gun of this type. In 1937 Bofors refused to sell the Army a gun for demonstration, but then two years later they changed their minds, at least entertaining the licensing of rights for manufacture of their gun. Guns evaluated separately by the Army and Navy proved their effectiveness, and also efforts were made by the services to adopt the superior Kerrison predictor. Navy production for shipboard guns and mounts starting in early 1941 and amounted to very large numbers procured.

The Army approved the gun as standard in February 1941, and production arrangements followed. The Army secured its gun needs mostly from Chrysler. The first guns were produced there in February 1942, peak production coming that December. Through the end of 1943 1,392 were made for the British and 23,685 for American services. The wheeled carriages came primarily from Firestone Tire & Rubber Co. and lessor quantities the J.I. Case Company and Koppers. The U.S. Navy used other exclusive producers for its common single, dual and quad shipboard mounts (over 40,000 tubes were fabricated for the U.S. Navy).

In Army usage twin mounts were carried on the M19 Gun Motor Carriage. During the war the 40mm was a common weapon supplied in lend-lease. Something over 9000 units were supplied, major recipients being the USSR (5,511), British Commonwealth (2,834), France (448) and China (180). In Army service the guns became standard in the mobile and semi-mobile anti-aircraft battalions and regiments, serving in all theatres from late 1942 to the end of the war. AFVs carrying the 40mm and the M2 field car-

40mm Automatic Gun M1 On Carriage M2 and M2A1

Gun Characteristics
Bore size: 40mm
Construction: Alloy Steel
Tube length overall: 88.58-inches
Nominal caliber: 56 caliber
Tube and breech weight: 1051-pounds
Usual carriage: 40mm Carriage M2
Weight of gun and carriage: 5549-pounds
Maximum elevation: 90-degrees
Loading: Fixed
Shell weight: 2-pounds
Maximum muzzle velocity: 2870-fps
Maximum range: vertical 7500-yards, horizontal 12,500-yards

Builders: Guns: Chrysler, Pontiac, Otis-Fensom (Canada), York Safe & Lock, Blaw-Knox Company
Carriages: Firestone Tire & Rubber Co., J. I. Case Company, Koppers Co.

Number built: 26,371 for army M2 mobile carriages through March 1944

Though not widely used, also developed was a 40mm Automatic gun M5 carriage of compact dimensions purportedly for issue to airborne units. It is not known how widely it was actually used.

riage were used for many years postwar and supplied through military assistance to other allies. The gun was never distributed in large numbers as a memorial piece, though inevitably because of the very large production units survive at military service clubs and in many public and private military museums.

A 40mm Automatic Gun M1 and carriage deployed in its defensive role in North Africa in 1943.

This is a 40mm M1 Automatic Gun and carriage on display at the outdoor ordnance collection at the Rock Island Museum in Rock Island, Illinois.

Railway Artillery

American practical experience with railway artillery began during the Civil War. While the Army gained some operational familiarity, the technical aspects remained unsophisticated. These early weapons were essentially heavy guns or mortars mounted on commercial flatcars, sometimes fitted with protective metal shields. No specialized carriages or purpose-built weapons emerged—these were expedients assembled from available ordnance. The small international conflicts that followed provided no impetus for further development. Only on the eve of the First World War did the U.S. Army begin developing dedicated railway artillery weapons.

Two projects commenced in 1916. The first involved modified flatcars carrying 4.7-inch howitzers for land defense of the Panama Canal Zone harbor defenses. Though successful, production totaled just four cars. The second, prompted by awareness of widespread railway artillery employment in Europe, aimed at creating an entirely new 14-inch gun and carriage. This carriage, designated Type "E," was not completed until after the war and remained a single prototype that never entered production.

The static nature of warfare on the Western Front generated an entire generation of American railway artillery. Beginning in 1915, the major combatants rapidly mounted their heaviest guns—often drawn from naval or fixed coast artillery installations—onto specialized railway carriages to provide mobile heavy firepower at the front. France produced a wide spectrum of railway types across multiple calibers. Their experience heavily influenced American development. When the United States entered the war, guns of 8-inch, 10-inch, and 12-inch caliber were selectively removed from seacoast defenses or drawn from reserve stocks. New carriages were quickly fabricated, though most were of simple design and thus limited capability. The U.S. Navy initiated its own program using reserve battleship gun barrels for 14-inch weapons and made available both 7-inch and 14-inch guns to the Army for prospective railway battery use.

American participation lasted only nineteen months, so little of this program reached fruition during the war. The Navy's five-gun 14-inch battery entered combat, but little Army materiel reached France, and none saw action. The interwar Army retained some 7-inch and 8-inch guns and carriages. The finished 12-inch railway guns lasted until the 1930s. A portion of 12-inch mortars on railway mounts also survived, though generally still in storage into the Second World War.

Several new projects initiated during the war aimed at providing more sophisticated mounts with greater stability and traverse than the wartime sliding carriages. These included a new 12-inch gun on a ground platform, both 12-inch and 16-inch howitzers, and two new carriages for the Navy's 14-inch/50-caliber tube. After years of development and testing, a very small number of the improved 14-inch gun carriages were produced and deployed in coast defense positions.

The final American railway artillery project to reach production was a new 8-inch railway carriage designed for surplus Navy 8-inch/45-caliber gun tubes. Developed through the 1920s, nearly thirty of these guns and carriages were produced by 1942 and deployed with coast artillery units during the war. No American railway guns saw combat in any theater, though many served coast defense roles in the continental United States and its possessions. Very little of this materiel survives. A single intact railway gun is displayed at the Naval Museum in Washington, D.C., and another former Army gun is displayed at Fort Lee.

12-inch mortar railway mount in 1937

4.7-inch Howitzer Model 1913
On 4.7-inch Railway Carriage Model E (M1917)

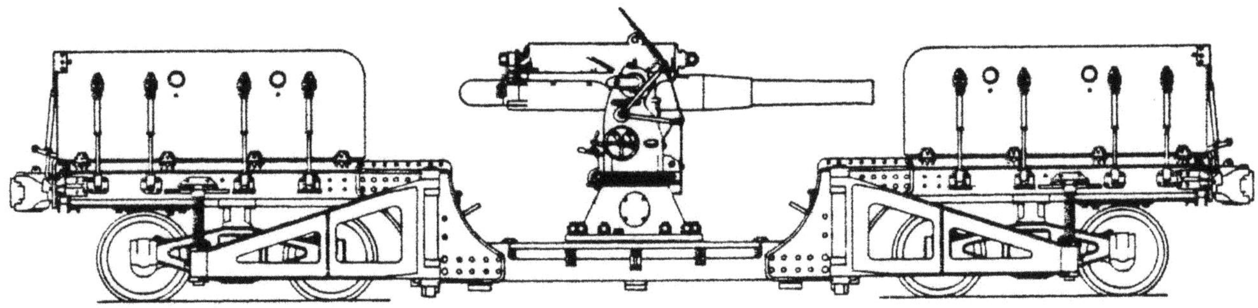

Army sketch of the 4.7-in howitzer M1913 on railway car M1917.

The 4.7-inch howitzer Model of 1913 on simple pedestal mount Model of 1915 had been developed to provide firepower for the land defense problems of defending seacoast defenses. The fear was that enemy boat parties could come ashore and neutralize the seacoast guns. While in some locations the defensive needs were met by using obsolete field pieces, for Panama a more dedicated solution was found. This new howitzer was emplaced on simple, open, fixed platforms—essentially a retaining bolt circle embedded in concrete. The emplaced howitzer had a high enough elevation (40-degrees) to get 10,000-yards of range.

The howitzer was a conventional wire-wrapped, alloy steel tube of just 22.5 caliber. It had an interrupted-screw breechblock with a mechanical firing mechanism. Recoil was controlled with a bulky, combined recoil and counter recoil cylinder mounted atop the tube. The gun fit in a yoke held by a pedestal. Elevation (-10 to +40-degrees) was with a screw hinged to the bottom of the cradle. Traverse was a full 360-degrees. It fired fixed, 60-pound HE rounds. The howitzer itself had a large defensive shield used both in the fixed and railway carriages.

Soon after entering service, railway carriages, eventually designated the Model 1917, were developed for this gun. The thought was that the guns with attached armored shield could be rapidly moved to any threatened area needing additional defenses. The railroad network on the isthmus was well developed with many spurs. The cars built could be rapidly converted from a 60-inch, broad gauge (at the time the standard of the Panama Canal railroads) to 56.5-inch standard military gauge. The Model 1917 railway car had a cast steel center platform and two structural steel plates riveted to each side. On each end of the car were hinged, three-pieced 0.5-inch armor plates with holes for machine guns. The plates could be raised to protect the crew or lowered if not in danger. Two other plates on the center section could be lowered to serve as a working platform.

In the center of the cast platform was a pedestal mounted to the base. Into this was inserted the yoke, cradle, recoil system and tube of the Model 1913 howitzer. The yoke retained its full traverse and elevation properties. The mount (car) was let down through screws and clamped to the track for firing. Hinged outriggers could be extended for additional stabilizing with a steel foot on the end of the arm. It was estimated that it would take at the most 5.5-minutes to prepare the gun for firing from traveling configuration. The car ran on commercial four-wheel trucks of 50,000-pounds capacity each. No special ammunition service jig or arrangements were required with the hand-service type of shell and powder used.

The first of four railway carriages (at the time designated Railway Mount Model E) was delivered by Watertown Arsenal and shipped for testing to Sandy Hook on October 1, 1916. The other three followed in 1917. The carriages were shipped to Panama by 1919 where they were taken in by the garrison for use. No new howitzers were built to go with these carriages, they were always thought to be substitute carriages for the fixed pedestals to use the same inventory of howitzers on hand. However, as the 4.7-inch howitzer was declared obsolete and slated for disposal in June if 1919, these carriages, at least in this role, did not last long.

4.7-inch Howitzer Model 1913
On 4.7-nch railway carriage Model 1917

Gun Characteristics
Bore size: 4.7-inch
Construction: Alloy steel, wire-wound
Tube length overall: 111.4-inches
Nominal caliber: 22.5-caliber
Tube and breech weight: 2,100-pounds
Usual carriage: 4.7-inch Railway Carriage Model 1917
Weight of gun and carriage: 45,495-pounds
Maximum elevation: 40-degrees
Loading: Fixed
Shell weight: 60-pounds
Maximum muzzle velocity: 1,300-fps
Maximum range: 10,000-yards

Builders: Guns: Watervliet Arsenal
Railway Carriages: Watertown Arsenal

Number built: 4 railway carriages which could be equipped with one howitzer each.

While a photograph taken in a shop, it clearly shows how the outrigger supports would help stabilize the 4.7-inch howitzer on railway car for firing.

First World War Expedient Railway Guns

The U.S. Army and U.S. Navy moved quickly after American entry into the First World War to identify available large-bore gun tubes for heavy artillery service on the Western Front. Domestic production of new tubes would require at least two years. The widespread, successful use of railway mounts by both Allied and German forces suggested that these large mobile platforms would prove most useful to the American Expeditionary Force. By 1917 the threat of German naval attack on the United States had become negligible, allowing a portion of seacoast defense guns to be temporarily dismounted for use as heavy field artillery. Orders were issued to remove fifty 8-inch and thirty 10-inch guns from seacoast batteries, supplemented by reserve and arsenal guns of the same calibers for mounting on railway carriages—paralleling the effort that produced over 120 guns of 5-inch and 6-inch caliber for new wheeled field artillery carriages.

An additional 150 12-inch mortars were withdrawn from coast defense emplacements. Ninety-one new railway carriages were ordered for these weapons to fulfill a request from General Pershing. The Navy contributed a dozen 7-inch guns that were likewise mounted on railway carriages. Later in the war a project was initiated to mount another dozen spare 12-inch tubes on the same type of carriage.

The U.S. Navy possessed a surplus of 14-inch/50-caliber Mark IV gun tubes—both the normal allotment of spares for the five battleships armed with this type and several dozen additional tubes in various stages of fabrication, originally ordered for battlecruisers no longer scheduled to receive guns of this caliber. In a rare example of effective interagency cooperation and rapid industrial mobilization, a new railway mount was designed, guns installed, and naval personnel organized to serve the batteries. Within months several of these mounts deployed to Europe and performed as expected. The program proved so successful that the U.S. Army ordered additional units of the same design. The mount selected was necessarily simple in design. A follow-on Type II railway mount incorporating numerous improvements was nearing completion when the war ended. With no strategic requirement for these weapons in peacetime, all materiel was transferred to the Army, which retained the gun tubes but over a number of years scrapped most of the carriages.

During the war the Army also initiated plans to produce large numbers of its own 14-inch gun, loosely based on the prewar Model E design. A new manufacturing plant was under construction when the armistice was signed. None of the guns or their carriages, though ordered, were ever compleed.

A final source of heavy tubes emerged from an order under construction by Bethlehem Steel for coast defense batteries in Chile. Contracted for delivery in 1917 were six powerful 12-inch/50-caliber tubes for a coast defense project. After acquisition, a simple sliding-mount railway carriage was designed and three carriages ordered. Only one was entirely completed, and not until after the war's end. As with the Navy guns, the tubes were retained but the expedient carriages soon scrapped.

8-inch Railway M1918 of the 41st Coast Artillery Regiment

7-inch Gun Mk II Mod3 On 7-inch Railway Carriage Model 1918M1

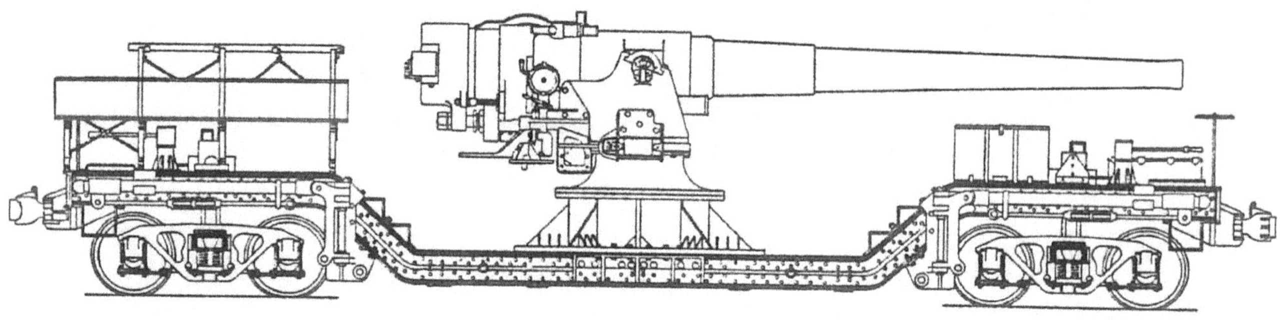

Sketch side view of the 7-inch Gun MK II Mod3 on 7-inch Railway carriage Model 1918MI.

At the start of the First World War for America, the U.S. Navy endeavored to transfer to the Army for use on mobile mounts a number of 6, 7, 8, 10, and even 14-inch gun tubes. A dozen 7-inch guns, previously tertiary broadside mounts from pre-Dreadnaught battleships were initially made available for transfer to the U.S. Army. The army quickly utilized a version of its standardized railcar, the Model 1918, for mounting these guns. They were always intended for mobile weapons used in American seacoast defense against enemy (German) submarine raids and not intended for deployment in Europe.

The gun was a typical steel naval built-up, 45-caliber type, with interrupted thread breech and mechanical firing mechanism. It had 42 grooves and uniform 1/25 rifling. Recoil was checked with one hydraulic cylinder on the bottom of the cradle and two spring recuperator cylinders on either side of the recoil cylinder but still mounted on the underside of the cradle. There was an elevation rack on the left side giving -5 to +15-degrees of elevation. Loading was achieved using gravity feed from a breech tray in the -5-degree elevation.

The gun, on its shipboard-type pedestal, was placed on the dropped center well of the Model 1918 railway carriage. That carriage was a basic design also slightly modified to carry the 8-inch gun and 12-inch mortar artillery. While this mounting still allowed a good depression useful for firing on low submarine targets offshore, an additional circular collar was used as a height extender for the pedestal so range could be increased as it allowed a higher elevation. Traverse of 360-degrees on the mount was achieved with a worm wheel on the pedestal. At the firing position the car was raised by four screw jacks and then lowered onto steel I-beams placed across the wood cross ties. The carriage could be stabilized during firing with its integral outriggers and jack screws. A loading platform was included behind the gun, receiving ammo from the separate ammunition car, the shell being hand-loaded from the tray at a slight depression. The car was carried on two 2-wheel trucks, one in front and one behind. Never intended for overseas deployment, they had only American-style coupling gear.

Twelve Model 1918 Railway carriages were fabricated, assembly being done by the American Car & Foundry Co., of Berwick, Pennsylvania. Upon delivery of completed units shortly after the war ended, they were immediately consigned to storage at Fort Eustis. They were never issued to an active unit. Unlike other materiel, the gun was retained after the war as reserve material with the U.S. Army. In 1931 consideration was given to declaring them obsolete and scrapping the guns and carriages, but the Coast Artillery Corps requested their retention for possible use in deployment to otherwise abandoned or undefended harbors. The twelve units were moved to Aberdeen for storage in 1932.

In September 1939 the whole inventory was considered for transfer to Brazil as part of an arms supply deal tied to U.S. access to new air bases in Natal. By March 1940 they were formally declared obsolete, enabling the transfer. Ultimately twelve guns and carriages, along with six spare tubes and 2400 rounds of ammunition were supplied to Brazil. The sale was finalized, and by year end 1941 the equipment was

7-inch Gun MK II Mod3
On 7-inch Railway carriage Model 1918MI

Gun Characteristics
Bore size: 7-inch
Construction: Alloy steel, built-up
Tube length overall: 330-inches
Nominal caliber: 45-caliber
Tube and breech weight: 28,450-pounds
Usual carriage: 7-inch Railway carriage M1918M1

Weight of gun and carriage: 171,000-pounds
Maximum elevation: 15-degrees
Loading: Separate
Shell weight: 165-pounds
Maximum muzzle velocity: 2,700-fps
Maximum range: 17,000-yards

Builders: Guns: Various builders
Carriages: American Car & Foundry Co.

Number built: 12 guns acquired by U.S. Army, placed on 12 railway carriages

overhauled at Baldwin Locomotive. The Brazilians had the guns inverted on their modified cradles to move the recoil assemblies on top—thus increasing the gun's elevation. They also were changed to one meter gauge. These served in Brazil until deleted in 1961. One of these 7-inch guns on its carriage remains on display at an army museum in Rio de Janeiro. There are more than a few of the 7-inch guns on fixed pedestals in various overseas locations.

No 7-inch guns on railway carriages survive in the U.S. But several of the twelve units supplied to Brazil at the start of the Second World War do. However, during restoration prior to transfer it was found possible to mount the tubes upside-down to provide higher elevation. Here is an example of one of the guns at the Ordnance Museum in Rio de Janeiro.

Photograph of a completed 7-inch gun Mk II on 7-inch railway carriage, from February 1919.

8-inch Gun Model 1888A1, 1888MIA1, and 1888MIIA1 On 8-inch Railway Carriage Model 1918M1

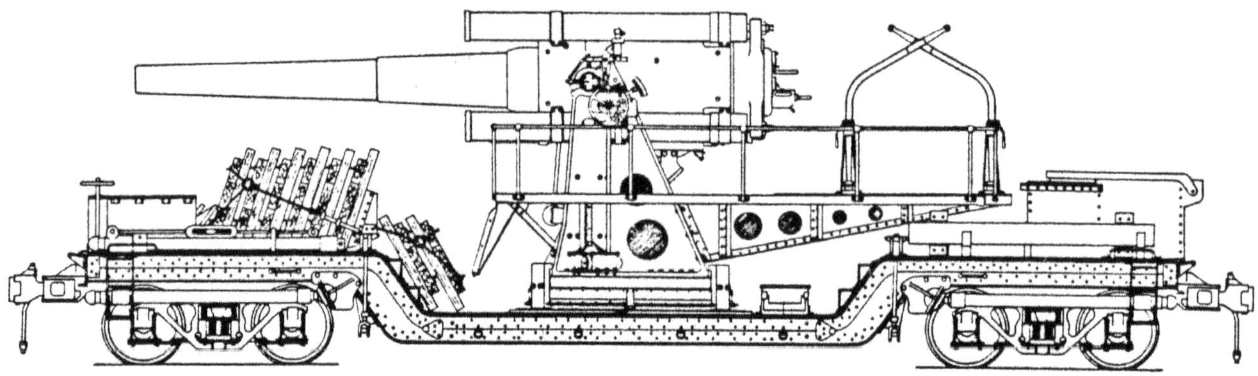

Army sketch of the 8-inch Model 1888 gun on Model 1918M1 railway carriage.

Almost immediately after America's entry into the First World War a survey was made to identify the number and type of heavy seacoast that could be removed from emplacements to use as heavy artillery for the proposed Expeditionary Force. By May 17, 1918 plans were cast for a total of 76 8-inch rail guns: with 50 army Model 1888 tubes from defenses, 12 of the same type of spare tubes, and 13 Navy, 35-caliber tubes. The Navy tubes ultimately weren't used in this application, but the project for the Army seacoast guns proved successful.

The 50 8-inch guns obtained were a mixture of Model 1888, 1888MI and 1888MII tubes, which had different hoops but were otherwise identical. This was a standard coast defense gun of 32-calibers length, with 72 grooves with 1/50 rifling increasing to 1/25. After addition of a new recoil band, the models were given an addition of "A1" on their designations. The gun was carried in an elevated top carriage (designated the Model 1918) with attached loading platform. It had one hydrospring recoil and four spring recuperator cylinders—the latter arranged symmetrically around the cradle. The gun could be elevated to 42-degrees with a segmented circular track attached to the bottom of the cradle with a Hindley worm and bevel gears.

A railway car mount (the Model 1918M1) utilizing aspects from previous design efforts for relatively light weight guns was designed for the new generation of 6 (ultimately not utilized), 7, 8-in guns and 12-inch mortars. In fact the first design specification, borrowing much from the 4.7-inch gun project, was dated September 1916. These cars had a cast steel dropped center with structural steel ends and a single 2-wheeled truck on either end. Firing was meant to be done by dropping the car with four onboard jack screws onto I-beams laid on the underlying cross-ties and stabilizing the car with its six outriggers anchored to floats. The 8-inch gun project used this car. The bottom carriage racer allowed a rack and pinion mechanism giving a full 360-degrees traverse. Two trucks with four wheels each moved the car.

Morgan Engineering, Harrisburg Manufacturing, and the American Car & Foundry Co. shared orders for the 8-inch railway carriages. The first one was completed for test evaluation in May 1918. A total of 24 were completed by year's end, three were actually sent to Europe, but did not see combat, before the armistice. Altogether 47 rail cars and mounts were produced.

After the war the guns and carriages were kept. While eventually a significant portion were kept at Aberdeen in storage, several coast artillery regiments used the gun as their primary weapon and were regularly deployed to several stateside posts. Fort Story, Eustis, and Hancock often hosted units with this railway gun. A total of 16 guns, barbettes, and railway carriages were sent to Oahu (the final eight as late as1940), where the gauge of the cars was changed to the island's narrow gauge standard. The type served well into the war. The Oahu guns were extensively deployed to a number of special spurs and sidings, and many were eventually mounted on their barbettes as fixed mounts. Guns were also supplied to Canada and four went to Bermuda. Seven guns and top carriages were sent to the Philippines as part of their Inland Seas Project in 1941 and subsequently lost there.

A proposal to scrap the remaining guns (except those still in Oahu) in 1943 was deferred as it was thought they still might be useful as reserve material. As expected, the ending of the war finally prompted their scrapping. No complete railway carriages still exist, but the gun and top carriage for one is on public display in a park in Tampa, Florida.

8-inch Gun Model 1888A1, 1888MIA1, and 1888MIIA1 On Railway Carriage Model 1918M1

Gun Characteristics
Bore size: 8-inch
Construction: Steel, built-up
Tube length overall: 295.3-inches
Nominal caliber: 32-calibers
Tube and breech weight: 32,218-pounds
Usual carriage: 8-inch Railway Carriage Model 1918M1
Weight of gun and carriage: 157,700-pounds
Maximum elevation: 42-degrees
Loading: Separate
Shell weight: 260-pounds
Maximum muzzle velocity: 2,600-fps
Maximum range: 23,900-yards

Builders: Guns: Watervliet, Bethlehem, West Point Foundry
Carriages: Morgan Engineering, Harrisburg Manufacturing, American Car and Foundry Co.

Number built: 47 barbette carriages and rail cars built. 54 existing guns Modified to A1 type for this carriage

8-inch guns on a M1918 railway mounts of the 52nd Coast Artilery Regiment at a shoot at Fort Story, Virginia

An early (1918) photograph of a complete 8-inch Model 1888 gun and M1918M1 railway carriage.

No complete 8-inch Model 1888 gun on railway carriage exists, but a complete barrel with top carriage is on display as a public memorial in Plant Park, Tampa, Florida.

10-inch Gun Model 1888A1, 1888MIA1, and 1888MIIA1 On 10-inch Railway Carriage Model 1919

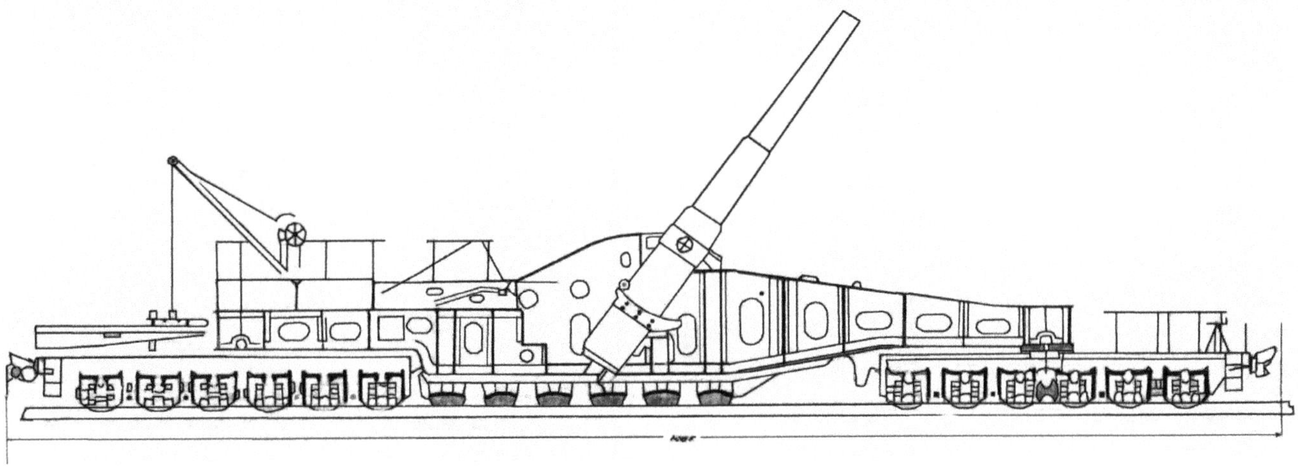

Army sketch of the 10-inch Model 1888 gun on Railway Carriage Model 1919.

In April 1917 the Army began securing forty 10-inch seacoast guns for use on railway carriages to send with the expeditionary force going to France. Thirty guns were removed from disappearing carriages at various defenses, and ten were of similar types kept as spares or at the proving ground or arsenals. They were a mix of Model 1888, 1888MI and 1888MII—all of similar dimensions and performance. It was intended to mount them on 36 railway carriages and hold four guns as replacement spares. The removals were done in the late summer of 1917; the original carriages being left in place as it was intended to replace the "borrowed" guns after the war (which usually never happened). Later plans in January 1918 proposed that additional 10-inch guns should be removed for use on an improved French Batignolles-type railway carriage, and in May the program was increased to 117 guns. While the guns were indeed removed from their seacoast disappearing carriages, none of these latter types of carriages were built for them.

The tubes were of typical built-up, gun steel construction, of 34 caliber length and with interrupted screw beach mechanism. There were 90 grooves with rifling of 1 in 50 increasing to 1 in 25 at the muzzle. The railway mount intended for these guns was a copy of the well-tried (and relatively easy to construct) French type of sliding mount. The simple mount consisted of placing the gun with trunnions in bearings on each side of the car side girder. The upper carriage was just a box of two steel side girders, deck plates and end transoms. There was no recoil mechanism, the gun recoil was constrained only by the friction of sliding along the sleepers by the rails. An elevating rack bolted to the right side provided a considerable -7 to +54-degree elevation. However, there was no traverse on the mount itself.

For operation steel I-beams were laid parallel to the tracks and wooden sleepers placed between the beam and the carriage which was lowered onto them by screw jacks. It was intended that this would be on a specially built curved section of track to optimize ability to move a short distance to get more traverse—as none was available from the carriage otherwise. After firing the entire car was to be physically moved back to its starting position. The railway car moved with two trucks of six axles each holding 12 36-inch wheels. There was a separate ammunition car and two jib cranes on the gun car to provide ammunition service.

The plan had been to fabricate the major parts of the 36 carriages in the U.S. and send them to France for final assembly. Fifteen gun tubes were sent to France in the fall of 1917. The Harrisburg Manufacturing & Boiler Co. made most of the gun car parts, with the Pullman Car Co. the trucks and the American Car & Foundry Co. the ammunition cars. By the armistice eight sets of parts had been produced, with four sets sent to France but not yet assembled when the war ended. Twenty-two sets of parts were reported ready by April 7, 1919. Apparently only 18 railway mounts were ultimately completed. At least one mount was completely assembled and tested at Aberdeen. None of the mounts ever was issued for service.

The 18 railway mounts with guns were declared obsolete and sold on February 25, 1932 to Michael Flynn Inc. of Brooklyn, New York. Several of the originally spare 10-inch M1888MIIA1 gun tubes were kept in the ordnance department inventory, presumably as spares for the few persisting coast artillery sites still in service, these too wcrc ncvcr again used and disappeared for scrap during the 1940s. No 10-inch railway carriage survives

10-inch Gun Model 1888, 1888MI, 1888MII, and 1895 On 10-inch Railway Carriage Model 1919

Gun Characteristics
Bore size: 10-inch
Construction: Steel, built-up
Tube length overall: 367.25-inches
Nominal caliber: 34-caliber
Tube and breech weight: 69,078-pounds
Usual carriage: 10-inch Railway Carriage Model 1919
Weight of gun and carriage: 387,100-pounds
Maximum elevation: 54-degrees
Loading: Separate
Shell weight: 617-pounds
Maximum muzzle velocity: 2,250-fps
Maximum range: 23,603-yards

Builders: Guns: Watervliet Arsenal and Bethlehem Steel
Carriages: Harrisburg Manufacturing and Boiler Co. and Pullman Car Co.

Number built: 18 completed carriages

A completed 10-inch Model 1888 Gun on Railway Carriage Model 1919. Few were completed, none got to Europe during the First World War, and they were discarded in the early 1920s.

12-inch Gun Model 1895A2 and 1895M1A2
On 12-inch Railway Carriage Model 1918M1

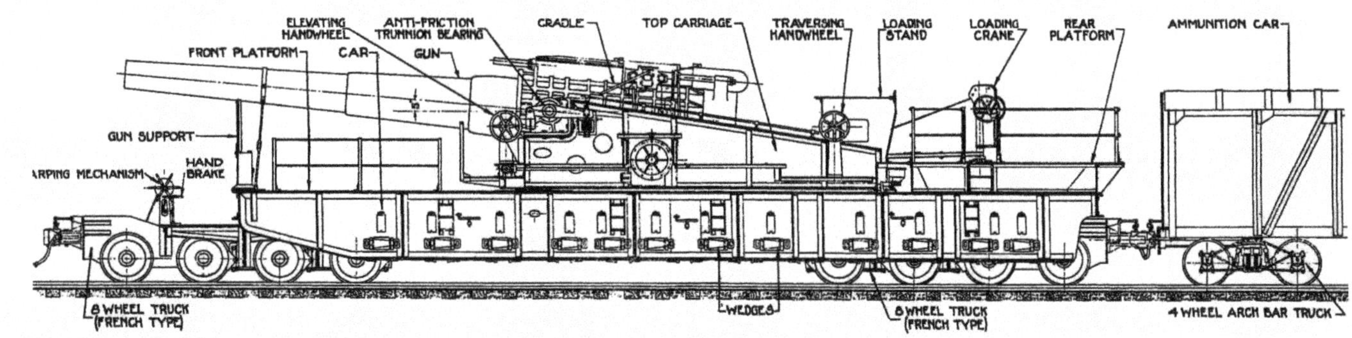

Army annotated sketch of the 12-inch Gun Model 1895A2 or 1895M1A2 on 12-inch Railway Carriage Model 1918.

During the First World War the Army approved a project for constructing twelve railway carriages for the 12-inch gun Model 1895. One of the best and most numerous of the Army's seacoast guns, there were guns available as operational spares—although they were also required for the new Model 1917 long-range barbette carriage just coming into service. By May 1918 twelve new railway carriages were slated for production and fourteen gun tubes identified and held for mounting—scheduled for completion in November and ready to be shipped to France. However, the project did not move quickly enough, the carriages were not finished until well into 1919 and 1920—so none ever reached Europe.

The Model 1895 12-inch gun was a development from the prior Model 1888. It had conventional tube and hoop arrangement and was of 35-calibers. They had 72 grooves and rifling of 1 in 50 increasing to 1 in 25. It had a simplified interrupted breech opening mechanism and mechanical firing device. Ninety of these guns were made by Watervliet and Bethlehem Steel. For use on rail carriages these guns were modified by removing the trunnions and installation of a new recoil band with heavy bronze keys or splines to hold the gun in its new cradle. So modified they became designated M1895A2 (or M1895M1A2). These changes added 4,300 pounds to the tube weight. Recoil system was two hydraulic cylinders on the bottom of the cradle and one pneumatic recuperator cylinder on the top. The clearance was such that no pit was required for firing at any elevation.

The carriage selected was the French Batignolles type. Unlike the all-around fire 8-inch type, and the fixed gun in simple sliding mount of the 10-inch type, the 12-inch was intended all along to be fired from a fixed ground carriage. The rail car was run over a prepared position with a six-section platform. The gun was lowered by means of wedges operated by ratchets onto the platform. No outriggers were required. A double elevation rack, one unit on each side of the cradle, could provide elevation from -5 to +38-degrees. A limited traverse of 5-degrees to each side of the centerline was possible through a worm gear operating between the two side frames. The car body was a simple box girder connected by suitable transoms. For operations the body was lowered on a special ground platform and attached through clips. This securely fixed the gun in place—though it was not a rotating base and no traverse other than the 10-degrees the carriage allowed was possible. It was estimated to take two hours to place the platform and gun. Each end of the carriage had an 8-wheeled (4 axle) truck. Ammunition supply was by a usual separate car with transfer being made by on-carriage jib.

Twelve carriages Model 1918 were fabricated at the Marion Steam Shovel Company (serial number 1-12). Twelve rail cars and 24 complementary ground platforms with cars were also constructed, along with the usual accompanying ammunition, machine shop and fire control cars. In the early 1920s a new all-around, full-traverse ground platform was developed and trialed using parts of an old 10-inch disappearing carriage racer. It was a success, but no additional units were produced.

Though retained in the postwar inventory, the call for the use of the gun was never issued. They were

12-inch Gun Model 1895A2 or 1895M1A2
On 12-inchRailway Carriage Model 1918

Gun Characteristics
Bore size: 12-inch
Construction: Gun steel, built-up
Tube length overall: 442.6-inches
Nominal caliber: 35 calibers
Tube and breech weight: 119,856-pounds
Usual carriage: 12-inch Railway Carriage Model 1918
Weight of gun and carriage: 334,900-pounds
Maximum elevation: 38-degrees
Loading: Separate
Shell weight: 700-pounds
Maximum muzzle velocity: 2,600-fps
Maximum range: 29,000-yards

Builders:Guns: Watervliet Arsenal and
Bethlehem Steel
Carriages: Marion Steam Shovel

Number built: 12 complete gun and carriage units

never issued to an active unit. Finally in March 1940 they were declared surplus and were put up for disposal. In 1942 two were supplied to the Navy facility at Dahlgren for experimental work. They were used for experimental firings and ballistic tests as late as the 1960s. One complete unit has been restored in recent years and is on display near the U.S. Army Ordnance Training Support Facility at Fort Lee, Virginia.

Veterans visiting the 12-inch Railway Carriage Model 1918 in 1931.

One 12-inch railway mount survives, this is 12-inch M1895M1A1 Watervliet No.19 on Railway Mount M1918 No. 9 Marion Steam Shovel now on display at Fort Lee, Virginia.

12-inch Gun Model 1918 (Chilean Gun)
On 12-inch Sliding Railway Carriage Model 1918

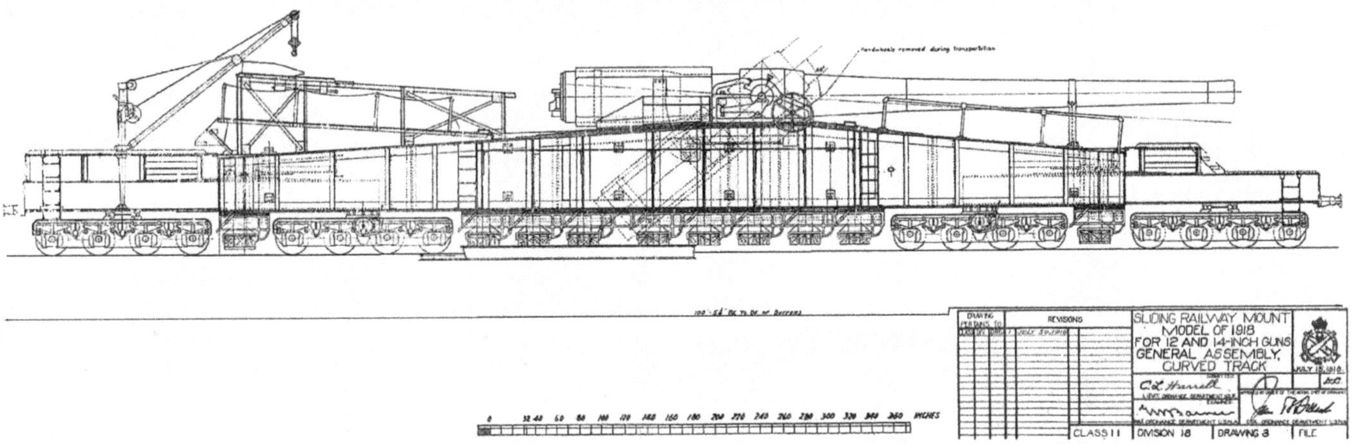

Army published sketch of the 12-inch Gun Model 1918 on 12-inch Sliding Railway Mount Model 1918. This was a mount specifically made to accommodate the Bethlehem-made 12-inch/50 caliber tubes taken over during the First World War from an order placed by the nation of Chile.

Bethlehem Steel was working on a large order of coast artillery materiel for the nation of Chile when the war intervened. The primary weapons were six very powerful 12-inch, 50-caliber, high-velocity guns on barbette carriage. After negotiations the guns and parts of their barbette carriages were purchased by the U.S. Army in early 1918. A project was quickly initiated to construct a simple sliding mount, not dissimilar to that being developed for the 12-inch guns removed from coast artillery disappearing emplacements. The guns and parts of the carriage continued to be completed at Bethlehem, the mount by Morgan Engineering the rail car carriages and girders were built at Baldwin Locomotive and the American Bridge Company. In June 1918 Bethlehem reported that five guns were done, and the sixth was just completing. All six guns were accepted by October 1918. Three carriages were ordered; it thought best to have one spare gun for each mount.

The Bethlehem guns were very powerful weapons. Not only were they a long 50-calibers, but they were also credited with a very high muzzle velocity of 3,200 feet/second. That was much in excess of other guns of this size. In fact, the Army had tried high muzzle velocity guns with their Model 1900 coast defense guns but bore erosion and gun life were so severe that it had to order reduced charges for most of these guns. How Bethlehem Steel was able to achieve this type of performance without impairing the life of the guns is unknown. The tube was built-up and was matched to a standard interrupted-screw type breech. The tubes were rifled with 72 grooves, with a twist increasing from 1/50 to 1/25.

The carriages were completed at Morgan Engineering in the last quarter of 1918. One went on a Liberty Loan train display, the other two went directly to Aberdeen. They followed the French Glissement form. The gun was essentially fixed on its carriage with only pivoting trunnions to allow elevation. An elevation of 40-degrees was possible. It had no lateral traverse capability outside of what the curvature of the track provided. Also, it had no recoil containment; simply rolling backward on the track after firing and needing to be restored to its firing position for a subsequent round. The carriage had two 4-axle/8-wheeled trucks forward and rear.

Tests of the first proof mount revealed several problems. The carriage proved of insufficient rigidity and had to be reinforced to prevent breakage of girders and transoms. The elevating gear also proved problematic and finding the right powder type and amount for a gun of such high velocity was difficult. After the test modifications to the other two mounts were delayed, but they were retained and classified only as war reserve.

The three carriages were finally authorized scrapped on July 29, 1931, and was duly accomplished. The gun tubes were kept in reserve by the Army for many years. Finally in 1941 they were authorized to be supplied to Brazil as part of a package of arms from the U.S. to its new ally. Coming full circle Brazil wished to use them in conventional coast defenses. While a major project was started to develop a new T-1 barbette carriage for ex-navy 12-inch 45-caliber gun tubes also transferred to Brazil; the old ex-Chilean guns may not ultimately have been sent. There are Army gun card notations that the guns were authorized for deletion on October 2, 1946.

12-inch Gun Model 1918
On 12-inch Sliding Railway Mount Model 1918

Gun Characteristics
Bore size: 12-inch (ex-Chilean)
Construction: Built-up Nickel Steel
Tube length overall: 622.1-inches
Nominal caliber: 50
Tube and breech weight: 172,000-pounds
Usual carriage: 12-inch Sliding Railway Mount Model 1918
Weight of gun and carriage: 550,000-pounds
Maximum elevation: 40-degrees
Loading: Separate
Shell weight: 700-pounds
Maximum muzzle velocity: 3,200-fps
Maximum range: 45,000-yards

Builders: Guns: Bethlehem Steel
Carriages: Morgan Engineering with Baldwin Locomotive

Number built: 6 guns, 3 carriages

Good photo of the ex-Chilean 12-inch gun on Model 1918 Railway mount in November, 1918.

12-inch Mortar Model 1890M1
On 12-inch Mortar Railway Carriage Model 1918MII

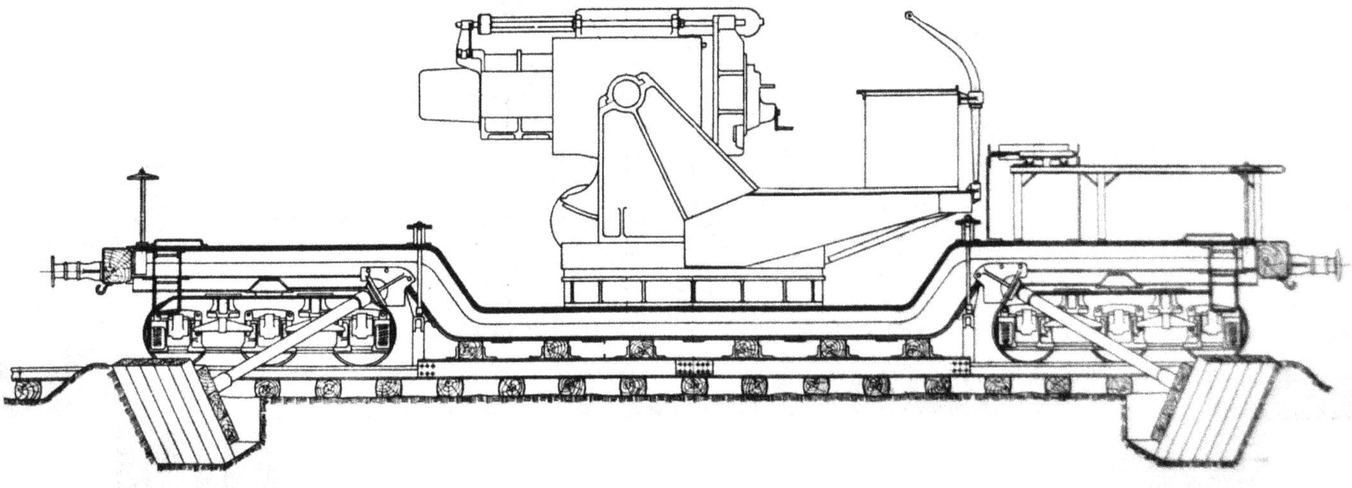

Army sketch of the 12-inch Mortar Model 1890M1 on 12-inch Mortar Railway Carriage Model 1918MII.

For the American Expeditionary Force, a large number of 12-inch seacoast mortars were available for use on railway carriages. Over 300 of the 10-caliber Model 1890 and 1890M1 mortars had been produced and emplaced in batteries in the continental U.S. and some insular forts. By removing two guns from pits normally containing four, 150 mortars could be spared. Fortuitously it had been found that the mortars were both more accurate than thought when originally emplaced, and that the emplacement pits were too crowded for efficient servicing anyway. Orders in September 1917 went out to reduce many of the pits. In some instances, entire defenses were abandoned anyway and the mortars made available, but for the territorial defenses it was not thought wise to make this reduction in the few batteries of this type actually emplaced.

The 12-inch bore tube was of built-up construction with numerous hoops or bands. An interrupted-screw breech mechanism supported by the mortar's jacket with the usual breech rotating crank. It had 72 grooves and rifling of 1 in 40 increasing to 1 in 20. The guns were produced from 1894 to 1902. Besides the first eight, all of the subsequent mortars were Model 1890M1 with a reduced number of bands. For the railway mounting it was fitted with a hydropneumatic recoil mechanism—with two cylinders on the bottom of the cradle. A pneumatic recuperator cylinder was mounted atop. The mortar was mounted in a sort of sleigh with two obvious, large rings and runners around the muzzle and breech ends. The sleigh was held in the cradle which fit in the side frames with trunnions. Elevation was by a segmented circular rack on the bottom of the cradle operated with a worm wheel. It could achieve -5 to +65-degrees of elevation.

The carriage was known as the Model 1918 and the railway car as the Model 1918MI or MII (only with minor differences). The gun car was the same basic type Model 1918 used for the 7-inch and 8-inch railway guns. The lower carriage sat on a barbette racer that allowed a full 360-degree traverse. An ammunition table was to the rear of a working platform, along with a jib crane for ammunition transfer. The car was lowered by four screw jacks onto steel I-beams laid across wooden supports on the rail crossties. The increased weight of the mortar did mean that the car had three axles and six wheels (of 28-inch diameter) on each of the two trucks. Narrow-gauge trucks could be substituted for standard gauge as needed. In firing position, it used the same type of outriggers and floats as the 8-inch railway car.

The production of both the carriages and rail cars was given to the Morgan Engineering Co. The firm had to build a special new plant for the work. The first production unit was finished in August 1918. By April 1919 45 units of the 91 ordered were delivered. None of the mounts ever made it to Europe. After the war the finished units were either stored (initially at

Fort Eustis, then after 1931 at Aberdeen) or issued to a few coast artillery regiments. One unit of 8 railway mounted mortars was sent to the Hawaiian Department in 1922. There they were converted to the 36-inch, narrow gauge used by the railways on Oahu. In 1936 there were still 13 carriages M1918M1 and 57 of M1918MII in Army inventory. During the Second World War the mortars on railway cars stayed in storage or at the few posts that maintained mobile coast artillery units—none were deployed to combat theatres.

In 1943 it was proposed to scrap the remaining units, but instead they were authorized for retention as reserve material until right after the war. No surviving examples of the railroad mount or car for these mortars persists.

12-inch Mortar Model 1890M1
On 12-inch Mortar Railway Carriage Model 1918MII

Gun Characteristics
Bore size: 12-inch
Construction: Built-up Gun Steel
Tube length overall: 160.32-inches
Nominal caliber: 10-caliber
Tube and breech weight: 29,120-pounds
Usual carriage: 12-inch Mortar Railway Carriage Model 1918MII
Weight of gun and carriage: 176,800-pounds
Maximum elevation: 65-degrees
Loading: Separate
Shell weight: 700-pounds
Maximum muzzle velocity: 1,500-fps
Maximum range: 15,291-yards

Builders: Mortars: Watervliet, Bethlehem Steel, Niles Tool, Builders Iron Foundry
Carriages: Morgan Engineering Co.

Number built: 91 railway carriages

The 12-inch M1918MII mount and mortar showing the extended supports for the firing configuration.

Railroad train of cars with the 12-inch mortar on M1918 rail cars at Cape Henlopen, Lewes, DE

12 inch mortar on M1918 railway mounts of the 52nd Coast Artillery Regiment at Fort Hancock

There was much anticipation of success for a new 12-inch howitzer mounted on railway carriage in the 1918-1920 period, though in the end only a single prototype was fabricated. This is an Army sketch of the proposed gun.

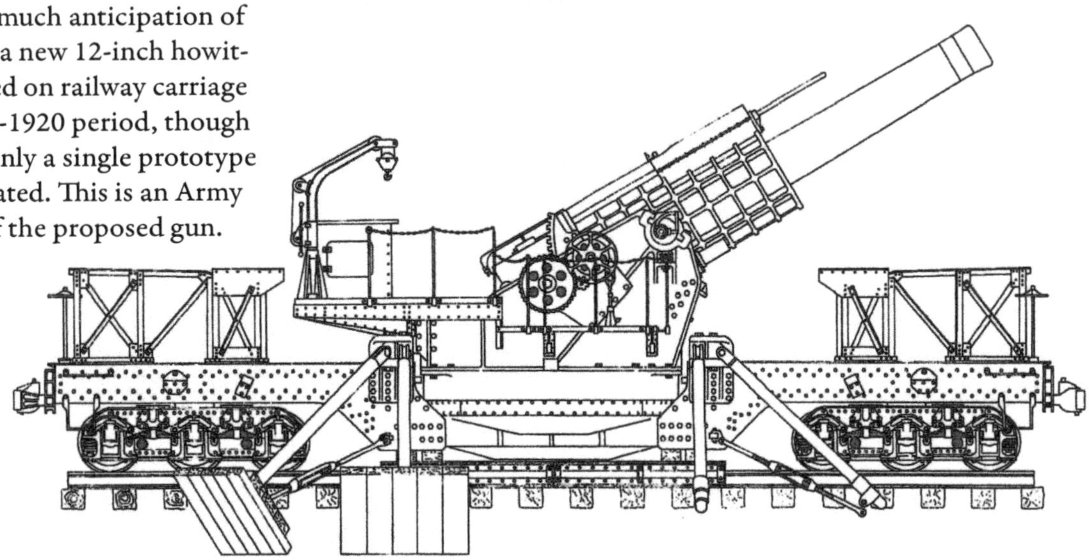

14-inch Gun Mk IV
On 14-inch Naval Railway Carriages Mk I and Mk II

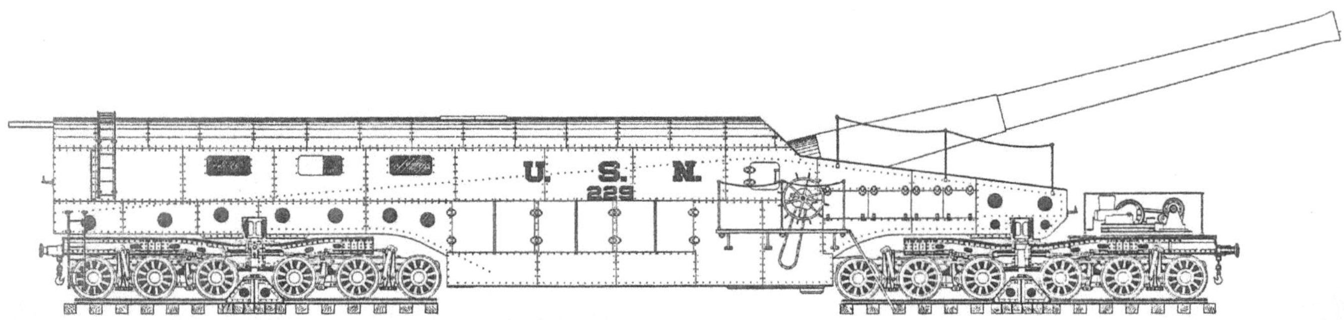

Sketch of the Navy's 14-inch Gun Mk IV on 14-inch Naval Railway Mount Mk I.

The U.S. Navy launched its own hurried project for a large-bore railway gun and mount after the entry of the country into the war. On November 12, 1917 the navy Bureau of Ordnance pointed out the advantages that could be obtained from the excess 14-inch/50-caliber tubes on hand. Several dozen of these guns, originally intended as main armament for the Lexington-class battlecruisers, were being completed by several arsenals and contractors. The Navy approved a project to design and construct railroad carriages for these on November 20th. With great urgency—and impressive results—this project was executed. The Washington Navy Yard completed the plans in January 1918, and Baldwin Locomotive selected to build the carriages and gun cars with Standard Steel Car Co. to supply the other train service cars.

Initially five trains, each with one gun car, were ordered. These mounts were delivered in April and May of 1918. They were shipped to France, assembled at St. Nazaire, and the first shot fired in combat in September 1918. The five Navy guns, commanded as a unit by Rear Admiral C.P. Plunkett were all in active service by the armistice. After the armistice the gun cars were returned to the U.S. and ultimately turned over to the Army. Most of the remaining train support cars remained in France to aid with the demobilization.

The tube was the navy Mk IV 14-inch, 50-caliber battleship gun. It was of conventional, built-up construction, with Welin drop breechblock. The tube had 126 grooves and a uniform twist of 1 in 32.

The carriage was a relatively simple rolling mount. Recoil mechanism consisted of one hydraulic cylinder and spring counter recoil cylinders. The gun could be elevated to its full 43-degrees only by excavating a recoil pit beneath the breech. Then the trucks could be removed and the carriage lowered with screw jacks on to a portable, cast steel foundation. Traverse was limited to just 5-degree total by shifting on the forward support of this foundation. Elevation was achieved with a large screw and oscillating net. The girder box-like rail car was covered with a curved, removable "cover shield". Transport on rail was facilitated with four trucks of three axles and six wheels each.

The Army was impressed with the quick availability, and after observing the proofing of the Navy gun in April 1918 placed two orders of three each for their own service. While completed, none were done by war's end. Eight of the 14-inch Mk I carriages still in Army hands at Fort Eustis were sold for scrap in May of 1931. One of these guns and railway carriages survives, now on display at the Navy museum at the Washington Navy Yard.

In October 1918 a new carriage design was begun by the Navy (actually at their contractor Baldwin Locomotive) to correct some of the Mk I deficiencies. In the new design the gun was carried on a lowered level and could be raised (a total of 61-inches) by hydraulic cylinders to a firing level without the necessity of digging a recoil pit. Maximum elevation was reduced to 40-degrees. The overhead cover shield was eliminated. Weight was better distributed with two trucks of ten axles each. Plans were made to construct five mounts, to be ready in early 1919. With the war's end

14-inch Gun Mk IV
On 14-inch Naval Railway Mount Mk I

Gun Characteristics
Bore size: 14-inch
Construction: Built-up Nickel Steel
Tube length overall: 724-inches
Nominal caliber: 50
Tube and breech weight: 192,500-pounds
Usual carriage: 14-inch Naval Railway Mount Mk I
Weight of gun and carriage: 539,760-pounds
Maximum elevation: 43-degrees
Loading: Separate
Shell weight: 1,560-pounds
Maximum muzzle velocity: 2,650-fps
Maximum range: 45,000-yards

Builders: Guns: Watervliet and Bethlehem Steel
Carriages: Baldwin Locomotive

Number built: 5 for U.S. Navy, 6 for U.S. Army

14-inch Gun Mk IV
On 14-inch Naval Railway Mount Mk II

Gun Characteristics
Bore size: 14-inch
Construction: Built-up Nickel Steel
Tube length overall: 724-inches
Nominal caliber: 50
Tube and breech weight: 192,500-pounds
Usual carriage: 14-inch Naval Railway Mount Mk II
Weight of gun and carriage: 610,000-pounds
Maximum elevation: 40-degrees
Loading: Separate
Shell weight: 1,560-pounds
Maximum muzzle velocity: 2,650-fps
Maximum range: 45,000-yards

Builders: Guns: Watervliet Bethlehem Steel
Carriages: Baldwin Locomotive

Number built: 5 begun, only 2 completed

three were cancelled, but two completed and evaluated in August 1919 at Dahlgren as the Mk II carriage. Soon they were also turned over to the Army. In Army hands the two 14-in railway carriages Mk II were authorized for scrapping in mid-1931.

A newly assembled 14-in Navy Railway Mk I at St Nazaire, France prior to being sent to the front.

Both the U.S. Navy and U. S. Army acquired the Navy-developed 14-inch Railway Mounts Mk I. This is a navy unit on February 11, 1919.

Five improved 14-inch Gun Mk IV on 14-inch Naval Railway Mount Mk II carriages were started, but only two were ever completed.

A nicely preserved Navy 14-inch gun Mk IV in its Mk I Railway mount exists and is on permanent display at the National Museum of the U. S. Navy in Washington DC

Interwar Railway Artillery Developments

Experience with the expedient 8-inch, 12-inch, and 14-inch Navy mounts demonstrated that significant improvements in railway artillery design were achievable once the war ended. With proper engineering resources and—most importantly—time, far more capable mounts could be developed. A series of experimental carriages were produced in the early 1920s. The severe financial restrictions that followed the war curtailed several projects, but not before a new, relatively sophisticated 14-inch mount could be standardized and a limited number produced.

Much of this development work should be credited to the Coast Artillery Corps. The postwar Army was quickly reduced to minimal size, though relatively well-equipped with the divisional and heavy weapons acquired during or immediately after the war. Facing similar cutbacks, the Coast Artillery recognized the utility of large, mobile weapons that could be rapidly deployed to threatened harbors. The service organized new regiments equipped with 155mm Model 1917-1918 field guns and 8-inch gun and 12-inch mortar railway equipment. It also advocated for a new, very heavy gun of exceptional range. The only available weapon meeting this requirement was the limited stock of Navy-type Mark IV 14-inch guns, so a new mount was developed around them.

Still utilizing the recently transferred 14-inch/50-caliber ex-Navy guns, a new carriage designed to be fired primarily from stable fixed platforms was developed for coast artillery service. In Army service the ex-Navy guns received new serial numbers and the designation Model 1920, which also became the model year designation for the new railway carriage.

The final production railway mount was a new 8-inch type developed throughout the 1920s. The program for a new railway carriage capable of carrying either a short 20-caliber 12-inch howitzer or a new long 50-caliber 8-inch gun began in 1919. Only a single example of the howitzer was produced, along with a strengthened, stable Model 1918 rail car featuring a fully traversing top carriage. The howitzer project was discontinued in the mid-1920s, but the carriage underwent multiple modifications to eventually become the 8-inch Railway Carriage M1 in 1938. Armament became available through the Army's acquisition of seventy-four ex-Navy Mark VI 8-inch/45-caliber gun tubes. Between 1940 and 1942, twenty-eight new carriages joined the original prototype for service in domestic coast defense during the war.

Though never reaching production or even prototype stage, one final project merits mention. In April 1944 the Army Ordnance Department initiated design of a railway carriage for a standard coast artillery 16-inch Mark II gun, to be transported on a simple railway carriage similar to the 14-inch M1920. The weapon was to be emplaced in a special pit with steel pintle and roller. The perceived requirement was for the heaviest possible gun for use against German fortifications during the anticipated campaign into Germany. For rapid production the design utilized existing seacoast guns, cradles, and recoil mechanisms already in service. As mounted, the weapon was expected to fire at 45-degrees elevation with 60-degrees traverse, using the special Mark XIII 2,240-pound projectile. Though design work progressed, the project was cancelled when it became clear the weapon would arrive too late for the European war.

14-inch railway gun at Fort MacArthur, CA

8-inch Gun Mk VI Mod 3A2 On 8-inch Railway Carriage M1A1

Technical Manual sketch of the 1940s 8-inch Gun Mark VI M3A2 on Railway Mount Model M1A1.

The 1919 project for a new railway mount that could carry either a short, 20-caliber 12-inch howitzer or a projected new 8-inch, 50-caliber gun was cancelled in the late 1920s without seeing production of more than a prototype. However, when this project ended, the promising carriage was retained. It was essentially a more robust version of the Model 1918 type used for the wartime 8-inch and 12-inch mortars. It had a strengthened well (of cast steel rather than structural steel parts) and improved outriggers for stability. In 1922 the Navy transferred in two increments a total of 74 excess 8-inch, 45-caliber Mk VI guns which had been taken from various pre-Dreadnaughts upon their postwar scrapping. In June 1930 the Army initiated a new project to adapt the previous Model 1918 12-inch howitzer railway carriage to the Navy gun. In June 1930 a separate project was started for a conventional barbette, fixed seacoast mount for these same type of guns, using some structural parts of the railway carriage.

The Navy gun was the Mark VI Mod 3 8-inch gun of conventional built-up type with an interrupted screw, Welin-type Mark V beech mechanism. It was fired with the usual electric and percussion firing mechanism. It had grooves and a uniform rifling twist of 1/25. In Army service the tubes were relined and modifications made to the powder chamber and tapering to fit army projectiles. Recoil was minimized with two hydraulic cylinders and one recuperator cylinder, all mounted under the gun tube.

The M1 carriage was a dropped-well railway car. In the well was a circular foundation for a fully rotating top carriage. That carriage had the usual side frames holding the gun in a cradle. To the rear of the breech was an elevated steel plate breech platform. Elevation and traversing were by worm and spar gear. To allow the gun to be mounted far forward and minimize the depth of recoil, a heavy (and obvious in photographs) counterweight was fitted to the breech end. This also eliminated any need for equilibrators. Two cranes were mounted to lift ammunition. The rail car moved on two 3-axle/6-wheeled trucks. One of the new features of this design were the eight tubular steel outriggers with floats that helped stabilize the car when in firing position.

The modifications to the Navy gun and revisions to the carriage were begun in earnest in 1934. The rebuilt M1 carriage (No. 1) was completed in 1938, and the combination was approved as standard in September 1939. After changes with some improvements and additional on-carriage fire control equipment, all subsequent carriages were designated M1A1. Two were ordered from Watertown in 1937 and four more in early 1939. The final 22 (making 29 in all, including the single M1 prototype) were ordered from Baldwin Locomotive on September 27, 1939. The completed guns were delivered from December 1940 to January 1942. They were all assigned to Coast Artillery mobile regiments (usually eight mounts per unit). A few went to Los Angeles and the Pacific Northwest, but most stayed at East Coast posts (Fort Custis, Hancock, and Miles) through the war. None saw combat usage.

The 8-inch railway mount was declared obsolete after the end of the war, and eliminated by selling for scrap in the late 1940s. No intact surviving carriages exist, though partial remains of one is at the Fort Miles Museum and parts of other gun tubes at some fixed barbette locations do still exist.

8-inch Gun Mark VI M3A2
On Railway Mount Model M1A1

Gun Characteristics
Bore size: 8-inch
Construction: Built-up, nickel steel
Tube length overall: 369-inches

Nominal caliber: 45-caliber
Tube and breech weight: 42,000-pounds
Usual carriage: 8-inch Railway Mount M1A1
Weight of gun and carriage: 230,000-pounds
Maximum elevation: 45-degrees
Loading: Separate
Shell weight: 260-pound (AP)
Maximum muzzle velocity: 2,840-fps
Maximum range: 32,000-yards

Builders: Guns: Washington Navy Yard, Watervliet Arsenal
Carriages: Watertown (No. 1-7), Baldwin Locomotive (No. 8-29)

Number built: 29 carriages with guns

An 8-inch Mark VI M3A2 gun on new type Model M1A1 Railroad Carriage at Watertown Arsenal.

14-inch Gun Model 1920 and 1920M1 On 14-inch Railway Mount Model 1920

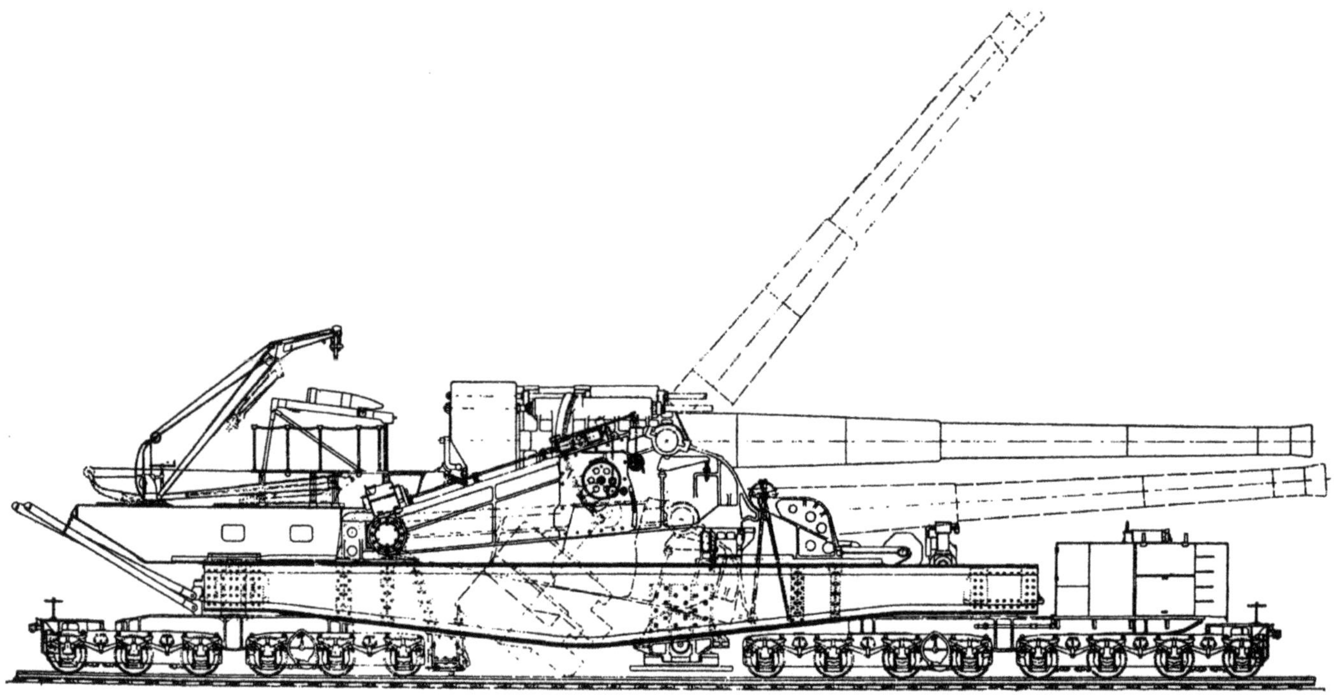

Army sketch of the 14-inch Gun Model 1920, 1920M1 and 1920MII on 14-inch Railway Mount Model 1920.

Several experimental 14-inch carriages were being developed when the First World War ended. The Army's own Model E prototype was undergoing evaluation, and a new Model 1918 sliding mount and a similar Schneider-inspired Model 1919 was scheduled for production (though cancelled soon after the war's end). The gun tube situation was fortuitously in a good position. The Navy possessed numerous 14-inch Mk IV 50-caliber tubes. Sixty of these modern guns had been used to arm the Mississippi and California classes a battleships. They were also earmarked for the six battlecruisers of 1916 until a design change switched the armament to 16-inch guns. Many guns were physically ordered and being fabricated.

On May 10, 1918 the Navy first offered a number of these guns to the Army. Some were diverted for use in the Navy's Mk I and Mk II railroad mounts, but eventually these and others done or partly finished (42 in all) were obtained by the Army. After completing to Army specifications most of these tubes emerged renumbered and reclassified as Model 1920 14-inch guns. The tube was a conventional built-up alloy gun, with navy-type drop block and step-cut interrupted breech mechanism. It had 126 grooves and the initial model a uniform 1 in 32 twist. (Note: A number of official reports credit the gun with just a rifling of 84 grooves, whether the guns were at some point relined with new rifling or this is due to some other recording error is unknown). At 50-calibers and a 2,650-fps muzzle velocity it was a very powerful weapon, though plagued with dispersion issues in its early career. Minor differences in liners and rifling differentiated three modifications (M1920, 1920M1 and 1920MII).

The railway carriage featured a gun in a heavy waffle-braced cradle that could travel in a lowered position but raised for firing to allow a higher elevation and deeper recoil. The top carriage was raised by the same lifting mechanism used on the Model E prototype. While primarily designed to fire from a prepared ground platform, it could be operated from tracks. In that case it utilized six outrigger struts for stabilization, the gun in this mode was capable of just 19-degree elevation, but it had 7-degrees of traverse moving the top carriage between the car body walls. Preferred firing was from a prepared platform, a forward pintle on the bottom of the carriage was lowered over the center of the base and a rear roller transom rotated around a circular ring. In this position it could attain a 50-degree elevation using a segmented rack and gears on the side of the cradle. Maximum range was rated at 45,000-yards.

Recoil was limited to just 35-37-inches by four hydraulic recoil cylinders and two hydraulic counter recoil cylinders. The carriage was supported for moving with two trucks forward and rear. Also forward was a self-contained power plant on a front span bolster with a Sterling gasoline engine and a 50-Kw electrical generator. With the mount was an M1 elevation quadrant, M8 panoramic telescope and M20 telescope mount.

The first unit was delivered in 1921 and demonstrated some teething problems. The design was adopted for limited procurement in September 1924, and three more carriages were ordered. Initially the Army projected obtaining 25 units, but funding for just four was appropriated. All four were eventually assigned to coast artillery roles. Two units went to the Panama Canal and fixed platforms were constructed on both sides of the isthmus in 1928-30 (Fort Grant and Fort Randolph). Another pair went to the Harbor Defenses of Los Angeles in 1925-26 and served at Fort MacArthur. These continued to serve during the war in this role. The four mounted guns were ordered salvaged in June 1946. The remaining unused gun tubes were scrapped in 1947 except for one M1920 MII tube recently confirmed as remaining at Dahlgren.

14-inch Gun Model 1920, 1920M1 and 1920MII On 14-inch Railway Mount Model 1920

Gun Characteristics
Bore size: 14-inch
Construction: Built-up Nickel Steel
Tube length overall: 724.-inches
Nominal caliber: 50
Tube and breech weight: 233,782-pounds
Usual carriage: 14-inch Railway Mount Model 1920
Weight of gun and carriage: 730,000-pounds
Maximum elevation: 50-degrees
Loading: Separate
Shell weight: 1,560-pounds
Maximum muzzle velocity: 2,650-fps
Maximum range: 45,000-yards

Builders: Guns: Watervliet Arsenal and Bethlehem Steel
Carriages: Watertown Arsenal

Number built: 26 guns acquired by the army, 4 used on 4 railway carriages

The completed 14-inch Model 1920 Gun and Railway Mount. Photograph taken at Watertown Arsenal in the early 1920s. Four carriages were completed, two were sent to Los Angeles and two to Panama.

One of the two 14-inch Model 1920 Railway Guns in Panama on October 27, 1932.
Prepared firing platforms were built on both the Pacific and Atlantic coasts.

A14-in Model 1920 Railway Gun at a 1937 remote shoot on the coast of Southern California,
photo taken by Col. George Ruhlen.

Surviving American Seacoast Artillery Weapons

Compiled by by the Coast Defense Study Group
March 2026

General Note: This is an attempt to list surviving weapons (or the same model/type) that were used by the American armed forces, either in the U.S. and overseas, in a coast defense role in the "modern era" (1890 to 1950). Items to be included in this list must retain at least the whole gun/mortar/howitzer. Surviving weapons from earlier periods, muzzleloading cannon (rifled and smoothbore), field artillery (except for 155mm), mobile anti-aircraft guns, and British, Canadian, or Mexican coast artillery are excluded from this list, even if used in North America. In a few cases, weapons have been included because they represent weapons similar to those used for coast defense and are sufficiently interesting to warrant inclusion. Every effort has been made to make this list as accurate as possible, but it will probably contain a number of errors and omissions.

The list is arranged by caliber (largest to smallest). The first line contains data about the weapon: the quantity at the site, the caliber of weapon (inches or millimeters), the model, serial number, place of manufacture, and carriage information, if known. The next lines contain information on where the weapon was previously located as coast defense weapon. The last line contains data about the location of the weapon: battery or site where the weapon is currently located, fort or installation, geographic location, and city and/or country.

Sources: Matt Adams, Tom Batha, Mark Berhow, Roger Davis, Colt Denfeld, Elliott Deutsch, Bill Dorrance, Clarence Drennon, Al Grobmeier, Greg Hagge, Ed Jerue, Nelson Lawry, Ray Lewis, Charles Kimball, David Kirchner, Danny Malone, John Martini, Dean Mayhew, Terry McGovern, Jim Parker, Dan Rowbottom, Bob Rutherford, Phil Sims, Peter Smith, Dick Weinert, Glen Williford, David Yeager, and Bob Zink.

16 Inch Guns

1. 16"/50 Mk III Mod 1, ______ No. 138 on M1919 Barbette Carriage Watertown Arsenal No. 1
   Location: at the Former Aberdeen Proving Ground, MD
2. 16"/50 Mk II, Naval Gun Factory No. 111
   Location: U.S. Navy Museum, Washington Navy Yard, Washington, DC
3. 16"/50 Mk II _____ No. 96
   Location: Naval Surface Warfare Center, Dahlgren, VA
4. 16"/50 Mk II_______ No. 100
   Location: Naval Surface Warfare Center, Dahlgren, VA
5. 16"/50 Mk II_______ No. ____
   Location: Naval Surface Warfare Center, Dahlgren, VA

14 Inch Guns

1. M1909, Watervliet No. 4 on M1909 Turret Mount, Newport News No. 1
   Location: Battery John M. Wilson, Fort Drum, El Fraile Island, PI
2. M1909, Watervliet No. 3 on M1909 Turret Mount, Newport News No. 1
   Location: Battery John M. Wilson, Fort Drum, El Fraile Island, PI
3. M1909, Watervliet No. 2 on M1909 Turret Mount, Newport News, No. 2
   Location: Battery William L. Marshall, Fort Drum, El Fraile Island, PI

4. M1909, Watervliet No. 1 on M1909 Turret Mount, Newport News, No. 2
   Location: Battery William L. Marshall, Fort Drum, El Fraile Island, PI
5. M1910, Watervliet No. 15 on M1907M1 Disappearing Carriage, Watervliet No. 20
   Location: Battery Gillespie, Fort Hughes, Caballo Island, PI
6. M1910, Watervliet No. 8 on M1907M1 Disappearing Carriage, Watervliet No. 17
   Location: Battery Woodruff, Fort Hughes, Caballo Island, PI
7. 14"/50 Mk II, _____ No. 119L2 on Mk 1 Railway Mount No. 148
   Location: U.S. Navy Museum, Washington Navy Yard, Washington, DC
8. 14"/50 M1920 MkII No. 26
   Location: Naval Surface Warfare Center, Dahlgren, VA

12 Inch Guns

1. M1895M1A4, Watervliet No. 1 on M1917 Barbette Carriage, Poole Engineering & Machine Co. No. 31
   Location: Battery Smith, Fort Mills, Corregidor, PI
2. M1895M1A4, Watervliet No. 6 on M1917 Barbette Carriage, Poole Engineering & Machine Co. No. 30
   Location: Battery Hearn, Fort Mills, Corregidor, PI
3. M1895M1A4, Watervliet No. 8 (spare tube) Battery Hearn, Fort Mills, Corregidor, PI
   Location: Baywalk, Dolomite Beach, Manila for mock turret display
4. M1895, Watervliet No. 13 on M1901 Disappearing Carriage, Watertown No. 14
   Location: Battery Crockett, Fort Mills, Corregidor, PI
5. M1895, Watervliet No. 27 on M1901 Disappearing Carriage, Watertown No. 15
   Location: Battery Crockett, Fort Mills, Corregidor, PI
6. M1895, Bethlehem No. 9 (spare tube) Battery Crockett, Fort Mills, Corregidor, PI,
   Location: Baywalk, Dolomite Beach, Manila for mock turret display
7. M1895, Watervliet No. 37 on M1901 Disappearing Carriage, Watertown No. 16
   Location: Battery Cheney, Fort Mills, Corregidor, PI
8. M1895, Watervliet No. 12 on M1901 Disappearing Carriage, Watertown No. 17
   Location: Battery Cheney, Fort Mills, Corregidor, PI
9. M1895, Watervliet No. 16 (spare tube) Battery Cheney, Fort Mills, Corregidor, PI,
   Location: Bottomside Area, Fort Mills, Corregidor, PI
10. M1895, Watervliet No. 36, with partially scrapped Disappearing Carriage
    Location: Battery Wheeler, Fort Mills, Corregidor, PI
11. M1895, Bethlehem No. 7 on M1901 Disappearing Carriage, Watertown No. 2
    Location: Battery Wheeler, Fort Mills, Corregidor, PI
12. M1895, Bethlehem No. 10 (spare tube)
    Location: Battery Wheeler, Fort Mills, Corregidor, PI
13. M1895 Watervliet No. 19 on M1918 Railway Mount, Marion Steam Shovel No. 9
    Location: US Army Ordnance School, Fort Lee, VA

12 Inch Mortars

1. M1890M1, Watervliet No. 173 on M1896M1 Carriage, Builders Iron Foundry No. 158
   Location: Battery Way, Fort Mills, Corregidor Island, PI
2. M1890M1, Watervliet No. 174 on M1896M1 Carriage, Builders Iron Foundry No. 151
   Location: Battery Way, Fort Mills, Corregidor Island, PI

3. M1890M1, Watervliet No. 170 on M1896M1 Carriage, Watertown No. 241
Location: Battery Way, Fort Mills, Corregidor Island, PI
4. M1890M1, Watervliet No. 172 on M1896M1 Carriage, Builders Iron Foundry No. 150
Location: Battery Way, Fort Mills, Corregidor Island, PI
5. M1890M1, Bethlehem No. 31 on M1896M1 Carriage, _____ No. 104
Location: Battery Geary Pit A, Fort Mills, Corregidor Island, PI
6. M1890M1, Bethlehem No. 40 on M1896M1 Carriage, ____ No. 210
Location: Battery Geary Pit A, Fort Mills, Corregidor Island, PI
7. M1890M1, Watervliet No. 22 on M1896M1 Carriage, American Hoist No. 183
Location: Battery Laidley Pit A, Fort Desoto, Mullet Key, FL
8. M1890M1, Watervliet No. 86 on M1896M1 Carriage, American Hoist No. 184
Location: Battery Laidley Pit A, Fort Desoto, Mullet Key, FL
9. M1890M1, Watervliet No. 132 on M1896M1 Carriage, American Hoist No. 185
Location: Battery Laidley Pit B, Fort Desoto, Mullet Key, FL
10. M1890M1, Watervliet No. 135 on M1896M1 Carriage, American Hoist No. 187
Location: Battery Laidley Pit B, Fort Desoto, Mullet Key, FL
11. M1908, Watervliet No. 2 on M1908 Carriage, No. 17
Location: Battery Geary Pit B, Fort Mills, Corregidor Island, PI
12. M1908, Watervliet No. 22 on M1908 Carriage No. 20
Location: Battery Geary Pit B, Fort Mills, Corregidor Island, PI
13. M1908, Watervliet No. 13 on M1908 Carriage No. 19
Location: Battery Geary Pit B, Fort Mills, Corregidor Island, PI
14. M1908, Watervliet No. 12on M1908 Carriage No. 18
Location: Battery Geary Pit B, Fort Mills, Corregidor Island, PI
15. M1912, Watervliet No. 40 on M1896MIII Carriage, Watertown No. 41
Location: Battery Craighill Pit A, Fort Mills, Corregidor Island, PI
16. M1912, Watervliet No. 39 on M1896MIII Carriage, Watertown No. 40
Location: Battery Craighill Pit A, Fort Mills, Corregidor Island, PI
17. M1912, Watervliet No. 41 on M1896MIII Carriage, Watertown No. 38
Location: Battery Craighill Pit B, Fort Hughes, Caballo Island, PI
18. M1912, Watervliet No. 38 on M1896MIII Carriage, Watertown No. 39
Location: Battery Craighill Pit B, Fort Hughes, Caballo Island, PI
19. M1911, ____ No. 8, no carriage (non- standard training dummy)
Location: Visitor Center, Fort Wadsworth, Staten Island, NY

10 Inch Guns

1. M1895M1, Watervliet No. 25 on M1901 Disappearing Carriage, Watertown No. 14
Location: Battery Grubbs, Fort Mills, Corregidor Island, PI
2. M1895M1, Watervliet No. 22 on M1901 Disappearing Carriage, Watertown No. 16
Location: Battery Grubbs, Fort Mills, Corregidor Island, PI
3. M1895M1, Watervliet No. 20 (spare tube only)
Location: Battery Grubbs, Fort Mills, Corregidor Island, PI
4. M1895M1, Watervliet No. 26 on M1901 Disappearing Carriage, Watertown No. 13
Formerly emplaced at Battery Warwick, Fort Wint, PI
Location: Battery Worth, Fort Casey, Coupeville, WA

5. M1895M1, Watervliet No. 28 on M1901 Disappearing Carriage, Watertown No. 15
Formerly emplaced at Battery Warwick, Fort Wint, PI
Location: Battery Worth, Fort Casey, Coupeville, WA
6. M1888, Watervliet No. 41
Formerly emplaced at Battery Harker, Fort Mott, NJ
Location: Fort Cape Spear, St John's, Newfoundland
7. M1888, Watervliet No. 3
Formerly emplaced at Battery Harker, Fort Mott, NJ
Location: Fort Cape Spear, St John's, Newfoundland
8. M1888, Watervliet No. 12, on M1893 Barbette Carriage, Watertown No. 11
Formerly emplaced at Battery Quarles, Fort Worden, WA
Location: Fort McNutt, McNutt Island, Nova Scotia
9. M1888, Watervliet No. 37, on M1893 Barbette Carriage, Watertown No. 1
Formerly emplaced at Battery Quarles, Fort Worden, WA
Location: Fort McNutt, McNutt Island, Nova Scotia (Note: Partially scrapped)

8 Inch Guns

1. M1888M1A1 Watervliet No. 32 on M1918M1 Barbette Carriage, Morgan Engineering No. 9
Formerly emplaced at Battery Bowyer, Fort Morgan, AL
Location: Plant Park, Tampa, FL
2. Navy 8"/45 Mk VI M3A2 _____No. 160L2
Location: Battery 404, Fort Abercrombie, Kodiak, AK
3. Navy 8"/45 Mk VI M3A2_____No. 154L2
Location: Battery 404, Fort Abercrombie, Kodiak, AK
4. Navy 8"/45 Mk VI M3A2_______No. 134L2
Formerly emplaced at Battery 403, Fort J.H. Smith, Kodiak, AK
Location: Kodiak Airport, Kodiak, AK
5. M1888MII, Bethlehem No. 8
Location: Bottomside artillery park, Fort Mills, Corregidor Island, PI
6. Navy 8"/45 Mk VI M3A2______No.______on M1 Railway Proof Mount Baldwin No.____
Note: bored out to 9.12" for experimental purposes, missing 8 ft. from muzzle
Location: Fort Miles Museum, Cape Henlopen State Park, DE

7 Inch Naval Guns

1. 7"/45 Mk II, Midvale No. 90 on Mk II Mod 4 Pedestal Mount No. ___
Formerly armed USS New Hampshire (BB-25)
Formerly emplaced at Harbor Battery, Sand Island, HI
Location: Battery Randolph, Fort DeRussy, Oahu, HI
2. 7"/45 Mk II, Midvale No. 92 on Mk II Mod 4 Pedestal Mount No. ___
Formerly armed USS New Hampshire (BB-25)
Formerly emplaced at Harbor Battery, Sand Island, HI
Location: Battery Randolph, Fort DeRussy, Oahu, HI
3. 7"/45 Mk II, Naval Gun Factory No. 100 on Mk II Mod 4 Pedestal Mount No. 43
Location: Battery North, Bora Bora, Society Island, French Polynesia, Also reported
As being relocated to Papeete, Tahiti.

4. 7"/45 Mk II, Bethlehem No. 97 on Mk II Mod 4 Pedestal Mount No. 55
Formerly emplaced at Battery North, Bora Bora, Society Island, French Polynesia
Location: private yard in Bora Bora.
5. 7"/45 Mk II, Watervliet No. 57 on Mk II Mod 4 Pedestal Mount No. 23
Formerly armed USS Louisiana (BB-19).
Location: Battery West, Bora Bora, Society Island, French Polynesia
6. 7"/45 Mk II, Naval Gun Factory No. 82 on Mk II Mod 4 Pedestal Mount No. 30
Location: Battery West, Bora Bora, Society Island, French Polynesia
7. 7"/45 Mk II, Naval Gun Factory No. 81 on Mk II Mod 4 Pedestal Mount No. 19
Location: Battery South, Bora Bora, Society Island, French Polynesia
8. 7"/45 Mk II, Naval Gun Factory No. 96 on Mk II Mod 4 Pedestal Mount No. 18 (?)
Location: Battery South, Bora Bora, Society Island, French Polynesia
9. 7"/45 Mk II, Watervliet No. 60 on Mk II Mod 4 Pedestal Mount No. 13
Formerly armed USS Louisiana (BB-19)
Location: Battery East, Bora Bora, Society Island, French Polynesia
10. 7"/45 Mk II, Midvale No. 69 on Mk II Mod 4 Pedestal Mount No. 14
Formerly armed USS Kansas (BB-21)
Location: Battery East, Bora Bora, Society Island, French Polynesia
11. 7"/45 Mk II, __
Location: Museo Naval Mexico, Vera Cruz, Mexico
12. 7"/45 Mk II, __
Location: Museo Naval Mexico, Vera Cruz, Mexico

6 Inch Guns

1. M1905 Watervliet No. 30 on M1903 Disappearing Carriage, Watertown No. 1
Formerly emplaced at Battery Schofield, West Point Military Academy, NY
Location: Battery Cooper, Fort Pickens, Pensacola, FL
2. M1905 Watervliet No. 9 on M1903 Disappearing Carriage, Watertown No. 2
Formerly emplaced at Battery Schofield, West Point Military Academy, NY
Location: Battery Chamberlin, Fort Scott, San Francisco, CA
3. M1905 Watervliet No. 31 on M1905M1 Disappearing Carriage, Watertown No. 12
Location: Battery Morrison, Fort Mills, Corregidor Island, PI
4. M1905 Watervliet No. 32 on M1905M1 Disappearing Carriage, Watertown No. 13
Location: Battery Morrison, Fort Mills, Corregidor Island, PI
5. M1905 Watervliet No. 4 on M1905M1 Disappearing Carriage, Watertown No. 9
Note: gun is in poor condition, cut up
Location: Battery Ramsay, Fort Mills, Corregidor Island, PI
6. M1905 Watervliet No. 33 on M1905M1 Disappearing Carriage, Watertown No. 11
Note: gun is in poor condition, cut up
Location: Battery Ramsey, Fort Mills, Corregidor Island, PI
7. M1905 Watervliet No. 27
Note: spare tube, no carriage
Location: Battery Morrison, Fort Mills, Corregidor Island, PI
8. M1905 Watervliet No. 6 on M1905M1 Disappearing Carriage, Bethlehem No. 6
Location: Battery Hall, Fort Wint, Grand Island, Subic Bay, PI

9. M1905 Watervliet No. 7 on M1905M1 Disappearing Carriage, Bethlehem No. 7
Location: Battery Hall, Fort Wint, Grande Island, Subic Bay, PI
10. M1908 Watervliet No. 6
Location: Battery Leach, Fort Hughes, Caballo Island, PI
11. M1908MII Watervliet No. 4
Note: with shield, no carriage
Location: Battery Roberts, Fort Drum, El Fraile Island, PI
12. M1900 Watervliet No. 22 on M1900 Barbette Carriage, Rock Island No. 12
Formerly emplaced at Battery Peck, Fort Hancock, Sandy Hook, NJ
Location: Battery Gunnison (Peck 2), Fort Hancock, Sandy Hook, NJ
13. M1900 Watervliet No. 23 on M1900 Barbette Carriage, Rock Island No. 17
Formerly emplaced at Battery Peck, Fort Hancock, Sandy Hook, NJ
Location: Battery Gunnison (Peck 2), Fort Hancock, Sandy Hook, NJ
14. M1905A2 Watervliet No. 16 on M1 Barbette Carriage, ________No. 58
Formerly emplaced at Battery 227, Fort John Custis, VA
Location: Battery 234, Fort Pickens, Pensacola, FL
15. M1905A2 Watervliet No. 21 on M1 Barbette Carriage, ______ No. 59
Formerly emplaced at Battery 227, Fort John Custis, VA
Location: Battery 234, Fort Pickens, Pensacola, FL
16. M1903A2 Watervliet No. 30 on M1 Barbette Carriage, ____No. 9
Formerly emplaced at Battery 281, Fort McAndrew, Argentia, Newfoundland
Location: Battery 246, Fort Columbia, Chinook Point, WA
17. M1903A2 Watervliet No. 61 on M1 Barbette Carriage, ____No. 10
Formerly emplaced at Battery 281, Fort McAndrew, Argentia, Newfoundland
Location: Battery 246, Fort Columbia, Chinook Point, WA
18. M1903A2 Watervliet No. 13 on M1 Barbette Carriage, ________No. 44
Location: Battery 282, Fort McAndrew, Argentia, Newfoundland
19. M1903A2 Watervliet No. 8 on M1 Barbette Carriage, ______No. 45
Location: Battery 282, Fort McAndrew, Argentia, Newfoundland
20. M1905A1 Watervliet No. 12 on M1917 carriage, Morgan Engineering No. 22
Formerly emplaced at Battery Parker, Fort Casey, WA, Note: with limber, Morgan No. 82
Location: U.S. Army Ordnance Training Support Facility, Fort Lee, VA
21. Rapid Fire BL Gun Mk 7, Armstrong No. 12139 on Mk 2 Barbette Carriage No. 11162
Formerly emplaced at Battery Burchsted, Fort Dade, FL
Location: Fort Desoto, Mullet Key, FL
22. Rapid Fire BL Gun Mk 7, Armstrong No. 12140 on Mk 2 Barbette Carriage No. 11157
Formerly emplaced at Battery Burchsted, Fort Dade, FL
Location: Fort Desoto, Mullet Key, FL

6 Inch Naval Guns

1. Navy 6"/50 Mk VIII Naval Gun Factory No. 368
Location: Near airport, Nome, AK
2. Navy 6"/50____________ on Pedestal Mount
Location: Cannon Beach, Yakutat, AK
3. Navy 6"/50__________ on Pedestal Mount
Location: Cannon Beach, Yakutat, AK

4. Navy 6"/50___________
   Location: City dump, Cold Bay, AK
5. Navy 6"/50 _________ on Mk X Pedestal Mount No.____
   Location: George Island, Cross Sound, AK
6. Navy 6"/50 Mk VIII Mod II Midvale No. 550 on Mk X Pedestal Mount
   Location: Blunts Point, Pago Pago, American Samoa
7. Navy 6"/50 Mk VIII Mod II Midvale No. 554 on Mk X Pedestal Mount
   Location: Blunts Point, Pago Pago, American Samoa
8. Navy 6"/50 Mk VIII Mod II _______No.____ on Mk X Pedestal Mount
   Location: Breakers Point, Pago Pago, American Samoa
9. Navy 6"/50 Mk VIII Mod II ______No.____ on ___ Pedestal Mount
   Location: Breakers Point, Pago Pago, American Samoa
10. Navy 6"/50 Mk VI Mod 1 No. 309L on Pedestal Mount
    Note: Ex-navy gun emplaced in 1942 at Fort New Amsterdam
    Location: Paramaribo, Suriname (formerly Dutch Guiana)
11. Navy 6"/50
    Location: Paramaribo, Suriname
12. Navy 6"/50 Mk VI Mod 2 No. 314L on Pedestal Mount
    Location: Paramaribo, Suriname
13. Navy 6"/50
    Location: Paramaribo, Suriname
14. Navy 6"/50 Mk VIII _______
    Location: Great Sitkin Island, AK
15. Navy 6"/50 Mk VIII _______
    Location: Great Sitkin Island, AK

155mm GPF Guns

1. M1918M1, Bullard Eng. Works No. 161; Carriage M3 No.___
   Location: Presidio Army Museum, Fort Scott, San Francisco, CA
2. M1918M1, Watervliet Arsenal No. 906; Carriage M2 No. 82
   Location: Fort Stevens State Park, Astoria, OR
3. M1917, Puteaux No. 629; Carriage M1917 No. 222, with limber
   Location: U.S. Army Ordnance Training Support Facility, Fort Lee, VA
4. M1918M1, Watervliet Arsenal No. 1073, Carriage M3 Minneapolis Steel Co. No. ___
   Location: Fort Morgan, Mobile Bay, AL
5. M1917, Puteaux No.918; Carriage M3 No.523
   Location: Fort MacArthur Military Museum, San Pedro, CA
6. M1918_______________No. ___; Carriage M1918 No.___
   Location: Shrine of Valor Memorial, Bataan, Philippines
7. M1918________________No. 810; Carriage M1918 No.___
   Location: Pulaski Park, Pulaski, VA
8. M1918_______________No. 75; Carriage M1918 No.___
   Location: Odlin County Park, Lopez Island, WA
9. M1917, Puteaux No. 1263; Carriage M1917 No. ___
   Location: Washelli-Evergreens Cemetery, North Aurora St., Seattle, WA

10. M1918M1, Bullard Eng. Works No. 314; Carriage M1918
Location: Memorial Park, Lansdale, PA

11. M1918M1, Bullard Eng. Works No. 209; Carriage M1918
Location: Memorial Park, Lansdale, PA

12. M1918M1, Bullard Eng. Works No. 149; Carriage M1918 No.___, with limber
Location: Barberton Veterans Memorial, Lake Anna, Barberton, OH

13. M1918M1, Watervliet Arsenal No. 218; Carriage M1918 No.___
Location: Cemetery on Rte. 40, Indianapolis, IN

14. M1917M1, Puteaux No.___; Carriage M3 No.473
Location: Fort Sill, OK

15. M1918M1, Watervliet Arsenal No. 920; Carriage M1918
Location: Lake St., Lancaster, NY

16. M1917, Puteaux No. 959; Carriage M1918 No. 208 (missing wheels & axel)
Location: Fort Miles Museum, Cape Henlopen State Park, Lewes, DE

17. M1917, Puteaux No.___; Carriage M1918 No.___
Location: Camp Shelby, MS

18. M1918M1, Bullard Eng. Works No. 376; Carriage M1918 No.___
Location: Cemetery on David Highway, Saranac, MI

19. M1917, Puteaux No. ___; Carriage M1918 No.___
Location: Fort Macon State Park, Atlantic Beach, NC

20. M1818M1, Watervliet Arsenal No. 806; Carriage __________No.___
Location: Cherryville, KS

21. M1918M1, Watervliet No. 989; Carriage M1918 No. ___, with limber
Location: Edgemont, SD

22. M1918M1, Watervliet No. 988; Carriage M1918 No. ___, with limber
Location: Edgemont, SD

23. M917, Bullard No. 427; Carriage M1918 No. ___
Location: Courthouse, Forsyth, MT

24. M1918M1, Bullard Eng. Works No. 171; Carriage _______No.___
Location: Mechanicsville, IA

25. M1918, ______________No. 967; Carriage __________No.___
Location: Marine Base, Quantico, VA

26. M1918M1, Watervliet Arsenal No. 976; Carriage M1918 No.___
Location: Quincy, IL

27. M1917A1, Puteaux No.765/USA No. 1158; Carriage M1917 No. 1158
Location: Scotland, SD

28. M1917, Puteaux No. 103; Carriage M3 No. ___
Location: American Legion Post, Seattle, WA

29. M1918,______________No.___; Carriage ___________No.___
Location: St. Louis, MO (possibly from Jefferson Barracks)

30. M1918M1, Puteaux No. 764; Carriage M1918A1 Rock Island Arsenal No. 633
Location: Lockport, NY (missing wheels & axel)

31. M1917, Puteaux No. 602; Carriage M1917 No. ___ Note: Nos. 31 & 32 believed never used as CA
Location: Fort Park, Walla Walla, WA

32. M1917, Puteaux No. 740; Carriage M1917 No. ___ Note: Nos. 31 & 32 believed never used as CA
Location: Fort Park, Walla Walla, WA

33. M1918M1,_____________No.____; Carriage ________No.__
Location: Grand Rapids, WI
34. M1917, Puteaux No. ___; Carriage M1918 No.___
Location: Fort Macon State Park, Atlantic Beach, NC

5 Inch Guns

1. M1897 Watervliet No. 6, Formerly emplaced at Battery Krayenbuhl, Fort Mott, NJ
Location: Chapman Park, Stanley, WI
2. M1897 Bethlehem No. 16, Formerly emplaced at Battery Shipp, Fort Caswell, NC
Location: Memorial Park, Woodstown, NJ
3. M1897 Bethlehem No. 21, Formerly emplaced at Battery Wagner, Fort Baker, San Francisco, CA
Location: Veterans Memorial Park, Mahoning Ave, Warren, OH
4. M1897 Bethlehem No. 13, Formerly emplaced at Battery Bell, Fort Schuyler, NY
Location: Valley Forge Military Academy, Wayne, PA
5. M1897 Bethlehem No. 14, Formerly emplaced at Battery McGrath, Fort Rosencrans, CA
Location: VFW Post 1989, Indiana, PA
6. M1897 Bethlehem No. 3, Formerly emplaced at Battery Boutelle, Fort Winfield Scott
Location: Virginia War Museum, Newport News, VA
7. M1897 Bethlehem No. 22, Formerly emplaced at Battery Vicars, Fort Worden, WA
Location: Chewsville, MD
8. M1897 Bethlehem No.____ Note: muzzle cut off, data inferred
Location: Boonsboro, MD

5 inch/51 Navy Guns

1. 5"/51 Mk VIII Four Lakes No. 1205
Location: NAS Trumbo Point, Key West, FL
2. 5"/51 __________No. 1093L2
Served on USS Indiana, USS Arizona
Location: Ropkey Museum, Crawfordville, IN
3. 5"/51 Mk VII Watervliet No. 774
Location: NAS Midway Island
4. 5"/51 Mk VII Watervliet No. _____
Location: NAS Midway Island
5. 5"/51 Mk VIII __ No. 415L
Location: NROTC, Tulane University, New Orleans, LA
6. 5"/51 Mk XV Naval Gun Factory No. 736L
Served on USS Idaho
Location: Naval Gun Factory, Washington, DC
7. 5"/51 __________No.__
Location: Fort Schuyler, Kings Point, NY
8. 5"/51__________
Location: Behind Museum, Treasure Island, CA
9. 5"/51 Mk XV _______No.___
Location: Park, Lewsiton, ME

10. 5"/51 Mk XV Bethlehem No.____
Location: NAS Brunswick, ME
11. 5"/51 __________No. ___
Location: Mitchell, IN
12. 5"/51 Mk IX Mod 3 Naval Gun Factory Mod. 3 No. 938L
Location: American Military Museum, S. El Monte, CA
13. 5"/51 _______ No. ______
Served on USS Colorado
Location: Veteran's Memorial Museum, Chehalis, WA

4.7 Inch Guns

1. QF Mk IV Armstrong No. 12123 on Central Pivot Mount Mk 1 No. 10981
Formerly emplaced at Battery Talbot, Fort Adams, RI
Location: Equality Park, Newport, RI
2. QF Mk IV Armstrong No. 12124 on Central Pivot Mount Mk 1 No. 10982
Formerly emplaced at Battery Talbot, Fort Adams, RI
Location: Battery Bingham, Fort Moultrie, Sullivan's Island, SC (shield missing)
3. QF Mk 3 Armstrong No. 11933
Formerly emplaced at Battery Dodge, Fort Ruger, HI
Location: State Armory, Wahiawa, Oahu, HI
4. QF Mk 3 Armstrong No. 11009
Formerly emplaced at Battery Dodge, Fort Ruger, HI
Location: State Armory, Wahiawa, Oahu, HI
5. QF Mk IV Armstrong No. 11856 on Central Pivot Mount Mk 1 No. 10842
Formerly emplaced at Battery Drum, Fort Strong, MA
Location: State Armory, Main Street, Ansonia, CT
6. QF Mk IV Armstrong No. 9718 on Central Pivot Mount Mk 1 No. 10841
Formerly emplaced at Battery Van Swearingen, Fort Pickens, FL
Location: County Courthouse, Danielsville, GA
7. QF Mk 3 Armstrong No. 11855 on Central Pivot Mount Mk 1 No. 10843
Formerly emplaced at Battery Drum, Fort Strong, MA
Location: American Legion Concord Post 431, Springville, NY
8. QF Mk 3 Armstrong No. 9724 on Central Pivot Mount Mk 1 No. 10885
Formerly emplaced at Battery Dodd, Fort Delaware, DE
Location: American Legion Post, Dravosburg, PA
9. QF Mk 3 40-caliber Armstrong No. 11857 on Central Pivot Mount Mk 1 No. 10835
Formerly emplaced at Battery Backus, Fort Screven, GA
Location: East Park, Girardville, PA

3 Inch Guns

1. M1898M1, Driggs-Seabury No. 11
Formerly emplaced at Battery Center, Fort McRee
Location: Schell Memorial Cemetery, Boyertown, PA
2. M1898M1 Driggs-Seabury No. 118
Formerly emplaced at Battery Terrett, Fort Gaines, AL

Locations: Schell Memorial Cemetery, Boyerstown, PA

3. M1898M1, Driggs-Seabury No. 25
Formerly emplaced at Battery Center, Fort McRee, FL
Location: Mount Pleasant Cemetery, Elbridge, NY
4. M1898M1, Driggs Seabury No. 27
Formerly emplaced at Battery Beecher, Fort Schuyler, NY
Location: Valley Forge Military Academy, Wayne, PA
5. M1898M1, Driggs Seabury No. 28
Formerly emplaced at Battery Baker, Fort Totten, NY
Location: Valley Forge Military Academy, Wayne, PA
6. M1898M1, Driggs-Seabury No. 60
Formerly emplaced at Battery Burchsted, Fort Dade
Location: Rte 22 Harrisburg, PA
7. M1898M1, Driggs-Seabury No. 75
Formerly emplaced at Battery Blaney, Fort Scott, CA
Location: Town Hall, Forked River, NJ
8. M1898M1, Driggs-Seabury No. 85
Formerly emplaced at Battery Irwin, Fort Monroe, VA
Location: Valley Forge Military Academy, Wayne, PA
9. M1898M1, Driggs-Seabury No. 88
Formerly emplaced at Battery Griffin, Fort Hamilton, NY
Location: Orange, MA
10. M1898M1, Driggs-Seabury No. 119
Formerly emplaced at Battery Terrett, Fort Gaines, FL
Location: Copper Street Cemetery, Vernon, NY
11. M1898M1, Driggs-Seabury No. 120
Formerly emplaced at Battery Adair, Fort Taylor, Key West, FL
Location: Central Park, West Rutland, VT
12. M1898M1, Driggs-Seabury No. 112
Formerly emplaced at Battery Mellon, Fort Dade, FL
Location: Pulaski Square, Cleveland, OH
13. M1898M1, Driggs-Seabury No.___
Formerly emplaced at Battery________
Location: Valhalla Firehouse, Valhalla, NY
14. M1898MI, _________, No. ________
Location: Oregon National Guard Museum storage, Camp Withycombe, Clackamas, OR
15. M1898M1, Driggs-Seabury No. 39
Formerly emplaced at Battery Dilworth Fort Taylor
Location: American Legion Post 397, Shrewsbury, MA
16. M1898M1, Driggs-Seabury No. 92
Formerly emplaced at Battery Irwin, Fort Monroe, VA
Location: Brighton Town Hall, State Route441, Brighton, N.Y.
17. M1898M1, Driggs-Seabury No. 37
Formerly emplaced at Battery Laval, Fort Crockett, TX
Location: American Legion Post, Penfield, N.Y.

18. M1898M1, Driggs-Seabury No. 38
Formerly emplaced at Battery Laval, Fort Crockett, TX
Location: American Legion Post, Penfield, N.Y.
19. M1898 Masking Parapet mounts (4ea)
Formerly emplaced at Battery Adair, Fort Taylor, Key West, FL
Location: Fort Taylor, Key West, FL
Note: cast-iron counter weights located in abandoned Fort St. Philip, New Orleans, LA
20. M1902M1, Bethlehem No. 6 on M1902 Pedestal Mount No. 6
Formerly emplaced at Battery Hindman, Fort Wool, VA
Location: Battery Irwin, Fort Monroe, VA
21. M1902M1, Bethlehem No. 7 on M1902 Pedestal Mount No. 7
Formerly emplaced at Battery Hindman, Fort Wool, VA
Location: Battery Irwin, Fort Monroc, VA
22. M1903, ______ No. 11 on M1903 Pedestal Mount No. 6
Formerly emplaced at Battery Flake, Fort Wint, Philippine Is.
Location: Battery Trevor, Fort Casey, Coupeville, WA
23. M1903, _____ No. 12 on M1903 Pedestal Mount No. 7
Formerly emplaced at Battery Flake, Fort Wint, Philippine Is.
Location: Battery Trevor, Fort Casey, Coupeville, WA
24. M1903, ____No. 17 on M1903 Barbette Carriage No. 10
Formerly emplaced at Battery Jewell, Fort Wint, Philippine Is.
Location: Battery Wansboro, Fort Flagler, WA
25. 3"/50 Mk VI, American Radiator Co. No. 4954, on American Ordnance Pedestal Mount Mk II
Formerly emplaced at Battery Jewell, Fort Wint PI as a Philippine Army practice battery
Location: Battery Wansboro, Fort Flagler, Nordland, WA
26. M1911 Watervliet No. 2 on M1912 Cowdrey Machine Co. Barbette Carriage No. 1
Note: training dummy (formerly at Aberdeen Proving Ground Museum)
Location: US Army Air Defense Training Support Facility Fort Sill, OK
27. M1911 Watervliet No.____ on M1912 Cowdrey Machine Co. Barbette Carriage No.__
Note: Training dummy
Location: Battery McCorkle, Fort Moultrie, Sullivan's Island, SC

90 mm Guns (Only Fixed Emplacement Carriages)

1. 90mm M1, Chevrolet No. 6931 on Carriage T3, Fisher Body No. 123
Formerly emplaced at Battery TAC 24, Fisherman Island, VA
Location: Battery Parrott, Fort Monroe, VA
2. 90 mm M1,_________ No.___ on Carriage T3, _______ No. ____
Location: Outside Building 600, Shemya Air Force Base, Shemya Island, AK (shield scrapped)
3. 90 mm Carriage T3,______No. ____
Formerly emplaced at Battery TAC 24, Fisherman Island, VA
Location: Fort MacArthur Military Museum, San Pedro, CA

Production Companies of American Seacoast, Anti-Aircraft, and Railway Artillery 1886-1955

The following is a list of all firms or government facilities that produced finished U.S. anti-aircraft, seacoast, or railway artillery cannons or carriages from 1886-1955. Space did not permit inclusion of other firms that might have supplied forgings, recuperators, wheels, limbers or caissons, optical equipment, or parts. Similarly full corporate histories, including acquisitions, mergers, corporate reorganizations, name changes, and final liquidations were not possible. The intent is to provide a brief overview of whom the major contributors were, their business, and their major products in the supply of modern American static artillery.

Government Facilities:

Rock Island Arsenal. Rock Island, Illinois. Created in 1862 by Congress as a major regional arsenal. With the construction of the Field Gun Carriage shop it became the primary Army facility for production of field artillery carriages from the 1890s through the Second World War. Most field carriage types were made here for the Army. The only seacoast carriage they produced was a limited run for the 6-inch Model 1900 pedestal carriage. They also produced the initial "type" carriage for the 75mm M1917 improvised AA mount and the 75mm truck-mount Model 1917 carriage. During the two world wars a large number of outside firms were engaged in carriage production, but often Rock Island was responsible for all early prototype construction and some ongoing service production. Also, between the wars Rock Island made or oversaw many of the conversions to high-speed capabilities and then after the Second World War both overhauled existing weapons and produced additional new field guns and howitzers.

Washington Navy Yard, Washington, D.C. This was one of the original ship construction yards funded at the start of the U.S. Navy, located on the north shore of the Anacostia River. In the 1840s it began to specialize in light foundry work, steam engine castings, anchor fabrication and gun production. The 1883 congressionally sanctioned Gun Foundry Board selected this yard to become the Navy's primary gun factory. Informally described as the "Naval Gun Factory" by the Navy's Bureau of Ordnance until 1945 when it officially adopted that name. Heavy ship ordnance was produced starting in 1854. Many of the Navy gun tubes used by the Marine Corps and U.S. Army on fixed platforms for temporary or permanent defenses were made by this factory. Some of the heavy ex-Navy guns used on railway carriages were originally products of the Navy Gun Factory. Establishment still exists, though mostly performing administrative functions.

Watertown Arsenal, Watertown, Massachusetts. A Boston-area ordnance facility acquired in 1816 and completed in 1819 beside the Charles River. During the Civil War it built field and seacoast carriages. While it saw some activity supplying early Army gun ordnance, in 1892 it was converted for the specialized manufacture of seacoast carriages. It had limited but useful forging and metal shops and participated in field artillery carriage production on a number of occasions. However, its real function from the 1890s to the First World War was as the Army's primary seacoast carriage producer. Generally, all government sponsored new carriage types had their first prototypes built here. Often production runs were also conducted by the arsenal, using commercial contractors to supplement when time or quantity demanded. Dozens of carriage types for mortars, disappearing guns, barbette and pedestal carriages numbering many hundreds in quantity were produced at Watertown Arsenal. During the First World War it also produced completed railway carriages, including the

Model 1920 14-inch type. It was also heavily used for anti-aircraft mounts, building meaningful amounts of five different types from 1917 to 1944. Participated in production of Second World War 6-in and 16-in barbette carriages. Facility operated postwar, not being closed until 1967. It is now a mixed use development of offices and residences.

Watervliet Arsenal, Watervliet, New York. Created as a government ordnance facility in 1813 in the village of Gibbonsville on the west bank of the Hudson River near Troy, New York. In 1883 the Gun Foundry Board recommended that this arsenal become the Army's dedicated gun factory. Subsequent funding fulfilled this commitment. From 1889 virtually all the Army's field and seacoast guns were developed at this arsenal. It had only limited forging capability and depended on commercially produced forgings. The shop's capacity was adequate for most of the following 30 years, although it was supplemented by commercial producers on occasion for large orders. By far it is the most mentioned source of U.S. Army guns in the modern period. Type guns and significant production runs occurred for most all the Army's seacoast, anti-aircraft, as well as field guns occurred at Watervliet. During both the First and Second World Wars the arsenal was expanded and increased production, but demand was so high that entirely new commercial sources were enlisted to meet the requirements. Gun carriages were almost never made by this arsenal. The facility served through the Second World War when it produced over 20 types of field, AA, seacoast (the T2 6-inch gun), and tank guns. Watervliet Arsenal is still an Army ordnance facility to this day, though cannon production is much reduced.

Private Facilities:

Aetna-Standard Engineering Company, Pittsburg, Pennsylvania. In 1902 organized by R. C. Stiefel and J. H. Nicholson, who bought the Ellwood City Gas Engine Co. and the National Machine and Tool Co. of Pittsburgh. Erected a new plant on Park Avenue to manufacture rolling and tube milling machinery along with high-grade iron castings. In 1926 merged with the Aetna Foundry of Youngstown, Ohio to become the Aetna Standard Engineering Company. During the Second World War they contracted to supply 37mm M3 anti-tank gun wheeled carriages (along with Watertown Arsenal and Koppers Co.). In the early 1950s they were the producer of the carriage and final assembler of the M51 75mm Skysweeper anti-aircraft gun and mount. Name changed postwar, but the plant still exists as part of the Edro Specialty Steel Corporation.

Alliance Machine Company, Alliance, Ohio. Firm founded in 1901 by employees formerly at nearby Morgan Engineering. President was W. H. Purcell, former general manager of the rival company. Specialized in cargo handling equipment, overhead cranes, and large structural steel components. In 1903-1908 bid on a number of contracts for Army seacoast carriages but secured only one successfully. Produced just two 12-inch Model 1901 disappearing carriages (No. 20 and No. 21) in 1909-1910. During the First World War it produced relining equipment for the Navy. In the Second World War continued to supply heavy cranes and plant equipment to the government but received no ordnance contracts. Firm existed until the early 2000s.

Allis-Chalmers Manufacturing Company, Milwaukee, Wisconsin. Company initially created in 1901 by the merger of the Edward P. Allis Company and Dickson Manufacturing Company. Had a tractor plant is West Allis, Wisconsin. Eventually also made steam engines, mill and mining equipment and compressors. Reorganized in 1912 as the Allis-Chalmers Manufacturing Company. During the Second World War was one of four new manufacturing companies selected to produce 90mm anti-aircraft gun carriages. Also, in the war produced parts for naval turbines, artillery tractors, and uranium processing equipment for the Manhattan Project. Firm existed until the 1990s when it was dissolved through divestitures.

American and British Manufacturing Company. See ***American Ordnance Company.***

American Bridge Company, Coraopolis, Pennsylvania. This company was created by famous financier J. P. Morgan by consolidating 28 separate businesses specializing in steel fabrication and contracting. It quickly became a major factor in supplying and erecting bridges and structural building steel components. It was selected to be one of the new contributors to ordnance production in the First World War. The firm erected a very large new gun forging plant to make 155mm guns (though not finished guns) in Gary, Indiana. Also, it produced the main girder parts for the three 12-inch Model 1919 (ex-Chilean) railway guns and parts for the Model 1918 16-inch howitzer railroad carriage. During the Second World War the company was a major landing craft contractor. It was spun off for U.S. Steel and went public in 1987 and still exists.

American Car and Foundry Company, New York, New York. This company was a classic conglomerate formed in 1899 by the merger or acquisition of thirteen separate companies involved in the rail equipment and car manufacturing business. Within a short period, several more were added. By the First World War consolidation had reduced primary manufacturing sites to less than a half dozen, mainly in Michigan, Indiana, and Pennsylvania. During the war the company delivered on major contracts for ammunition (Detroit), artillery vehicles (Detroit), sub chasers (Wilmington), and specialized railroad cars for artillery (Berwick, Pennsylvania). It delivered 77 rail cars Model 1918 for mounts (7, 8, 10, 12-inch) and the Model E experimental railway mount. During the Second World War it produced the majority of the American M3 and M5 Stuart tanks. The firm acquired majority control of the J. G. Brill Company of Philadelphia in 1926, fully absorbing it in 1944. Brill was the primary contractor for the 75mm pack and field howitzers. Postwar it became ACF Industries with headquarters in St. Charles, Missouri, and still exists today with its major plant in Milton, Pennsylvania.

American Hoist and Derrick Company, St. Paul, Minnesota. Company founded in 1883 making heavy mining and lumbering equipment. Soon added hand and horse hoisting gear, overhead and traveling cranes. Eventually heavily involved in producing ship deck equipment and shipyard cranes and construction machinery. Initially named the American Manufacturing Co., in 1892 became the American Hoist and Derrick Co. Bid on several seacoast carriage offers starting in 1894 but only secured one order. Produced for the Army Ordnance Department a total of 35 Model 1896 12-inch mortar carriages in 1899. Apparently, the firm overestimated the ease of making these and ran into trouble during construction. Deliveries were late, and five assigned carriages had to be transferred to another firm. The firm did not return to ordnance manufacture again, though its production of ship and dockyard equipment was important during both world wars. The firm closed operations in 1985 in St. Paul, and was renamed the American Crane Corp. in 1985 and acquired by Terex Corp. in 1987.

American Oerlikon Gazda Corporation, Providence, Rhode Island. Antoine Gazda was an Austrian inventor and businessman who had acquired a power of attorney to act on behalf of the Swiss Oerlikon company. He initially worked on arrangements in 1940 to have the 20mm Swiss gun adopted by the British Admiralty and produced with a royalty license in a U.S. manufacturing site. After getting backing he formed the American Oerlikon Gazda Corporation in Providence. For manufacture he used a former textile factory in that city. In mid-1941 a firm order for production split by the British and Americans was received and production begun. At that time the Swiss government was unable to confirm the license, but war needs pushed the project forward leaving the final negotiations for license fees to be settle postwar. Also, late in 1941 the needs easily outstripped the capabilities of the firm, and both Pontiac and Hudson car companies were brought in to supplement production. Ultimately the American Oerlikon Gazda Corporation built only a small percentage of the total American 20mm guns during the war. The firm was dissolved postwar.

American Ordnance Corporation, Bridgeport, Connecticut. The firm was founded in 1895 by combining or acquiring financial interest of the Driggs Ordnance Company, the ordnance works of William Cramp & Sons and the American plant of French Hotchkiss Ordnance—which it represented as both exclusive agent and domestic manufacturer. Plant was located in Bridgeport, Connecticut where it made guns, ammunition, lathes and other machine tools. They produced light guns (up to 3-inch), carriages, and ammunition for both the Army and Navy. Provided the four 4-inch Driggs-Schroeder type Navy guns and carriages emplaced by the coast artillery during the Spanish American War emergency. Completed an order for 23 3-inch Model 1903 pedestal mounts (No. 22-44) for the Model 1903 seacoast gun. In 1902 reorganized to become the American and British Manufacturing Corp. Firm secured a considerable amount of foreign business, particularly in Latin America. The company existed through the First World War primarily as an ammunition producer but went into receivership in 1921. The plant building itself was razed in 1940.

The Baldwin Locomotive Works, Inc., Eddystone, Pennsylvania. Established in 1831, this was a long-standing producer of railroad locomotives. The firm incorporated in 1909. During the First World War it began production of ordnance. It formed the Eddystone Ammunition Corp. subsidiary. The locomotive division itself contracted to build the Mk V tracked 7-inch gun mounts for both the U.S. Navy and Army. It produced the trucks for the three ex-Chilean 12-inch railway cars and assembled all the Navy-type 14-inch railway gun cars and carriages (both Mk I and Mk II for both Navy and Army requirements). In the Second World War it contributed another piece of railroad armament—22 of the new 8-inch Mk I railway mount and car. Additionally during this conflict, it participated in Army tank production and made 3-in/50 AA guns for the Navy.

Bethlehem Steel Corp., South Bethlehem, Pennsylvania. A large American iron and steel company started in the 1830s to make rails and wheels for the railway industry. It became the Bethlehem Iron Company and then finally incorporated as the Bethlehem Steel Corporation in 1904. Purchased technical expertise from the British Whitworth Company in 1886 to make ordnance-grade forged steel for armor plate and guns. In an effort to encourage domestic gun forging capability, the War Department awarded Bethlehem a 100-gun contract in 1890. Over eight years the firm produced 25 8-inch, 25 12-inch, and 50 10-inch guns in accordance to this award. Other gun forgings followed to both services. Later (1910s-1920s) added facilities for assembling guns, building carriages and artillery vehicles. Provided about half of all Army gun forgings and a large number to the Navy. Produced 55, finished 12-inch steel Model 1890 mortars. Supplied many finished 8, 10, and 12-inch seacoast guns as part of a "Hundred-Gun Contract" to the U.S. Army. Also supplied almost 70 large seacoast gun carriages and all 60 of the 3-inch Model 1902 RF gun and pedestal for coast defense. Was equally active in satisfying contracts for field artillery material. Produced 22 Model 1917 fixed AA pedestal mounts. During the First World War additionally produced ordnance for several allied nations. Provided the six 12-inch "Chile guns" and the 14-inch Mk IV Navy guns for First World War railway carriages. Many naval gun forgings and finished guns were supplied to that service through the Second World War, including many of the 14-inch and 16-inch ex-Navy guns later used on seacoast and railway mounts. Purchased many of the assets of Midvale Steel and Ordnance Company in 1923 and continued to supply ordnance through the Second World War and modern age. Finally shut down in 1998.

Blaw-Knox Company, Pittsburgh, Pennsylvania. Formed in 1917 by the merger of the Blaw Collapsible Steel Casting Company and the Knox Pressed and Welded Steel Company. Firm specialized in manufacturing steel and asphalt forms for the construction and transportation industries. In 1927 they branched into making radio towers and then paving equipment. During the Second World War shared responsibility with York Safe and Lock for the majority of the 40mm anti-aircraft gun production for the United States military (both Army and Navy). Were a primary contributor of finished 40mm guns made primarily from forgings supplied by Midvale and Crucible Steel Cos. The major production plant at Martins Ferry, Ohio was returned to U.S. Navy control after the war and not closed until 1965. After several changes in ownership and organization, the brand still exists as part of Volvo Construction Products.

Builders Iron Foundry, Providence, Rhode Island. Established in 1820 in Providence. Was an early producer of iron cannon for both the Army (13-inch mortars) and Navy (11-inch guns) since 1861. By the 1890s had a good business volume in castings for pumps, air compressors, rope drives, structural iron and steel members, grinding and polishing machines. Was the sole producer of all 73 Model 1886 cast iron 12-inch mortar tubes. With these were half (36) of the Model 1891 mortar carriages. The firm successfully transitioned into steel products and was a major supplier of the next generation all-steel mortars, building 55 12-inch mortars Model 1890M1 and 20 Model 1896 carriages for them. All of this was accomplished by 1900. They also produced 13 Model 1900 6-inch barbette carriages and 16 Model 1896MIII carriages for the Model 1912 12-inch mortar in the 1910s. In 1898 the firm spun off its steam engine business as the Providence Engineering Works. During the First World War produced the 50 carriages for the improvised 3-inch anti-aircraft gun Model 1917. In the 1930s it increasingly concentrated its business in water pumps and water filtration businesses. After the Second World War it changed its name to BIF Industries. The original works in Providence were closed and destroyed in the 1950s, but the business itself became a division of the New York Air Brake Co. in 1961.

Chalkis Manufacturing Company, Detroit, Michigan. This was one of the businesses specifically organized to use government funds to erect a plant and accept orders for ordnance production during the First World War. Two businessmen from Detroit's Chalmers Motor Company (Hugh Chalmers and William Kiser) created the firm and erected a new plant. The company accepted and fulfilled a contract during the First World War for producing 3-inch anti-aircraft guns Model 1917. All 120 such guns accepted by the Ordnance Department came from this order. This appears to be the only ordnance order performed by this company. An attempt was made to enter the car part business, and in 1919 they began manufacturing lock washers. No further records for the company have been found, presumably they went out of business in the early 1920s.

Chevrolet Division of General Motors, Detroit, Michigan. Firm started in 1911 by Louis Chevrolet and William Durant to produce motor cars. Acquired by General Motors in 1917. Possessed many plants for assembly of automobiles and parts in the Midwest. Large producer of military vehicles during the Second World War. Also, during that war made 2000 90mm anti-aircraft gun tubes, breech rings and recoil mechanisms. Firm still exists as a division of General Motors.

Chrysler Corporation, Auburn Hills, Michigan. Automobile company created by Walter Chrysler in June 1925, using some of the assets of the prior Maxwell Motor Company. Like other major industrial firms, they contributed to the wartime production during the Second World War. A new plant on Lynch Road in Detroit was built in 1940 for the assembly of 40mm anti-aircraft guns for both the U.S. Navy and Army. Parts and subassemblies came from other Chrysler plants in Indiana and Ohio. Many of these guns went onto carriages built by Firestone Tire and Rubber Company. After the war Chrysler reverted to automobile production. Eventually the firm merged with Fiat of Italy.

Colt's Patent Fire Arms Manufacturing Company, Hartford, Connecticut. Organized since 1855 in Hartford, Colt's is best known for its pistols and other small arms. However, at times it delved into the business of supplying services for light artillery production. They were the subcontractors making most all of Gatling's machineguns with field and shipboard mounts. During the Second World War they were the primary producer of the Army's 37mm anti-aircraft gun. While eventually split into two firms, the company still exists.

Columbus Machine Company, Columbus, Ohio. A long-established (since at least 1849) metal fabricating firm located in Columbus. Specialized in large commercial boilers, rollers, mechanical shears and gas engines. Was one of numerous firms producing custom heavy steel and iron equipment that occasionally bid on complicated seacoast carriages in the late 19th and early 20th Century. Secured in 1898 a bid for 15 12-inch Model 1896 mortar carriages and completed and delivered these on time. These were serial numbers No. 193-207. While the firm subsequently bid on 3-inch and 6-inch carriages in the early 1900s, they were not successful. There are indications that the firm failed, and may have been guilty of fraudulent practices, on a subcontract of work for the Kilby Manufacturing Co. on other carriages. Company no longer exists.

Detrick & Harvey Machine Company, Baltimore, Maryland. A specialized manufacturer of metal-working machinery and precision-made tools such as planers, drilling and boring machines. Company founded in 1881 by Jacob Detrick and Alex Harvey in Baltimore. From 1898 to 1909 they secured bids for several runs of coast defense gun carriages. They made 12 Model 1897 5-inch balanced pillar mounts (No. 21-32), 25 Model 1903 6-inch disappearing carriages (No. 46-60 and No. 71-80) and 13 3-inch Model 1903 pedestal carriages (No. 83-95). Slightly later they also boasted of ordnance accomplishments in machining hoops and jackets and breech parts for gun tubes made in government arsenals. The company went out of business in 1927.

Driggs-Seabury Gun & Ammunition Company, Derby, Connecticut. Incorporated in May 1897 by Louis Driggs, who had just departed from nearby American Ordnance Company. The new firm included the Seabury Gun Company (around since the mid-1890s to commercialize Lt. Seabury's breech patents). The plant was in Derby, formerly used by the Brady Manufacturing Co. It made light 1, 3, and 6-pounder and 3-inch RF guns, carriages, and ammunition for the Army and Navy. Made all of the Army's seacoast 3-inch Model 1898 balanced pillar guns and carriages to their proprietary design. In 1904 the company moved to a new plant in Sharon, Pennsylvania. The old plant was taken over by the U.S. Rapid Fire Gun and Power Company. Driggs-Seabury purchased Savage Arms in 1915. Company struggled to maintain ordnance business, but in 1917 was awarded a contract by the Navy for 3-in and 6-pdr guns. After the First World War the company was reorganized as Savage Arms Corp. and relocated to Utica, New York.

Erie Forge and Steel, Erie, Pennsylvania. Firm established in 1903 as a relatively small forge and machine shop as the Erie Forge Company. Renamed in 1912 as the Erie Forge & Steel Company. Produced gun forgings for the Army in the First World War and helped build and staff a plant for the Navy on adjacent property for naval 4 and 5-inch gun tube forgings and destroyer shafting in 1917. Purchased this plant in 1920 from the U.S. Navy. During the Second World War was one of the producers of the seacoast 6-in T2 gun and new M1-M4 barbette carriage. Also continued as a prominent Navy supplier of marine shafts, gearing, and rotors. Acquired in the 1960s by the National Forge Co. Still exists today on the same property as a subsidiary of Whemco, Inc.

Firestone Tire and Rubber Company, Akron, Ohio. Company established by Harvey Samuel Firestone in Akron in 1900 to produce pneumatic tires. Expanded with the growth of the automotive industries to include various manufacturing plants and even natural rubber plantations in West Africa. In February 1941 began negotiations to produce Bofors 40mm four-wheeled carriages. Firm returned to its tire business postwar, eventually being purchased by the Japanese company Bridgestone Corporation, though the brand trademark still is marketed.

Fisher Body Division, General Motors Corporation, Detroit, Michigan. Firm founded by Fred and Charles Fisher in Detroit in 1908. Converted early from producing wagons to automotive bodies. Incorporated in 1916 and purchased by General Motors in 1919 who continued to operate the Detroit plant as a subsidiary. Several of the division's plants received wartime contracts for ordnance manufacture. Made 2105 mobile and 254 fixed 90mm anti-aircraft mounts at several Detroit-area plants. The Grand Rapids plant made 550 120mm mounts in 1943-1944. Other contracts were for Navy 5-inch gun housings, 3-inch AA breech housings, and cradles and yokes for 155mm and 8-inch howitzers. Postwar reverted to its automotive production until it was discontinued as a separate business in 1984 by its parent corporation.

Harrisburg Manufacturing and Boiler Company, Harrisburg, Pennsylvania. This firm was founded in May 1895 as a producer of commercial and railway boilers. They were new to ordnance production at the time of the First World War. On March 31, 1918 they got a contract to build many of the new 8-inch Model 1918 railway carriages and authorized a new $1,000.000 plant and equipment. They were to build 31 of the 8-inch carriages and parts for use in the new 10-inch carriage being assembled by Marion Steam Shovel. The latter project was cancelled before any deliveries. The firm no longer exists.

J. I. Case Company, Racine, Wisconsin. Firm established by Jerome I. Case in 1844 as the J. I. Case Threshing Machine Co. Eventually became a major manufacturer of tractors, threshers and combines, and light construction vehicles. In the Second World War produced shells, tanks, and airplane parts. Was one of several prime producers of the Army's M2 wheeled trailer mount for the 40mm Bofors-licensed anti-aircraft gun. They manufactured 1434 carriages on a contract originally signed for 2000 units. Also made the twin 40mm mount used in the gun motor carriage M19. Postwar returned to the farm and construction equipment business. In 1984 became Case IH with a merger with Tenneco's International Harvester division. There was an additional merger in 1999 with New Holland. While the Case IH brand still exists, the independent manufacturer does not.

Kilby Manufacturing Company, Cleveland, Ohio. Firm established in 1888, succeeding the Prospect Engine and Machine Company. Initial products were steam engines, including the Cumner engine. Also, through most of its history specialized in producing heavy processing equipment for the sugar industry, both cane and beet sugar. In 1894, with the active involvement of the Secretary of War, secured with a low bid a contract for the first ten 10-inch Model 1894 disappearing carriages (No. 1-10). Obviously underestimating the skill and precision needed, along with the failure of a subcontractor they ran into problems. Deliveries were over a year and a half late, and substantial penalties were applied. Eventually all ten carriages passed inspection and were accepted. They did make a bid in 1897 on the 10-inch Model 1896 carriage but were not successful. They did not again participate in ordnance production. The firm continued in its core business for many more years, before being going bankrupt and being liquidated in 1948.

Koppers Company, Pittsburgh, Pennsylvania. This was a diversified manufacturing firm established in 1893 in Pittsburgh by Dr. Heinrich Koppers. It was incorporated in 1912 with main offices in Chicago, had several plants in the Pittsburgh area, and specialized in equipment for steel and coke oven plants. Later had a considerable specialty chemical and explosives product line. During the Second World War made wheeled carriages for the model M3 37mm and the M2 40mm mobile anti-aircraft mounts. Firm was absorbed in 1988 in a managed corporate buy-out.

Lake Erie Engineering Works, Buffalo New York. Company established in 1888 in Buffalo. Initially made steam engines, electric power machinery, iron castings for water works and blast furnaces and rolling mills. Secured a bid for 25 12-inch Model 1896 mortar carriages, later reduced to a total of 15 (No. 168 through No. 183) in 1898. In the same year also awarded a contract for four 8-inch Model 1896 disappearing carriages (No. 32 – 35). Struggled fulfilling these, apparently lacked adequate working capital and was unable to get an extension of a line of credit from their bank. However, all 19 carriages were eventually delivered. In 1923 became the Lake Erie Engineering Corporation. Before and during the Second World War made large hydraulic presses for aluminum airplane parts. Also produced equipment for making 8-inch artillery shells. Company was acquired by Bell Aircraft in 1957.

Link-Belt Company, Chicago, Illinois. A firm founded by William D. Ewart to manufacture detachable-link train drives in 1880. By 1906 it was known as the Link-Belt Company. During the Second World War the company operated plants in Chicago, Philadelphia, and Indianapolis. It received a contract to operate and manufacture a plant to produce the 90mm M2 anti-aircraft gun carriage, and then slightly later all of the M8 anti-tank gun carriages. In later years company was sold to FMC Corp.

Marion Steam Shovel Company, Marion, Ohio. Firm founded in August 1884 in Marion Ohio by Henry Barnhart and George King. Grew in both volume and by acquisition over the next years. Specialized in making steam shovels, power shovels, and excavators used in construction and mining. Had a major share of this business, was widely recognized for the contributions to the digging of the Panama Canal. During the First World War received contracts for making 12 12-inch railway carriages and would have also built the 10-inch Batignolles railway mount if the project wasn't cancelled at war's end. In 1946 became the Marion Power Shovel Company. Sold to Dresser Industries in 1977 and then absorbed and closed down in 1997 by competitor Bucyrus International.

Maryland Steel Company, Sparrow's Point, Maryland (Baltimore). This was a wholly owned, independent subsidiary of Pennsylvania Steel Company of Harrisburg. A new plant site was purchased in 1887 on the Patapsco River at Sparrow's Point, Maryland—to be convenient to Pennsylvania coal but also to use Cuban iron ore. The plant, initially intended to build railroad rails, was constructed in 1888-1890. A well-equipped steel fabricating shop was later added. In 1898 the firm won a contract from the Army for ten Model 1896 12-inch mortar carriages (No. 208-217). This apparently was the only ordnance bid the firm performed. In 1916 Pennsylvania Steel and its Maryland Steel Company were acquired by Bethlehem Steel. Bethlehem was so impressed with the site that they transformed it into one of the largest integrated steel and shipbuilding plants in the nation.

Mesta Machine Company, Homestead, Pennsylvania. Firm founded in 1898 by George Mesta with the amalgamation of Leechburg Foundry and Robinson-Rea Manufacturing Companies. It specialized in manufacturing machinery for steel mills. In the First World War made steel recuperators for 155mm field guns and howitzers. During the 1940s it received funding and a contract to operate a plant for production of 155mm guns M1A1. Produced a number of carriages for the heavy seacoast 16-inch barbette gun during the Second World War. Firm failed in 1983 with the decline of much of the American railroad industry.

Midvale Steel Company, Philadelphia, Pennsylvania. Main plant located in Nicetown, Pennsylvania from 1866. In 1872 became the Midvale Steel Works and 1880 the Midvale Steel Company. In 1915 became the Midvale Steel and Ordnance Company. Experts in specialized forged and alloy steels. The first domestic American steel company capable of producing steel forgings for both services. Produced forgings for the Navy's 3-in breechloading howitzer (steel) in the 1870s and the inserts for the Army's 3.2-inch conversion of ordnance rifles. Thereafter they became the major supplier of forgings for the U.S. Navy and a substantial part of the Army's. In the late 1890s shop facilities were created to produce a limited number finished guns and fixed coast artillery carriages. Produced 15 finished 12-inch and four 10-inch disappearing carriages. They manufactured complete 8-inch howitzers during the First World War for British, Russian, and American orders. Bought Cambrian Steel in 1906. Produced forgings for 16-inch Navy guns later used in seacoast defenses (along with Bethlehem Steel). In 1923 most assets were sold to Bethlehem Steel although the Nicetown plant continued in the forged gun business through the Second World War. Midvale was acquired and the plant closed in the 1970s.

Morgan Engineering Company, Alliance, Ohio. Firm founded in 1868 by Welsh immigrant Thomas Morgan, Sr., re-located to Alliance in 1871. In 1877 it became Morgan-Williams & Co. The Morgan Engineering Company was incorporated in 1900. It produced steam hammers and pneumatic equipment but found a special niche in making overhead traveling cranes from 1878. One of the primary suppliers of disappearing and barbette carriages for seacoast material—fabricated over seventy in total from 1894 to 1908 for 5, 6, 8, 10, and 12-inch sized guns. Was a subcontractor to build the experimental Gordon disappearing carriage in 1891. During the First World War 120 special wheeled carriages and limbers for seacoast 5 and 6-inch guns Model 1917 were produced. They also built many 8, and 12-inch railway gun carriages. Assembled the 12-inch railway carriages for ex-Chilean guns and the railway carriage for the Model 1918 16-inch howitzer. A large contract for numerous (91) 12-inch seacoast mortars on railway carriages necessitated the construction of an entirely new assembly shop in Alliance. Morgan was a major supplier of naval ordnance during the Second World War. The firm, after several acquisitions and separations, still exists.

New Britain Machine Company, New Britain, Connecticut. Founded in the 1890s as a tool producer in New Britain. They made mostly chainsaw parts of sockets and drive tools; also supplied small amount of specialized carbon tool steel. Contracted by the Army in the First World War to produce several types of anti-aircraft gun carriages. Manufactured under their name 51 carriages (all but the pilot example) of the Model 1917 75mm truck carriage, 120 of the 3-inch Model 1917 fixed AA mount pedestal and the four Model 1918 4.7-inch AA gun pedestals. Company did not return to ordnance production after the war. Was acquired in 1972 by the Litton Tool Corporation.

Niles Tool Works Company, Hamilton, Ohio. This firm was created in 1833 in Cincinnati as Niles Works by brothers James and Jonathan Niles. It relocated to Hamilton, Ohio in 1866, and the name changed to Niles Tool Works in 1874. As implied by the name, the firm specialized in large, custom machine tools. In 1897 it successfully bid for 25 Model 1890M1 12-inch mortars. It acquired the steel forgings from Midvale Steel, but fabricated and finished the guns itself, they being assigned a dedicated serial run of No. 1-25. Soon after it also won bids for two series totaling 14 of the 10-inch Model 1896 disappearing carriages, making serials No. 1-8 and No. 43-48. The firm built a special building as a gun shop for this work, though not subsequently winning any additional Army orders. Also, in the 1890s the firm was one of the players in a major merger involving Niles, the Bement Company, Pond Machine Tools, the Philadelphia Engineering Works and Pratt and Whitney. By 1902 a new firm named Niles-Bement-Pond emerged. That company eventually migrated to the Pratt and Whitney aircraft engine company. The tool and fabricating operation in Hamilton eventually closed in 1996.

Oldsmobile Division, General Motors Corporation, Lansing, Michigan. An automobile manufacturing firm established by Ransom Olds in 1897. The firm was acquired by General Motors in 1907 but continued to operate as a division with its own plant for many years. During the Second World War the plant in Lansing became a major supplier of Army and Navy artillery. Among many products were 20mm Navy AA gun mounts and numerous tank cannon. It also produced large numbers of 37mm M4 aircraft cannons and relatively few 37mm anti-aircraft M1A2 guns. For field artillery Oldsmobile completed orders for 105mm M2A1 gun barrel assemblies. The Oldsmobile brand and plant were discontinued by General Motors in 2010.

Otis-Fensom Elevator Company Limited, Hamilton, Canada. Company founded in 1905 with the amalgamation of the Toronto firm the Fenson Elevator Company and the Hamilton, Ontario Canadian subsidiary of the Otis Elevator Company. Subsequently marketed and produced elevators throughout Canada. In the Second World War constructed and operated a new plant for the Department of Munitions and Supply for production of 40mm anti-aircraft guns and carriages. Successfully produced many of these for the British, Canadian, and American military services. Postwar in 1949 became simply the Otis Elevator Co. Ltd for Canada and is a total subsidiary of United Technologies.

Philadelphia Engineering Works, ltd., Philadelphia, Pennsylvania. Company in existence at least from the late 1880s. Produced engines (Corliss, triple and quadruple expansion marine types) and heavy equipment for furnaces and boilers. In 1894 bid unsuccessfully on 10-inch disappearing carriages. Secured a bid and subsequent contract for six 6-inch Model 1898 disappearing carriages (No. 21-25). Acquired by Niles Tool Works in 1899 and consolidated in the new Niles-Bement-Pond heavy engineering company.

Pneumatic Dynamite Gun Company, New York, New York. Following the demonstration in 1884 of an early experimental, compressed air dynamite gun, Army Lt. Edmund Zalinski personally helped organize this firm to commercialize the concept. Formed with investors in New York City, it was originally named the Pneumatic Dynamite Gun Company. Subsequently after an 1894 reorganization it emerged as the Pneumatic Torpedo and Construction Company. The company received orders for a number of 15-inch and 8-inch pneumatic guns from both the U.S. Army and U.S. Navy (and indeed interest from, and at least one purchase from a foreign government). Model 1890 and 1893 15-inch guns were supplied to arm four Army coast defense batteries. The firm exercised financial and contractual responsibilities for producing the guns and even for constructing the batteries but relied on subcontractors for physical gun and construction work as it did not operate its own manufacturing plants. In the case of the pneumatic dynamite guns it used the West Point Foundry in Cold Spring, New York for most of the gun fabrication. Carriages were assembled at the New Jersey Steel and Iron Company of Trenton, New Jersey. After the failure of the concept to deliver to service expectations, the firm was dissolved in the early 1900s. The company should not be confused with the Pneumatic Gun Carriage and Power Company of Washington D.C. that produced under contract several seacoast experimental carriages for guns and mortars to the Army in the same period of time.

Pond Machine Tool, Plainfield, New Jersey. Company founded in 1872 by David W. Pond. Relocated to a new plant in Plainfield, New Jersey in 1887. Manufacturer of machine tools for the railway, boiler, and machine shop trades. In 1890 was one of the primary suppliers of new metal-working machine tools for the Army's gun shop at Watervliet Arsenal. From 1894 to 1898 secured several bids for Army seacoast gun carriage production. Manufactured eight 10-inch Model 1894 disappearing carriages (No. 11-18) and 21 8-inch Model 1896 disappearing carriages (No. 1-21). Also contracted to rebuild the experimental disappearing carriage for 8-inch gun to a Model 1896 standard in 1898. In 1898 the firm was acquired by Niles Tool Works, eventually becoming one of the operating plants of the new Niles-Bement-Pond Company.

Pontiac Division of General Motors, Pontiac, Michigan. Automobile company with a long history. Founded in 1893 as the Pontiac Buggy Company. Incorporated in 1899, began manufacturing cars in 1907 as the Oakland Motor Car Company. Purchased in 1909 by General Motors, brand revived as a separate business unit of General Motor's Oakland Division, and then the Pontiac Division. After the start of the Second World War received a government contract to manufacture licensed 40mm tube assemblies using existing factory floor space. Deliveries began in October 1942. Slightly later also began producing 20mm guns for Navy use. Was one of the largest producers of both of these effective gun types. Converted back to automobile production postwar, which continued until the brand was discontinued in 2009.

Providence Engineering Works, Providence, Rhode Island. The firm of Providence Steam Engine Company was founded in the 1830s on the waterfront in Providence. They represented and produced the Green steam engine, widely used in maritime applications. The firm provided the steam machinery plant for at least two naval vessels in the Civil War. The firm incorporated in 1863. In late 1897 the company acquired Rice and Sargent Engine Company, causing a reorganization and renaming as the Providence Engineering Works. Also starting in 1897 the company produced several lots of seacoast gun carriages for the U.S. Army. The first order was for six 8-inch Model 1894 disappearing carriages (No. 20-25). Later they built 15 Model 1896 12-inch mortar carriages (No. 103-117). Subsequent attempts to get more such bids in 1903 and 1905 were not successful. The company lasted for many more years, not being finally liquidated until 1955.

Pullman-Standard Car Manufacturing Company, Hammond, Indiana. Pullman was established in 1862 by George Pullman to both manufacture and operate the famous Pullman sleeper railroad cars. Pullman purchased the Standard Steel Car Co. in 1934 to become Pullman-Standard. It operated plants in both Butler, Pennsylvania and Hammond, Indiana. The Indiana plant was substantially expanded under contract during the Second World War to produce tanks and artillery. It produced carriages for the 3-inch anti-tank gun, 155mm M1 guns, and 8-in howitzer. The firm ceased to exist and sold off in separate parts in the early 1980s.

Rarig Engineering Company, Columbus, Ohio. Firm founded as a large machine shop in Columbus in 1867. Produced hoisting engines, blast furnace equipment, boilers and pressurized tanks, and structural iron bridge components. Aggressively pursued Army ordnance bids in 1897-1903, succeeding in completing a comparatively large number of orders. In 1898 began producing, under a series of bids and extensions, a total of 69 Model 1896 12-inch, spring-return, mortar carriages (No. 218-227, No. 244-259, No. 281-288, and No. 292-306). In 1903 they won a bid for 20 6-inch Model 1903 disappearing gun carriages (No. 26-45) but soon ran into trouble. Initially they had problems getting forged steel parts needed for the contract; both Bethlehem and Midvale Steel were unable to schedule additional orders, and two or three smaller producers defaulted on their deliveries. Then the firm hit financial difficulties, and by mid-1904 had to cut back to just three workers in the shop. Work on the order came to a standstill, leading Rarig to propose that it be moved to another contractor. That was done in 1905, Wellman, Seaver, and Morgan being assigned the order as a subcontractor to Rarig, who was still held accountable for deliveries and financial penalties. These troubles ended the firm; in 1905 the plant and assets were bought by the Ralston Rail Car Company.

Robert Poole & Son Company, Baltimore, Maryland. Created in the 1850s as a foundry and machine shop. They furnished the iron columns for the U.S. capitol building. Company went through several reorganizations and name changes. During the First World War were known as the Poole Engineering & Machine Co. They supplied the U.S. Army 36 Models 1891, and 97 Model 1896 seacoast mortar carriages in the 1890-1900s. An additional five of the latter were completed when they took over the contract from the American Hoist and Derrick Company. Produced the 21 sets of parts to modify older 15-inch Rodman carriages to carry modern 8-inch breechloading rifles for emergency emplacements at the start of the Spanish American War. Also furnished 12 Model 1897 12-inch disappearing carriages (No. 1 through No. 12). During the First World War was assigned a contract to supply the American-made 37mm infantry gun Model 1916 barrels and breeches. Also during the war, they had large Navy contracts for 3-inch/23 dual-purpose guns (contracts for 800) and 4-inch deck guns. Late in and after the war produced 11 Model 1917 12-inch barbette carriages (No. 12-20 and No. 30-31). The Firm no longer exists.

South Boston Iron Works, Boston, Massachusetts. A long-established iron and brass works in Boston, founded by Cyrus Alger in 1809. Produced cannons and ammunition during the War of 1812. One of the major suppliers of ordnance to the North during the American Civil War, including large numbers of Rodman cannon (along with West Point Foundry). A major proponent of cast iron work, the firm strenuously tried to interest the government in purchasing cast iron guns or even building it a gun factory throughout the 1880s and 1890s. Was given an order for four 12-inch rifled seacoast guns in 1880 of an early type which could not be fulfilled and eventually was cancelled. A final test cast-iron 12-inch mortar exploded on trial with the Army, leading to an end of this effort. Eventually the firm diversified into other heavy iron work and became the Hunt-Spiller Manufacturing Corporation in 1904; though it never did subsequently produce any ordnance.

Southwark Foundry and Machine Company, Philadelphia, Pennsylvania. This was a long-standing base metal working company established in 1836 in Philadelphia. By the end of the 19th Century, they produced steam and pumping engines, centrifugal pumps, blast furnaces, hydraulic presses and accumulators. They made large structural steel and iron pieces for building and bridge construction, like the iron lighthouses in Florida. In the 1890s they successfully pursued a number of bid orders for seacoast ordnance carriages. In 1891 they contracted to build the first two prototype Model 1894 disappearing carriage types—one each for 8 and 10-inch guns. In 1897 they built ten 12-inch mortar Model 1896 carriages (No. 58-67), and then a decade later built eight Model 1908 mortar carriages (No. 9-16). They also were responsible for 13 Model 1896 10-inch disappearing gun carriages (No. 9-18 and No. 62-64). They did not subsequently participate in ordnance fabrication and were acquired by Baldwin Locomotive Works in 1929 and became part of its Eddystone Division. In both wars their products were extensively used in naval ships and in the re-equipping of the nearby old Cramp shipyard.

Walker Company, Cleveland, Ohio. Firm established in 1883 by inventor/machinist John Walker in Cleveland. Specialized in producing heavy equipment for the power transmission industry, such as cable, transmitters, accumulators, and other electrical devices. After a prolonged legal battle with the industry giants started in 1894, the Walker Company was forced to reorganize and eventually had to sell itself to Westinghouse in 1898 for $1 million dollars. Continued to operate as a wholly owned subsidiary and became a casting plant for bronze and aluminum structures. Eventually became a plant for producing lighting devices until closed in 1979. In the late-1890s the firm bid on and won contracts for two series of seacoast disappearing carriages for the Army. Produced five 8-inch Model 1896 disappearing carriages (No. 27-31) and six 10-inch Model 1896 carriages (No. 65-70).

Wellman-Seaver and Morgan, Cleveland, Ohio. Steelman Samuel Thomas Wellman started the Wellman Steel Company in Ohio in 1890. After the failure of that business, he created an iron and steel manufacturing consulting business in 1896 known as the Wellman-Seaver Engineering Company in Akron. Through other mergers and acquisitions, it again became a manufacturing firm, known from 1903 as the Wellman-Seaver-Morgan Company. In early 1904 it bid unsuccessfully for 5-inch pedestal mounts, but subsequent orders came as the result of two separate firms defaulting on their contracts. Both Mansfield Engineering and Raring Engineering encountered financial difficulties in 1904 and had to forfeit their contracts. The firm's first ordnance order was for taking over and finishing an order of five 6-inch Model 1903 disappearing carriages. These had originally been contracted to The Mansfield Engineering Company of Mansfield, Ohio. When that company entered bankruptcy in April 1904 and was unable to fulfill its order, the work in progress was transferred to Wellman-Seaver-Morgan. They completed carriage No. 61-65. Similarly, they soon also were assigned to finish carriages No. 26-45 that had originally been contracted with Rarig. Wellman-Seaver-Morgan was able to fulfill both of these assignments. The firm was renamed the Wellman Engineering Company in 1930. During the Second World War they were brought in to build both the new 6-inch and large 16-inch barbette carriages for fixed seacoast emplacements. After numerous mergers and buy-outs the firm exited its Ohio location in about 1990.

West Point Foundry, Cold Spring, New York. The West Point Foundry was one of four private cannon foundries subsidized by the U.S. Government following the War of 1812. The foundry was fully established by 1817. It specialized in large iron castings such as those used in sugar mills and cannon production. Famous ordnance engineer and designer Robert Parrott joined the firm in the 1830s, and subsequently invented the "Parrott Rifle" of Civil War fame. This was the largest American cannon producer of the mid-19th Century. However the firm could not transition to the steel age. In 1893 it built the type mortar carriage Model 1891, following it in 1894 with seven Model 1891 seacoast mortar carriages (No. 73-79) It finished the assembly of 11 8-inch steel seacoast guns (No. 1-11)--though needing to get the steel forgings from subcontractor Midvale. For field guns it assembled the converted 3.2-inch guns for the Army and secured just one contract for 20 3.2-inch General Service field guns in the 1880s. It also produced the 15-inch dynamite guns for contract-holder the Pneumatic Dynamite Gun Company. The foundry went out of business and closed in 1911. The original site is abandoned and park land today.

The Wheland Company, Chattanooga, Tennessee. An early 20th Century manufacturing firm making steam engines, saw mill equipment, oil transmission systems, and drilling equipment. During the Second World War was able to manufacture for the government under contract a number of gun tubes for the 90mm anti-aircraft gun. A total of five firms and arsenals made this type of gun tube and breech mechanism. After the war converted to production of brake drums. Later renamed the Wheland Foundry, the company went into bankruptcy and was liquidated in 2003.

The William Cramp & Sons Ship & Engine Building Company, Philadelphia, Pennsylvania. This was one of the major shipbuilders in Philadelphia founded in the 1830s. Incorporated in 1872. Transitioned from wooden ships to iron and steel by the 1890s. Became the leading commercial supplier of vessels to the U.S. Navy by the time of the First World War. Cramp had an interest in the gun business, if nothing else to help supply foreign orders that were being pursued. The firm established a gun shop in 1891 and acquired a small order in 1894 for Army seacoast carriages. They produced two Model 1894 10-inch disappearing carriages (No. 23 and 24). The firm also assisted in constructing the experimental Pneumatic Disappearing Carriage. They helped launch the Driggs Ordnance Company; with a financial investment, plant site property, and a seat in management. Some of the original Driggs orders were fulfilled in the Cramp shops. Contributed its ordnance assets to the formation of American Ordnance Company in 1895. The shipyard failed in 1927, though revived for a short time during the Second World War.

Worthington Pump & Machinery Corporation, Buffalo, New York. A machinery company founded in 1845 as Worthington and Baker by Henry R. Worthington near the New York Navy Yard. Manufactured hydraulic machinery, pumps, compressors, and other machinery. Moved in 1904 to Harrison, New Jersey. Reorganized in 1916 and the Worthington Pump & Machinery Corporation with its primary manufacturing plants in Buffalo and Cincinnati. During the Second World War was one of several producers of the carriage for the 90mm anti-aircraft guns. After several postwar mergers, was acquired by McGraw Edison in 1979.

York Ice Machinery, York, Pennsylvania. Founded in 1872 as the York Machinery Company, the company specialized in ice handling equipment and then migrated to refrigeration units in later years. During the war, the firm expanded to manufacture air conditioning and refrigeration equipment for the Navy, and operated training schools for the maintenance of such equipment for service personnel. Under the leadership of York president William Shipley, York and numerous smaller industrial shops in York, Pennsylvania banded together to participate on ordnance contracts during the war. In fact the alliance was so successful as to be termed the "York Plan" and reproduced elsewhere. For heavy ordnance York Ice Machinery and its allies produced some of the carriages for the new 6-inch seacoast barbette gun in the Second World War.

York Safe & Lock Company, York, Pennsylvania. An industrial company founded in 1882 by S. Farry Lauks at York. The firm specialized in safes, locks, vaults, and other precision steel parts. It became an important ordnance producer for both the Army and Navy during the Second World War and was ultimately the operator of four government-funded plants. It produced license-built 40mm Bofors AA guns for the Navy and Army. For the U.S. Army it contracted for and made 37mm anti-tank guns, and produced a number of 90mm anti-aircraft gun carriages. Following the war the firm failed and its plants were all disposed of by 1947.

Foreign Facilities:

French Arsenals. Two French government arsenals were producers of 75mm guns Model 1897 subsequently supplied to U.S. Army A.E.F. in France. Arsenal Bourges (gun markings "ABS") was the primary producer of such guns during the war and by far the most common source of these cannon. Some of the earlier guns were produced by Arsenal Tarbes ("ATS") and carry that mark as manufacturer. Such French-produced 75mm guns were used on the Model 1917 Improvised anti-aircraft mount used by American forces in Europc.

Joseph Whitworth and Company, Manchester, England. Whitworth was a British steel and arms firm founded by inventor and engineer Joseph Whitworth in 1833. It was incorporated in 1874, producing over the years weapons from small arms to heavy ordnance of high quality. In particular they pioneered the use of a process known as fluid compression of steel that produced forgings particularly suited to armor and heavy guns. The firm supplied arms to America during the Civil War and were one of the entities that helped supply critical, large forgings to the Ordnance Department pending the development of the country's own steel industry. In the mid-1880s Whitworth provided five steel tubes (8, 10, and 12-inch), three jackets and five trunnion hoops for the first series of experimental seacoast type guns. In 1886 they reached a deal with Bethlehem Steel to sell technology (especially heavy forging presses and technical expertise) to allow that firm to produce large steel forgings in the U.S. Further orders were curtailed when both Bethlehem and Midvale demonstrated the ability to satisfy cannon orders domestically. Whitworth and Company merged with W. G. Armstrong, Mitchell and Company in 1897 and ceased independent operations.

Schneider et. Comp., Le Creusot, France. A large, private French cannon producer since 1782. Founded by Joseph-Eugène Schneider. An early adopter of steel forgings in ordnance, the very first American 12-inch Model 1888 seacoast gun was imported from Schneider, along with a 12-inch gun lift carriage. Also, a purchased example of a Schneider 4.7-inch seacoast gun on pedestal mount was emplaced at Fort Hancock in the emergency of the Spanish American War. Developed the 75mm M1897 field gun, though produced at government arsenal at Puteaux and other locations and used in the American Model 1917 improvised anti-aircraft mount.

Vickers Sons and Maxim, Ltd., London, England. A British steel and shipbuilding firm. Began design and sale of cannon with the acquisition of the Sheffield Steel and Gun Works at the River Don Works. In 1897 Vickers acquired the Maxim Nordenfelt Guns and Ammunition Co. Ltd. of London, changing the corporate name as above. Maxim's plant at Erith was kept for the manufacture of light guns and machine guns. The United States purchased a number of small guns from Maxim Nordenfelt starting in 1894. At the start of the Spanish American War the United States acquired in England eight 6-inch pedestal carriages made by Vickers, on which were mounted gun tubes made by Armstrong. Vickers merged with Armstrong in the 1920s.

Sir W. G. Armstrong Whitworth & Co. Ltd., Newcastle upon Tyne, England. Firm founded by Sir William G. Armstrong in the mid-19th Century. A key subsidiary was the Elswick Ordnance Company, which specialized in designing and production of guns, starting as early as 1855. Merged in 1897 with Joseph Whitworth Co. and then in 1927 with Vickers. Operated extensive naval ship construction facilities at Newcastle on Tyne for many years. Armstrong was the source of 34 4.7-inch pedestal guns and carriages, and eight 6-inch gun tubes purchased in April 1898 for use in the Spanish American War. Also, additional 6-inch guns and mounts were obtained by the U.S. Navy when it bought two cruisers under construction also in 1898. These were eventually dismounted as used as coast defense by the Navy in a couple of locations. The company survived as an ordnance and aircraft manufacturer until late in the 20th Century.

Notes to Sources and Bibliography

Primary Sources:

U.S. Army

Annual Report of the Chief of Ordnance to the Secretary of War, various years 1880-1919, Government Printing Office, Washington.

Army Ordnance Department Publications:

FM 4-120, Coast Artillery Field Manual, Antiaircraft Artillery (1940)

FM 4-140, Coast Artillery Field Manual, Antiaircraft Artillery Service of the Piece, 37-mm Antiaircraft Gun (1940)

Description of 12-inch Mortar Carriage Model 1891 (July 1899)

No. 1060 Service Handbook of the 3-inch Antiaircraft Gun Mount Model of 1917 (July 1921)

No. 1075 Service Handbook of the Mortar Carriage Model of 1890M1 and MII (May 1921)

No. 1665 Description of 8, 10, 12, 14, and 16-inch Seacoast Guns (July 1912)

No. 1676 Table of United States Army Cannon and Carriages (January 1924)

No. 1683 Instructions for 5-inch Barbette Carriages, Model 1896 (1909)

No. 1684 Instructions for 5-inch Barbette Carriage, Model of 1903 (May 1908)

No. 1685 Instructions for 8-inch Barbette Carriage Model of 1892 (1903)

No. 1686 Instructions for 6-inch Disappearing Carriage LF Model 1898 (1904)

No. 1687 Instructions for 6-inch Disappearing Carriage LF Model 1903 (1917)

No. 1688 Instructions for 6-inch Barbette Carriage Model 1900 (December 1907)

No. 1689 Instructions for Disappearing Carriages LF Model of 1894 for 8-inch Guns (October 1916)

No. 1690 Instructions for Disappearing Carriage LF Model 1896 for 8-inch Gun (March 1912)

No. 1691 Instructions for Disappearing Carriage LF Model 1894M1 for 10-inch Gun (October 1913)

No. 1692 Instructions for Disappearing Carriage LF Model 1896 for 10-inch Gun (June 1916)

No. 1693 Instructions for Disappearing Carriage ARF Model 1896 for 10-inch Gun (December 1907)

No. 1694 Instructions for Disappearing Carriage LF Model 1901 for 10-inch Gun (October 1907)

No. 1695 Instructions for Disappearing Carriage LF Model 1896 for 12-inch B.L. Rifle (June 1898)

No. 1696 Instructions for Disappearing Carriage LF Model 1897 for 12-inch Gun Model 1895 (November 1909)

No. 1697 Instructions for Disappearing Carriage LF Model 1901 for 12-inch Guns Model 1895 and 1900 (July 1911)

No. 1698 Instructions for Mortar Carriages Model of 1891 (April 1917)

No. 1699 Instructions for Mortar Carriages Model of 1896 (March 1905)

No. 1700 Instructions for Barbette Carriage Model of 1893 for 10-inch guns Model of 1888 (October 1916)
No. 1701 Instructions for 3-inch 15-pdr. Barbette Carriage Model of 1903 (May 1916)
No. 1702 Instructions for Barbette Carriage Model of 1892 for 12-inch guns Model of 1888 (February 1908)
No. 1703 Instructions for 6-inch Disappearing Carriage LF Model 1905 (October 1908)
No. 1704 Instructions for 6-inch Disappearing Carriage LF Model 1905M1 (April 1917)
No. 1707 Instructions for Mortar Carriages Model of 1908 (October 1912)
No. 1709 Instructions for Mortar Carriages Model of 1896MIII (December 1913)
No. 1711 Instructions for 6-inch Disappearing Carriage LF Model 1905MII (January 1914)
No. 1712 Instructions for Disappearing Carriages LF Models of 1907 and 1907M1 for 14-inch Guns (1917)
No. 1713 Instructions for 6 inch Barbette Carriage Model of 1910 (May 1914)
No. 1716 Instructions for Disappearing Carriage L.F. Model of 1912 for 16-inch Guns (July 1917)
No. 1722 Instructions for Barbette Carriage Model of 1917 for 12-inch Gun (August 1917)
No. 1749 Instructions for 4.72-inch gun, Armstrong (January 1911)
No. 1752 Instructions for 6-inch Gun Armstrong Mounted on 6-inch Barbette Carriage Armstrong (1903)
No. 1756 Instructions for Driggs-Seabury 15-pounder Rapid-Fire Guns and Masking Parapet Mounts (1904)
No. 1756 Instructions for 3-inch (15-pounder) Gun Model of 1898M1 (May 1916)
No. 1765 Description for 5-inch and 6-inch Rapid-Fire Guns (1903)
No. 1766 Instructions for 15-pounder R.F. Gun and Barbette Carriage Model of 1902 (January 1908)
No. 1816 Description of Railway Cars, 8-inch Barbette Carriage Models of 1918 M1 and MII (April, 1918)
No. 1820 Description of Twelve-inch Mortars (October 1913)
No. 2033. Handbook of Artillery. May, 1920, 1920.
No. 2033. Handbook of Artillery. July, 1921, Revised May, 1924, 1925.
No. 2034. Railway Artillery, 1921
No. 2042. American Coast Artillery Matériel, June 1922, 1923.

TM 4-210 Seacoast Artillery Weapons (October 1944)
TM 9-227 20-mm Technical Manual 20-mm Automatic Gun M1
TM 9-235 37-mm AA Gun Materiel (January 1944)
TM 9-252 40-mm Automatic Gun M1 (January 1944)
TM 9-361 75mm Gun T83E1 and Gun Mount T69 (November 1951)
TM 9-456 12-inch Seacoast Materiel (October 1942)
TM 9-471 Technical Manual 16-inch Seacoast Gun Materiel (November 1942)
TM 9-2005 Ordnance Materiel – General, Volume 4, Railway and Seacoast Artillery (December, 1942)
TM 9-2300 Standard Artillery and Fire Control Materiel (February 1944)
TR-1310-3B, Antiaircraft Artillery Matériel 3-inch M1918 (November, 1928)

Ordnance Files at the National Archives and Records Administration:

Record Group 156, U.S. Army Chief of Ordnance:

Entry 21, Chief of Ordnance Correspondence 1812-1894

Entry 28, Chief of Ordnance Correspondence 1894-1913

Entry 29, Chief of Ordnance Correspondence 1910-1915

Entry 36A, Chief of Ordnance Correspondence, 1915-1931

Entry 36B, Chief of Ordnance Correspondence, 1931-1941

Entry 712, Record Cards of Seacoast Guns and Carriages

Entry 819, Historical Files:

Design, Development and Production of 37mm and 40mm Guns (15 May 1945)

Design, Development and Production of 90mm and 120mm AA Guns (February 1945)

History of the Americanization of the Bofors 40mm Antiaircraft Gun (May 1944)

Entry 894, Report of The Coast Artillery Board on Antiaircraft Armament (February 1938)

Entry 1451, Registers of Guns Manufactured and Modified at Watervliet Arsenal 1891-1920

U.S. Navy

Annual Report of the Bureau of Ordnance to the Secretary of the Navy, various years 1870-1919, Government Printing Office, Washington.

Navy Bureau of Ordnance Publications:

Ordnance Pamphlet No. 77, 1-Pounder Case Guns, January 1903
Ordnance Pamphlet No. 212, Deck Mounts 6-inch Mk III, October 1902
Ordnance Pamphlet No. 217, 6-inch Mount, Mark VIII Armstrong Pedestal Mount, October 1903

Ordnance Files at the National Archives and Records Administration,

Record Group 74, U.S. Navy Bureau of Ordnance:

Entry 25, Chief of Bureau of Ordnance Correspondence 1888-1941

Entry 111, Record of Armament of Naval Vessels 1841-1903

Entry 112, Registers of Naval Guns 1842-1900

Entry 118, Record of Guns and Mounts Afloat 1860-1942

Secondary Sources:

Anonymous. *History of Watervliet Arsenal 1813 to Modernization 1982.* (Watervliet, New York: U.S. Army, Watervliet Arsenal, n.d.).

Anonymous. *Naval Ordnance, A Textbook.* (Annapolis, Maryland: The United States Naval Institute, 1937).

Anonymous, *Navy Ordnance Activities World War 1918-1918.* (Washington: Government Printing Office, 1920).

Barnes, G. M. *Weapons of World War II.* (New York: D. Van Nostrand Company, Inc., 1947).

Bennett, John D. "The 5-inch Emergency Naval-Gun Batteries on Oahu, T.H. 1942-1943", *The Coast Defense Study Group Journal* Vol. 17, No. 1 (February 2003): 81-99.

Berhow, Mark A. editor, *American Seacoast Defenses, A Reference Guide.* (McLean, VA: The CDSG Press, third edition 2015).

Berhow, Mark A., "U.S. Army Coast Artillery Corps and Antiaircraft Defense," *The Coast Defense Study Group Journal* Vol. 29 No. 2 (May 2014): 27-69.

Birnie, Capt. Rogers, Jr. *Gun Making in the United States.* (Washington: Government Printing Office, 1907).

Boyd, William B. and Buford Rowland. *U.S. Navy Bureau of Ordnance in World War II.* (Washington: Bureau of Ordnance Department of the Navy, n.d.).

Brauer, F., "14-inch Railway Mount Model 1920", *Army Ordnance,* 1921.

Brown, Sevellon. *The Story of Ordnance in the World War.* (Washington: James William Bryan Press, 1920).

Bruff, Captain Lawrence L., U.S. Army. *A Text Book of Ordnance and Gunnery.* (New York: John Wiley & Sons, 1896).

Buckner, Maj. David N. "WWI Tractor Gun Acquired by Museum." *Fortitudine* IX No. 4 (Spring 1980): 16-17.

Campbell, John, *Naval Weapons of World War Two.* (Annapolis: Naval Institute Press 1985).

Crozier, Maj. Gen William. *Ordnance and the World War.* (New York: Charles Scribner's Sons, 1920).

Crowell, Benedict. *America's Munitions 1917-1918.* (Washington: Government Printing Office, 1919).

Friedman, Norman. *Naval Weapons of World War One.* (S. Yorkshire, UK: Seaforth Publishing, 2011).

Fritz, Robert D. "Development of America's Large Bore Mobile Guns" *Warship International,* (No. 2. 1973): pp. 188-198.

Hansen, David M., "The Gun-Lift Battery in the Defenses of the United States", *The Coast Defense Study Group Journal* Vol. 10 No. 4 (November 1996): 4-16.

Hodges, Maj. LeRoy. *Notes of Post-War Ordnance Development.* (Richmond, Virginia: Richmond Press Inc., 1923).

Hagge, Greg, "Mortar Carriage Notes" *The Coast Defense Study Group News* Vol. 6 No. 2 (May 1992): 23-31.

Kirchner, D. P. and E .R .Lewis, "The Oahu Turrets", *Warship International* (No. 3, 1992): 273-301.

Lawry, Nelson H., "Another Five Degrees: WWI Alterations to the Disappearing Carriage" *The Coast Defense Study Group Journal* Vol. 10 No. 2 (May 1996): 4-10.

Lissak, Lieutenant-Colonel Ormond M. U.S. Army. *Ordnance and Gunnery.* (New York: John Wiley & Sons, 1910).

Marshall, Peter A., "History of the U.S. Navy 7-in/45 Gun", *Warship International* (No. 2, 2010): 153-179.

Mueller, Chester. *The New York Ordnance District in World War II.* (New York: The New York Post, 1947).

Peck, Taylor. *Round Shot to Rockets, A History of the Washington Navy Yard and U.S. Naval Gun Factory.* (Annapolis, Maryland: United States Naval Institute, 1949).

Robbins, Charles B. and E.R. Lewis, "The Chilean-American 12-inch Guns", *Warship International,* (Vol. No. 2, 2000): pp 184-190.

Smith, Bolling W., "Antiaircraft Weapons of the Coast Artillery Corps." *The Coast Defense Study Group Journal* Vol. 12 No. 2 (May 1998): 93-112.

Smith, Bolling W., "Seacoast Guns and Carriages", *The Coast Defense Study Group Journal* Vol. 12 No. 4 (November 1998): 73-118

Smith, Bolling W., "12-Inch Breech Loading Mortars", T*he Coast Defense Study Group Journal* Vol. 7 No. 4 (November 1993): 57-59.

Thomson, Harry C. and Lida Mayo. *United States Army in World War II, The Ordnance Department: Procurement and Supply.* (Washington: Center of Military History, 1991).

Tschappat, Lt. Col. William H., U.S. Army. *Text-Book of Ordnance and Gunnery.* (New York: John Wiley & Sons, Inc., 1917).

Williford, Glen M., *American Breechloading Mobile Artillery, 1897-1953.* (Atglen, PA, Schiffer Publishing Ltd., 2016).

Index to gun types - Seacoast Guns

Index to gun types - Anti-aircraft Guns

Index to gun types - Railway Guns

The Coast Defense Study Group, Inc. (https://CDSG,org) is a tax-exempt corporation dedicated to study of seacoast fortifications. CDSG's purpose is to promote and encourage the study of coastal defenses, primarily but not exclusively those of the United States of America. The study of coast defenses and fortifications includes their history, architecture, technology, strategic and tactical employment and evolution. The primary goals of the CDSG are the following:

- Educational study of coast defenses
- Technical research and documentation of coast defenses
- Preservation of coast defense sites, equipment and records for current and future generations
- Accurate coast defense site interpretations
- Assistance to groups interested in preservation and interpretation of coast defense sites
- Charitable activities which promote the goals of the CDSG

Membership is open to any person or organization interested in the study or history of the coast defenses and fortifications. Membership in the CDSG will allow you to attend the annual conference, special tours and receive the CDSG's quarterly journal and newsletter. For more information on the CDSG, please visit the CDSG website at cdsg.org or contact us at 22150 Pondsville Road, Smithsburg MD 21783 USA, Attn: Glen Milam Membership.

The CDSG Fund supports the efforts of the Coast Defense Study Group by raising funds for preservation and interpretation of American seacoast defenses. The CDSG Fund is seeking donations for projects sup- porting its goals. Donations are tax-deductible for federal tax purposes as the CDSG is a 501(c)(3) organization, and 100% of your gift will go to project grants. Major contributions are acknowledged annually. The Fund is always seeking proposals for the monetary support of preservation and interpretation projects at former coast defense sites and museums. A one-page proposal briefly describing the site, the organization doing the work, and the proposed work or outcome should be sent to the address below. Successful proposals are usually distinct projects rather than general requests for donations. Upon conclusion of a project a short report suitable for publication in the CDSG Newsletter is requested. The trustees shall review requests and pass their recommendation onto the CDSG Board of Directors for final approval. Send donations and grant requests to: CDSG Fund c/o Terry McGovern 1700 Oak Lane McLean, VA 22101- 3326 USA or use your credit card via PayPal on the cdsg.org website.

The CDSG Press

CDSG Books and CDSD Gear ($ domestic / $ International) prices include domestic/international postage $US currency only (cash, check, money order or credit card via PayPal at the cdsg.org store), allow 6-8 weeks for delivery.

- *Notes on Seacoast Fortification Construction* by Col. Eben E. Winslow, 1920, 1994 reprint HC with drawings $45
- *Seacoast Artillery Weapons Technical Manual (TM) 9-210* by U.S. War Dept. 1944, 202 pp. 1995 reprint PB $25
- *The Service of Coast Artillery* by F. Hines & F. Ward, 1910, 736 pp. 1997 reprint HC $40
- *Permanent Fortifications & Sea-Coast Defenses by U.S. Congress,* 1862, 544 pp. 1998 reprint HC $30
- *American Coast Artillery Matérial* Ordnance Dept. Doc#2042 by U.S. War Dept., 1922, 2001 reprint HC $45
- *American Seacoast Defenses: A Reference Guide* (3rd Edition) by Mark A. Berhow, (2015) 732 pp. HC $45
- *The Endicott & Taft Board Reports of 1886 and 1905* by U.S. Congress, 525 pp. 2007 reprint HC $45
- *Artillerists and Engineers: The Beginnings of US Fortifications 1794-1815* by Col. Wade, U.S. Army. PB, $25
- *WWII Harbor Defenses of San Diego,* 2021 by H.R. Everett, CDSG Press, 2022, HC 538 pp. $50
- *American Seacoast Forts - A Directory,* 2025, by Terry McGovern, Mark Berhow and Glen Williford
  Vol. 1 The North Atlantic Coast – Portland to New York – 2025, 376 pp HC $40;
  Vol. 2 The Mid-Atlantic, South Atlantic and Gulf Coasts – Delaware to Galveston – 2025, 316 pp HC $40;
  Vol. 3 The Pacific Coast – San Diego to Puget Sound - 2025, 264 pp HC $40;
  Vol. 4 Alaska, Hawaii and the Overseas Bases – Phillipines, Panama, others – 2025, 336 pp HC $40.
- Hole in the Head Press books : *Rings of Supersonic Steel* $25; *To Defend and Dete*r $25, *The Last Missile Site* $15, *Fort Baker Through the Ages* $15, *Artillery at the Golden Gat*e $15

Send order to: CDSG Press Attn: Terry McGovern 1700 Oak Lane, McLean, VA 22101-3326
Or order via our online store at cdsg.org.

The CDSG Digital Library

The CDSG has digitized an extensive collection of historic manuals, reports, records and documents on the harbor defenses of the United States Army.

- The CDSG provides back issues of the CDSG Publications (from 1985) in electronic format.
- The US Seacoast Defense Documents (CDSG Doc) covers a range of historical material related to seacoast defenses -- most are from the National Archives. Included are the annual reports of the chief of coast artillery and chief of engineers; several board proceedings and reports; army directories; text books; tables of organization and equipment; WWII command histories; drill, field, training manuals and regulations; ordnance department documents; ordnance tables and compilations; and the ordnance gun and carriage cards.
- Harbor Defense Document Collection (CDSG HD Doc) related to specific harbor defenses. These PDF documents form the basis of the conference and special tour handouts that have been held at these locations. They include RCBs/RCWs; maps; annexes to defense projects; CD engineer notebooks; quartermaster building records; and aerial photos taken by the signal corps 1920-40. Please consult cdsg.org for more details.

Information on the CDSG ePress items can be obtained from Mark Berhow at berhowma@cdsg.org or by post to PO Box 6124, Peoria, IL 61601 USA, or at cdsg.org.

MCGOVERN PUBLISHING PRESENTS:

McGovern Publishing is comprised of two divisions: Redoubt Press and Three Sisters Press offering a range of military titles. McGovern Publishing is interested in new titles, especially those dealing with fortifications and artillery, please contact Terry McGovern at 703/538-5403 or at tcmcgovern@att.net or post to 1700 Oak Lane, McLean, Virginia 22101 USA, if you have a title that you are seeking to have published.

The Chesapeake Bay at War!
The Coastal Defenses of Chesapeake Bay During World War II
by Terrance McGovern

Given the importance of the U.S. naval bases around Norfolk, Virginia and the shipyards of Hampton Roads, the seacoast defenses protecting Chesapeake Bay contained the largest collection of firepower in the continental United States as they reached their apex during World War II. This book tells the story of preparing the coastal defenses of the Chesapeake Bay for the coming of World War II and their operations from 1941 to 1945. Over a hundred rare black & white U.S. Army photographs and plans help document our nation's extensive efforts to defend against naval attacks and raids from Nazi Germany. A product of extensive research this book brings together rare images and the little-known military history of the Chesapeake Bay for the first time. 70-page softbound book for $30 plus $5 for domestic shipping.

The Delaware Bay at War!
The Coastal Defenses of Delaware Bay During World War II
by Terrance McGovern

Given the importance of the military- industrial complex along the banks of the Delaware River, including the large Philadelphia Naval Shipyard, the seacoast defenses protecting Delaware Bay had declined dramatically since the turn of the century resulting in a whole program of modern coast artillery batteries and other defenses to be constructed starting in the late 1930's and reaching their reached their apex during the middle of World War II. This book tells the story of preparing the coastal defenses of the Delaware Bay for the coming of World War Two and their operations from 1941 to 1945. Over a hundred rare black & white U.S. Army photographs and plans help document our nation's extensive efforts to defend against naval attacks and raids from Nazi Germany. A product of extensive research this book brings together rare images and the little-known military history of the Delaware Bay for the first time. 65-page softbound book for $30 plus $5 for domestic shipping.

Seacoast Cannon Coloring Book
By Brian B. Chin

This detailed coloring book shows how coast artillery once protected American shores from enemy attack. This second edition brings to life the coast defenses that today remain as empty forts and batteries with our national, state, and local parks. The author, Brian B. Chin, is a well-known author and artist that has produce several works on military fortifications. This coloring book allows a new generation to learn the history of American seacoast defenses while having fun coloring in this 52-page softbound book. $10 plus $5 for domestic shipping.

American Defenses of the Panama Canal
By Terrance McGovern.

This book documents the thirty-five years of fortification construction and use in the Panama Canal Zone. The first part of this book discusses how the various fortification boards decided, with the help of Congress, on what defenses were needed and how evolving threats resulted in changes to these defenses. The second part of the book describes in detail the construction, service life, and current status of each major battery. Assisting this narrative are many current and period photographs, along with engineering drawings and site maps detailing the construction of these impressive fortifications. This 116-page softbound book has 45 maps and drawings, 43 historical B&W photographs, and 40 color photographs. $55 plus $5 for domestic shipping.

The Concrete Battleship
Fort Drum, El Fraile Island, Manila Bay
by Francis J. Allen and Terrance McGovern

Fort Drum on El Fraile Island in the Philippines is unique in the development of United States coastal fortifications. Fort Drum is part of a chain of forts built across the entrance of Manila Bay to defend the Bay from naval attack. The design of the concrete and steel fort followed a naval pattern with turrets, a cage mast, and secondary armament in side casemates. Due to these characteristics, Fort Drum became known as the "Concrete Battleship". During World War II, Fort Drum became an important weapon during the Japanese siege of Corregidor and the other island forts during 1942 but only play a minor role during the American retaking of these islands in 1945. The battles of World War II would transform Fort Drum from an American-manned, fully operating fort to a burned-out hulk inhabited by lifeless Japanese sailors. This revised and enlarged 64-page softbound volume tells the story of The Concrete Battleship in words, diagrams, and photographs from its inception to the present day. $30 plus $5 for domestic shipping.

A Legacy in Brick and Stone – 2nd Edition
American Coastal Defense Forts of the
Third System 1816-1867
By John R. Weaver II

The definitive history of the American Third System of Fortifications that defended America's coastline for more than half of century, these architectural wonders were built from 1816 through 1867 from Maine through the Florida Keys to New Orleans, with two forts in San Francisco Bay. Most of these 42 masonry forts still stand guard along our shores, and open to the public. A Legacy in Brick and Stone provides the background of these famous Civil War forts – why they were built where they are, who built them, and how they functioned – as well as descriptions of each of the forts. This revised and expanded edition has grown to 340 pages with 400 new photographs and drawings. John Weaver II, a nationally known expert on masonry coastal fortifications, has invested over 30 years of research into A Legacy in Brick and Stone to produce the only a full treatment of the magnificent American Third System. Deluxe hardcover edition with color illustrations $65 plus a $5 fee for domestic shipping. Paperback edition with black and white illustrations $40 plus a $5 fee for domestic shipping.

Pacific Rampart:
A History of Corregidor and the Harbor Defenses
of Manila and Subic Bays,
By Glen M. Williford

This is the definitive history of the American-built forts and harbor defenses of Manila and Subic Bay in the Philippines. This a heavily illustrated work tells the history of the fortified islands (the most famous being the island of Corregidor) from the fortifications built by the Spanish and the Americans to what remains today. Compiling research from the National Archives and many primary sources this work describes in great detail the defensive plans as well as the fortifications built between 1904 and 1940. The 470-page text is supported with thorough referenced endnotes, a bibliography, and six appendices. It contains over 340 illustrations—black and white photographs, maps, and plat diagrams for many of the fort structures. ISBN 978-1-7323916-3-5 Hardcover edition is $50 plus $5 for domestic shipping for single orders.

Pacific Fortress
A History of the Seacoast Defenses of Hawaii
By Glen M. Williford

This is a revealing historical account of the U.S. Army's coastal defenses of Oahu. This work provides a thorough account of both the seacoast and anti-aircraft branches of the Coast Artillery Corps in this important American Pacific base. Compiling research from the National Archives and many primary sources this work describes in great detail the defensive plans as well as the fortifications built between 1904 and 1945. The 450-page text is supported with thorough referenced endnotes, a bibliography, and eleven appendices listing all fortification sites. Heavily illustrated with over 460 maps, plans, and period photographs, this work is the most comprehensive coverage of these defenses ever offered in print. ISBN 978-1-7323916-3-5 Hardcover edition is $50 plus# $5 for domestic shipping for single orders.

www.ingramcontent.com/pod-product-compliance
Lightning Source LLC
LaVergne TN
LVHW071343210926
848486LV00003B/219